CPSIA information can be obtained
at www.ICGtesting.com
Printed in the USA
LVHW082040090221
678838LV00001B/26

DIGITAL TRANSFORMATION
Evaluating Emerging Technologies

World Scientific Series in R&D Management

Print ISSN: 2591-7498
Online ISSN: 2591-7501

Series Editor: Tugrul U Daim *(Portland State University, USA)*

Published

Vol. 6 *Digital Transformation: Evaluating Emerging Technologies*
edited by Tugrul U Daim

Vol. 5 *Managing Medical Technological Innovations: Exploring Multiple Perspectives*
edited by Tugrul U Daim and Alexander Brem

Vol. 4 *Managing Mobile Technologies: An Analysis from Multiple Perspectives*
edited by Tugrul U Daim and Alexander Brem

Vol. 3 *Cooperative Innovation: Science and Technology Policy*
by Frederick Betz

Vol. 2 *Technology Roadmapping*
edited by Tugrul U Daim, Terry Oliver and Rob Phaal

Vol. 1 *Managing Technological Innovation: Tools and Methods*
edited by Tugrul U Daim

World Scientific Series in R&D Management – Vol. 6

DIGITAL TRANSFORMATION
Evaluating Emerging Technologies

Editor

Tugrul U Daim

Portland State University, Portland, Oregon, USA
Higher School of Economics, Moscow, Russia
Chaoyang University of Technology, Taiwan

NEW JERSEY · LONDON · SINGAPORE · BEIJING · SHANGHAI · HONG KONG · TAIPEI · CHENNAI · TOKYO

Published by

World Scientific Publishing Co. Pte. Ltd.
5 Toh Tuck Link, Singapore 596224
USA office: 27 Warren Street, Suite 401-402, Hackensack, NJ 07601
UK office: 57 Shelton Street, Covent Garden, London WC2H 9HE

Library of Congress Control Number: 2020006572

British Library Cataloguing-in-Publication Data
A catalogue record for this book is available from the British Library.

World Scientific Series in R&D Management — Vol. 6
DIGITAL TRANSFORMATION
Evaluating Emerging Technologies

Copyright © 2020 by World Scientific Publishing Co. Pte. Ltd.

All rights reserved. This book, or parts thereof, may not be reproduced in any form or by any means, electronic or mechanical, including photocopying, recording or any information storage and retrieval system now known or to be invented, without written permission from the publisher.

For photocopying of material in this volume, please pay a copying fee through the Copyright Clearance Center, Inc., 222 Rosewood Drive, Danvers, MA 01923, USA. In this case permission to photocopy is not required from the publisher.

ISBN 978-981-121-462-2 (hardcover)
ISBN 978-981-121-463-9 (ebook for institutions)
ISBN 978-981-121-464-6 (ebook for individuals)

For any available supplementary material, please visit
https://www.worldscientific.com/worldscibooks/10.1142/11675#t=suppl

Desk Editors: Anthony Alexander/Amanda Yun

Typeset by Stallion Press
Email: enquiries@stallionpress.com

To My Family, Yonca, Tolga and Eda

About the Editor

Tugrul U. Daim is a Professor and the Director of the Technology Management Doctoral Program in the Maseeh College of Engineering and Computer Science at Portland State University, USA (PSU). He is also the Director of the Research Group on Infrastructure and Technology Management. He is a Faculty Fellow at the Institute for Sustainable Solutions.

The US Department of Energy, National Science Foundation and National Cooperative Highway Research Program are some of the many regional, national and international organizations that have funded his research work.

He has published over 200 refereed papers in various journals and conference proceedings, and has edited more than 20 books and conference proceedings. He is the Editor-in-Chief of *IEEE Transactions on Engineering Management* and has held various editorial roles in other journals, such as the *International Journal of Innovation and Technology Management, Technological Forecasting and Social Change, Technology in Society, Foresight, Journal of Knowledge Economy* and *International Journal of Innovation and Entrepreneurship.*

He was the advisor for 11 PhD graduates who are now in leading positions in government, industry and academia.

Acknowledgements

The time of Tugrul Daim in this book is partially funded by the Basic Research Program of the National Research University Higher School of Economics (HSE) and by the Russian Academic Excellence Project '5–100'.

Contents

About the Editor   vii
Acknowledgements   ix

Part 1   Technical Transformation   1

Chapter 1   Technical Transformation: Transportation Technologies   3
Joshua Binus, Barrett Lewis, Horatiu Corban, Fayez Alsoubaie, Rasnia Tabpla and Tugrul Daim

Chapter 2   Technical Transformation: Cloud Services   25
Cody Miller, Wendy Lally, Liyan Xiao, David Burchfield, Shihab Hanayneh and Tugrul Daim

Chapter 3   Technical Transformation: Cloud Computing   55
Amit Pingle and Tugrul Daim

Chapter 4   Technical Transformation: Internet of Things   83
Surekha Rani Chanamolu and Tugrul Daim

Chapter 5   Technical Transformation: IT in Disaster Management   141
Namitha Shetty and Tugrul Daim

Part 2	**Personal Transformation**	**211**
Chapter 6	Personal Transformation: Evaluation of Smart Home Hubs *Ahmed Alzahrani, Majed Alshamlani, Wei-Chen Hsu, Shreyas Harish and Tugrul Daim*	213
Chapter 7	Personal Transformation: Protocols for Home Automation Application *Ahmed Alzahrani and Tugrul Daim*	245
Chapter 8	Personal Transformation: Smart House *Ahlam Alsuwiada, Ahmed Al-Shareef, Zuhair Alheayk, You Hong Yong, Wei Ming Jang, Kenny Phan and Tugrul Daim*	269
Chapter 9	Personal Transformation: Wearable GPS Device for Children *Bhawinee Banchongraksa, Jessie Truong, Lu Chuan Chieh, Mufeed Yacoub, Papit Meteekotchadet and Tugrul Daim*	299
Chapter 10	Personal Transformation: Smartwatches *Alexander Blank, João Ricardo Lavoie, Felix Maier, Kenny Phan and Tugrul Daim*	329
Chapter 11	Personal Transformation: Drones *Donavon Nigg, Sarah Alobaidi, Rushikesh Jirage, Tejas Deshpande, Haitham Alkharboosh and Tugrul Daim*	367
Chapter 12	Personal Transformation: Electric Scooter *Esraa Bukhari, Dana Bakry, Farshad, Mert Tonkal and Tugrul Daim*	405
Chapter 13	Personal Transformation: Wireless Services *Asma Razavi, Prajakta Patil, Ritu Chaturvedi, Pallavi Sandanshiv, Kenny Phan and Tugrul Daim*	419

Part 3	**Organizational Transformation**	**443**
Chapter 14	Organizational Transformation: Semiconductors *Tejas Deshpande and Tugrul Daim*	445
Chapter 15	Organizational Transformation: Universities *Ahmed Bohliqa, Corey White, Srujana Penmetsa, Sara Bahreini, Zeina Boulos and Tugrul Daim*	479
Chapter 16	Organizational Transformation: Consumer Goods *Yogi Hamdani and Tugrul Daim*	523

Part 1
Technical Transformation

Chapter 1

Technical Transformation: Transportation Technologies

Joshua Binus*, Barrett Lewis*, Horatiu Corban*, Fayez Alsoubaie*, Rasnia Tabpla* and Tugrul Daim*,[†],[‡]

*Portland State University, Portland, Oregon, USA
[†]Higher School of Economics, Moscow, Russia
[‡]Chaoyang University of Technology, Taiwan

Abstract

The power grid is an incredibly complex and important system, and is one of the most impressive engineering works of modern times. Previous research has confirmed that electric transport is now ready to move from traditional and complex uses to be more beneficial from social, economic, political and environmental perspectives. These perspectives, in the use of intelligent transportation, contribute significantly to the provision of energy, cost and time.

In our research in this paper, we offer many assessments of transportation technology, which included evaluating a range of market-emerging Electric Vehicles (EVs) and Electric Vehicle Service Equipment (EVSE) options. We did so in order to craft a recommendation for future grid-integration programs that will be capable of providing realistic and affordable assistance to electric utilities during summer peak periods (typically occurring about 20 days/year). This research also discusses the most opportune behind-the-meter transportation technologies and products to use for future summer peak Vehicle-to-Grid (V2G) programs in California, Oregon and/or Washington.

This paper applied a multicriteria decision methodology known as the Hierarchical Decision Model (HDM). This model assessed current transportation technology to determine the technology options based on the judgments of experts who selected multiple criterions.

Keywords: Technology assessment, transportation, electric vehicles.

1. Background

Electrical grids across the world are undergoing a period of prolonged transformation, from centralized, utility-controlled systems with unidirectional power flows (from generators to end-users) and captive customers, to grids that are increasingly integrating Distributed Energy Resources (DERs) at the "grid edge". The grids taking shape in the 21st century are subsequently becoming more decentralized/distributed, with bi-directional flows (of energy and data) and an ever-increasing number of "prosumers" that are capable of exporting power to the grid from their homes and/or electric vehicles.

At the same time, local- and state-level policies are increasing the presence of renewable energy sources (especially wind and solar), which is having an impact on both wholesale and retail markets, systems and reliability requirements [1]. This growth in renewables has, in turn, created new challenges and considerations for electric utilities as they make determinations for the most cost-effective strategies to modernize their distribution and transmission grids through traditional resource, transmission and distribution planning efforts [2].

One of the key drivers shaping the grid of tomorrow is the threat of climate change—in particular, the need for stakeholders of all kinds to reduce their carbon emissions. The Pacific United States' states of Washington, Oregon and California have been relatively aggressive in addressing the challenges associated with carbon-emission reduction (compared to other state and federal parties). Having already taken some substantive steps to clean their power generation portfolios, each of the three Pacific states have now begun to target emissions from the transportation sector, which has become the lead sector in emissions in each state (for California, see [3]; for Washington, see [4]; for Oregon, see [5]). Toward this end, Electric Vehicles (EVs) are not the only solution being pursued, but they play a significant role in each state's climate action plans. However, while these states (along with many of their larger cities and electric utilities) are developing and promoting policies meant to increase their constituents' adoption of EVs, there are issues that must be addressed to maintain reliability and cost-effective services in light of the increasing likelihood of a scenario that will see rapid and significant market adoption of EVs over the coming decades.

From a utility (or even transmission operator) perspective, the nightmare scenario of EV penetration involves the specter of uncoordinated charging. In particular, there is concern that if all EV owners charge their vehicles at the end of a work-day, the aggregate demand could dramatically increase evening peak loads. Ultimately, uncoordinated charging introduces a dual threat of higher costs (to build more peak-serving generation) and diminished reliability of the grid (at overtaxed portions within distribution systems) [6].

1.1. *Objective*

This research project endeavored to shed some light on what utilities might (or should) do to effectively integrate EVs into the grid in ways that reduce market barriers (for EV adoption) and maintain reliability at the lowest cost to ratepayers. As the project team began disaggregating the range of issues and decision-making factors informing the business challenge posed to utilities, it became obvious that two tiers of decision-making needed addressing.

The first decision point is defined by the current state of the market and utility strategy needs. The EV market adoption is already underway, but options for Vehicle-to-Grid (V2G) tactics are limited by the lack of commercial availability of bi-directional charging equipment. Specifically, outside of pilot projects, most consumers only have cost-effective access to unidirectional charging equipment; they cannot export power from their vehicles to the grid yet. This means that utilities witnessing significant EV growth in their distribution territories are limited to grid-support tactics that rely more on behavior change. Until bi-directional chargers become widely available, utilities must find ways to incentivize their customers to avoid charging during summer peak hours, especially during heat waves. At present, there is a viable solution: dynamic pricing tariffs that vary the cost for power during peak and off-peak hours, respectively. Today, utilities in twenty states (including Oregon and California) have made dynamic pricing programs available to customers with EVs.

Having determined that applying the hierarchical decision-making methodology would be unnecessary to address the first decision-point, this research team decided to focus its efforts on supporting a near-future decision point that will emerge with the commercial availability of bi-directional chargers. Additionally, while there are many potential services that could be provided through V2G approaches, our team focused on one use case: the summer peak grid support.

1.2. *Problem definition*

Research problem: What are the most opportune behind-the-meter transportation technologies/products to use for future summer peak V2G programs in California, Oregon and/or Washington?

To explore this question, we investigated a range of potential EV applications (see Table 1) and challenged ourselves to view the availability of these options over time (as they might emerge in the market place as viable resources for future V2G programs). For this evaluation, we defined summer peak periods as those generally experienced in Washington, Oregon and California, that is, from 1 June to 30 September, between 4 pm and 9 pm. Of course, while there is a daily evening peak, there

Table 1. Potential V2G applications.

Technology/Application	Availability During Peak	Likely to be commercially adopted (and usable) by			Sufficient SOC During Peak	Adaptable to Business Ops
		2020–2025	2025–2030	2030–2035		
Municipal Bus Fleets	✗	✗/✓	✓	✓	✗	✗
Municipal Non-Bus Fleets	✓	✓	✓	✓	✓	✓
School Bus Fleets	✓	✗	✗/✓	✓	✓	✓
Police Fleets	✗	✗/✓	✓	✓	✗/✓	✗
Taxi Fleets	✗	✗/✓	✓	✓	✗	✗
Military Fleets	✓	✓	✓	✓	✓	✓
Garbage Truck Fleets	✓	✗/✓	✗/✓	✓	✗/✓	✓
Delivery Fleets	✗/✓	✗/✓	✓	✓	✗	✗
Individually Owned EVs	✓	✓	✓	✓	✓	✓
High-Speed EVSE	✗/✓	✓	✓	✓	✗/✓	✗
Off-Road EVs	✗/✓	✗/✓	✗/✓	✓	✗/✓	✗/✓

Note: An "X" means that the technology/product is not a good fit; a "✓" mark indicates a potentially good fit; and the use of both indicators means that the application depends on local considerations to determine whether it can be appropriately enrolled in a summer peak grid support program.

are typically only about 20 days per year where utilities need additional resources to successfully serve peak loads (most often during summer heat waves or days of highest summer temperatures).

1.3. *Gap analysis*

Following brainstorming potential options, the research team identified a total of eleven EV products/technologies that could potentially provide exports to distribution grids during periods of summer peak stress. The initial eleven options identified included eight fleet options and three non-fleet options. Municipal buses, municipal non-bus vehicles, school buses, police vehicles, taxis, military vehicles, garbage vehicles and delivery vehicles made up the fleet options, while individual electric vehicles, off-road vehicles and Electrical Vehicle Supply Equipment (EVSE) comprised the non-fleet options.

As pointed out, these technologies/products were considered with the assumption that bi-directional charging equipment will become available to EV owners in the near-future. With this first enabling capability in mind, the next task required us to evaluate those options to determine which option, if any, would likely be available and capable enough to participate in a summer peak V2G program. Key gaps that needed to be filled, included: availability during summer peak periods, commercial availability, the likelihood of having sufficient export capability (determined by the State of Charge (SoC) available in the battery system during peak hours) and the capability of the EV owners (that is, their ability to participate without negatively impacting their primary use requirements/needs). Table 1 summarizes our findings; the prioritized options are shaded.

1.4. *Perspectives and criteria*

The Transportation Technology Assessment conducted for this report considered three overarching perspectives: *Availability*, *Readiness Status*, and the *Likelihood of Owner Participation*. These three perspectives are important for analyzing the adoption rate of Vehicle Grid Integration (VGI) technology in the near future. Each perspective includes criteria that inform the decision-making model and the options it provides.

1. *Availability* essentially considers potential EV types as options based on time and power. There are three criterions that build out this perspective:

    i. The likelihood of being connected to a charger during the summer peak. Certain types of EVs, such as those that are individually owned, may be used more during peak times than others, e.g., garbage truck fleets.

ii. Whether the existing SoC is high enough to provide exportable power during peak hours, which depends on an EV's charging load and speed. The key factor is whether an EV can be charged fast enough during pre-peak or current peak times, so that it can export power to help support the grid during such peak times.

iii. Whether EVs are capable of being scheduled for pre-charging prior to peak times. Some EVs, such as school buses, are not constantly in use and may be scheduled more easily than other EVs that are regularly in use, such as police vehicles.

2. The *Readiness Status* is the product of a combination of two elements—technology and market.

    i. *Technological readiness*, the first criterion, is a measure of product maturation. What this means for the EVs considered for connecting to and supporting the electrical grid, is the level to which VGI EV technology is ready for but may differ between EV types.

    ii. Technological readiness measures the technology in its current capabilities, while the second criterion—*market adoption and existing market conditions*—measures it based on the market. This criterion seeks to measure the existing adoption of EV types and the potential for their market growth. This takes into consideration existing market conditions, more specifically, consumer interest and demand for VGI EVs.

3. The last perspective is the *Likelihood of Owner Participation*, which is influenced by two criterions.

    i. The first criterion analyzes the incentives and benefits for owners who participate in bi-directional grid support programs, as well as existing and planned incentives for transmission and distribution construction.

ii. The second criterion is the likelihood of an owner's willingness to invest in bi-directional charging equipment that is likely needed to implement wide-scale VGI EV infrastructure.

1.5. *Relevant application alternatives*

As we assessed the technologies/products in question, certain options were deemed unlikely to help utilities serve summer peak needs, although we did see opportunities for these options to serve other service requirements, for example, ancillary services, renewables integration, volt/VAR support, etc. Municipal buses, for example, were likely to be in use during summer peak hours, but they could also be very helpful in integrating wind power at night (off-peak hours). Police fleets were cut from the list because emergency responders would likely need to keep their SoCs as high as possible, but they could also support ancillary services while plugged in. Taxi fleets were a mismatch in the same way as municipal buses, but could also help with renewables integration. Delivery fleets were cut from the list for the same reason, although some types of delivery fleets could provide some export power if they finished their routes early enough in the day.

High-speed charging equipment (especially on-route chargers backed up with stationary storage batteries) might offer some potential for program participation; however, we determined that it would be best to include them with their corresponding fleets and not treat them as a stand-alone option. While off-road EVs offered promise, the most significant potential involved the use of large-load vehicles associated with airports and seaports (e.g., tugs, ferries, cranes, electric rail, etc.). However, these technologies/products are still being developed and tested in early pilots and demonstrations. After performing the gap analysis, we were left with five application alternatives—municipal non-bus fleets, school bus fleets, military fleets, garbage truck fleets and individually owned EVs.

2. Municipal Non-Bus and Non-Emergency Fleets

More and more municipalities and cities are beginning to electrify their fleets, as moving towards EVs can help reduce emissions, lead to better public health and lower government spending [7]. Municipal non-bus fleets include a whole host of vehicles that cities use every day, from sanitation inspections to water meter readings and building inspections. Most non-bus fleets are made up of the same make and model that individuals can purchase for everyday use. While these are much smaller than a city bus, for example, a large city would have a number of smaller EVs as part of the fleet.

A key factor in the selection of non-bus fleets as an alternative for study was their immediate market availability, as well as their availability to be plugged in and ready during peak periods. Assume a 2019 Nissan Leaf as a case example of a municipal fleet vehicle: the Leaf has a maximum range of 150 miles [8], which can more than accommodate the average daily needs to fulfill municipality duties and still have an available SoC for the 4 pm peak hour. Additionally, since the Leaf has the ability to quick charge up to 90 miles in 30 minutes via a DC quick-charger, it can still be drawn down to a low SoC and then quickly charged again once the peak ends and still be ready to resume duties the following morning.

3. School Bus Fleets

School buses were selected for their large stored potential energy, as well as for their significant downtime during the summer months. As school buses normally sit idle during the summer months, they would be able to support a peak V2G program with a larger portion of their SoC. As a whole, the school bus system is the largest form of public transportation system in the country [9]. With more than 480,000 buses in service across the United States, an entire electric fleet capable of V2G interconnection could be a very attractive alternative.

However, only a fraction of the over 480,000 available school buses are electric, as of 2018 [10]. The largest barrier to entry is the significant cost associated with electric models; a full electric school bus typically costs three times a conventional diesel or propane model. Even so, some school districts are beginning to convert or supplement their fleets with electric models. For example, the White Plains of the New York School District, in partnership with their bus contractor—National Express and Con Edison—has purchased five Lion Electric bus models, each with a battery capacity of 88 kWh [11], specifically with V2G in mind. As part of the agreement, Con Edison contributed US$100,000 [10, 12] towards the purchase of each bus and plans to use all five buses during summer peak periods, which will provide an additional 75 kW of energy to the grid [13].

With such pilot programs, we feel that V2G electric school bus integration could be a reality by 2025, especially if more school districts and operators partner with utilities to help offset the initial, higher costs.

4. Garbage Truck Fleets

Like school buses, garbage trucks were selected as a possible alternative due to their flexible operating schedules as well as their potential for adoption. To date, most of the garbage trucks in service across the United States use internal combustion engines fueled with either diesel or Compressed Natural Gas (CNG). However, the segment shows large potential towards electrification, with large, heavy duty trucks (e.g., electrified garbage trucks) that start and stop every 200 feet through the use of regenerative braking. This type of braking can help recover energy consistently [14]. With the use of a regenerative brake system, it is estimated that an electric garbage truck can save up to US$35,000 per year in operating costs when compared to a traditional diesel/CNG model [15].

There are a number of manufacturers that are either building or developing electric garbage truck models, with models currently in use in California. Manufacturers of electric garbage trucks include Chinese Build Your Dreams (BYD), Swedish Volvo and Peterbilt, Mack and Wrightspeed. Battery capacities currently range from 60 to 300 kWh [16, 17].

Depending on its battery capacity, a garbage truck could be a good option for V2G integration. Larger capacity models should have a sufficient SoC in the peak period during the summer months, and flexible routes/schedules can be integrated as well. Ultimately, our research indicates that the factor for electric garbage truck adoption will result in savings for fleet operators. Electric truck models are currently estimated to save US$35,000 per year in operating costs, while partnerships with utilities for V2G integration could result in further operational savings. Ultimately, we estimate electric garbage trucks to be a tested and available alternative by 2025–2030.

5. Individually Owned EVs

Individually owned vehicles are the most widely adopted electric vehicles in the United States. However, these vehicles as well as their chargers are not currently enabled for V2G participation there. Fortunately, a trial V2G program that is a partnership between Enel and Nissan of Europe has been underway in Denmark since 2015 [18], which allows owners of Nissan Leaf models to supply energy to the grid. If this program is successful, a similar partnership is scheduled to begin in Italy and Germany [19].

For 2018, the new Nissan Leaf model has been approved for V2G integration. It is the first EV to gain such an approval [20]. Based on our assumption that bi-directional charging equipment will become widely available in the near-future to interested EV owners in the United States, we believe that individual EVs will be a good alternative for peak power V2G integration between 2020 and 2025. Furthermore, with battery capacities increasing with each EV announced, individual EVs should have a sufficient SoC to support peak demand integration during the peak hours of 4 to 9 pm.

6. Military Fleets

The final alternative or candidate for V2G integration we selected for analysis were military non-combat vehicles. We decided to restrict military EVs to only non-combat or non-tactical vehicles, since combat vehicles need to keep their SoCs as high as possible for operational readiness. We do acknowledge that combat vehicles could likely support ancillary services while plugged in, but that scenario would require a separate analysis outside our scope.

In 2013, the US Department of Defense (DoD) acquired 500 alternative fueled vehicles [21]. It is predicted that the DoD will own or lease 92,000 hybrid and electric vehicles through 2020 to help lower its fuel consumption, and reduce the risk and associated impact of fuel price volatility [22]. Furthermore, through Phase 2 of the Smart Power Infrastructure Demonstration for Energy Reliability and Security (SPIDERS) program, the DoD along with the Department of Energy and the US Army ran the first V2G test at Fort Carson, Colorado. The test integrated a 1 MW solar microgrid with five electric vehicles coupled "with advanced bi-directional vehicle chargers to integrate the battery capacity of electric vehicles in both microgrid and normal operations" [23].

With the current DoD experience in mind, we believe that future non-tactical EVs could be used during summer peak times. Such vehicles, if coupled with PV installations already in place on many military installations should also have sufficient SoCs during the 4 to 9 pm window selected for study.

6.1. *Model building*

At the early stage of the technology development, this study adopted the Hierarchical Decision Model (HDM), which was created and developed by Dundar Kocaoglu and Tugrul Daim [24] to better understand and track decision. The HDM is a methodology to analyze and evaluate best fitting alternatives in order to accomplish a specific objective. It uses a multicriterion that flows into alternatives selection process. Our study applied the HDM into four levels, which were Objective, Perspectives, Criteria and Alternatives, that contributed to the best option for the main objective. The HDM used the judgment of experts to prioritize the important perspectives, criteria and alternatives through the pairwise comparison technique [24, 25]. Those perspectives and criteria were then weighed by these experts, who then evaluated and estimated the complex and complicated system to gain the best decision strategically. However, the results from the HDM provided inconsistent and disagreement ratios, which indicated how much their responses did not agree with each other.

As discussed above, the objective was to determine the best opportunity behind the meter transportation technologies to use for future summer peak V2G programs. As Figure 1 shows,

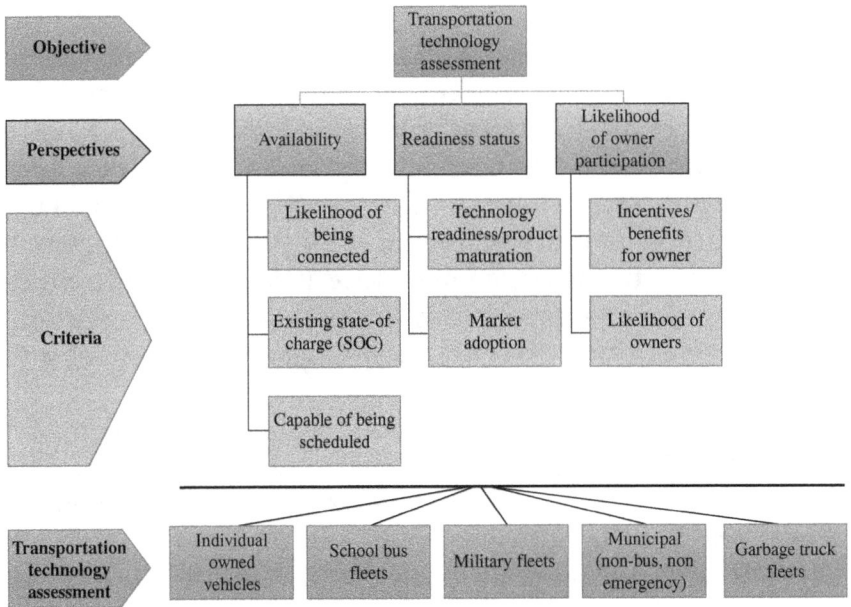

Figure 1. The HDM in four levels.

the model was created based on the HDM analysis to accomplish the goal. The decision model is illustrated in Figure 1. This model was created through the HDM link website to collect data from the experts. The respondents did pairwise comparison through the link for all three perspectives and a separate comparison in each node among the criteria is seen in Figure 1.

Finally, the experts completed weighing the pairwise comparison of all the perspectives, criteria and potential alternatives. Then, their opinion is submitted to the model and contributed to the result as the best opportunity of technology options.

6.2. Data analysis and results

The HDM results showed the best option of potential alternatives through the highest score. Moreover, there was a critical statistic result which is the inconsistency that explores how consistent and careful the experts weighted different factors. The standard acceptable rate for inconsistency was less than 0.1. If it had been more than 0.1, the quality of judgment should not be considered [26]. However, it also depends on the variety of perspectives, criteria and different tolerance levels. In this study, the inconsistency for each expert was less than 0.1, as shown in Table 2, therefore the results from all the experts can be considered as consistent judgment. Moreover, it shows that the disagreement rate is less than 0.1, which means that all the experts are in the same agreement with regard to weighting the criteria and perspectives relating to the objective.

The F-test value was calculated through pairwise comparison in the HDM model from all the participating experts, as shown in Table 3. The value indicated a degree of agreement due to the benchmark value of 2.33 at 0.1 level (90% confidence level) and the final value of 2.61, which is over 2.33. Therefore, it proves that the HDM weights from the selected experts were in agreement with a 90% confidence.

Table 2. HDM results based on the alternatives.

Expert	Individually Owned Vehicles	School Bus Fleets	Military Fleets	Municipal Fleets (Non-Bus and Non-Emergency)	Garbage Truck Fleets	Inconsistency
1	0.25	0.22	0.16	0.16	0.21	0.01
2	0.28	0.21	0.15	0.18	0.18	0.02
3	0.22	0.23	0.22	0.24	0.1	0.01
4	0.66	0.06	0.12	0.1	0.06	0.03
5	0.27	0.23	0.13	0.19	0.17	0
Mean	0.34	0.19	0.16	0.17	0.14	
Disagreement						0.071

Table 3. HDM statistical results.

Source of Variation	Sum of Square	Deg. of Freedom	Mean Square	F-test Value
Between Subjects	0.12	4	0.03	2.61
Between Conditions	0.00	4	0.000	
Residual	0.19	16	0.012	
Total	0.31	24		
Critical F-value with degrees of freedom 4 & 16 at 0.01 level				4.77
Critical F-value with degrees of freedom 4 & 16 at 0.025 level				3.73
Critical F-value with degrees of freedom 4 & 16 at 0.05 level				3.01
Critical F-value with degrees of freedom 4 & 16 at 0.1 level				2.33

Through the pairwise comparison in HDM methodology, the important perspectives and criteria reveal overall scores from all the experts. Figure 2 illustrates the score of each element in all levels calculated through the mean score of all the experts. In the perspective level, the *Likelihood of Owner Participation* tends to be the most important, which can also

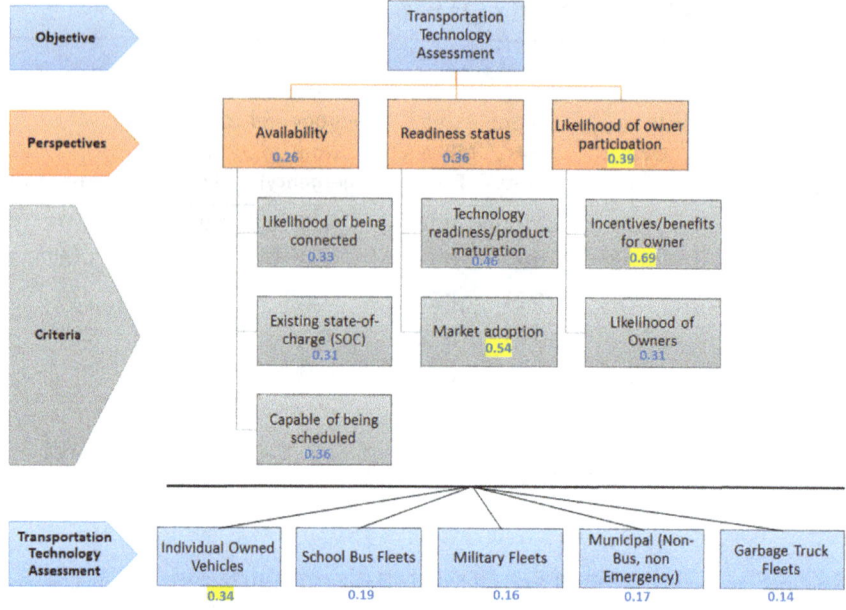

Figure 2. The model with HDM results in all levels.

potentially influence decision making. This perspective is important because the power of owner participation can persuade them to participate in the program. This perspective is influenced by two criterions, which includes the highest impactable criterion—*incentives and benefits for owners* to participate in a bi-directional grid support program. Therefore, to make the best decision with regard to the objective, the critical, important perspective and criteria need to be first considered carefully.

Considering relevant application alternatives, the *Individual Owned Vehicles* alternative, for example, electric vehicles in the United States, has the highest score based on many contributions. Such a huge market adoption and technology readiness factor could make this alternative become the first option for this opportunity to become the objective. Moreover, the direct power a vehicle owner could introduce to the program through incentives and benefits could potentially be key to making this the best option.

The *School Bus Fleets* is the second option because buses have large stored potential energy and significant downtime during the summer's peak, which highly contributes to all the three criteria in the *Availability* perspective. This can be the best option to support the peak V2G program; however, there is a barrier—the cost to transform these buses into electric models. Therefore, the lack of market adoption has a huge impact on this alternative.

The remaining three options—*Municipal (Non-Bus and Non-Emergency)*, *Military* and *Garbage Truck Fleets*—have similar scores. They have immediate market availability and market adoption, which give them opportunities to participate in the program. These options could help the program to manage the energy consumption time. However, it also depends on their availability, readiness and market adoption factor with regard to the V2G integration. Nevertheless, the *Garbage Truck Fleets* has the lowest score because of the transformation cost to electric models in the first period of time, which is related to the market adoption and incentive/benefit of the owner criterion.

Conclusion

This paper analyzed the most opportune behind-the-meter transportation technologies and products to use for future summer peak V2G programs in California, Oregon and/or Washington. All five options have potential as V2G technologies but are in varying degrees of readiness. Through this research, the perspective of the *Likelihood of Owner Participation* was the most influential to the decision-making process. Within this perspective, the *incentives/benefits for owner* criterion was weighted most highly. *Readiness Status,* the second-highest weighted perspective, had *market adoption* as its highest-ranking criterion.

The third perspective, *Availability*, had a fair equal distribution among its three criteria.

Of the five technological options, individually owned vehicles were chosen by all the experts to be the closest in delivering V2G integration capabilities. While there are still many hurdles and challenges for this technology type, the direct power and decent adoption rate of EVs in the United States indicate great promise. Research should be undertaken to examine how federal or state policy might stimulate this technology option through incentivization schemes, either in the form of government subsidies to consumers or VGI services, or regulatory prioritization.

Because school buses are common in almost every locality of the United States and that they are already often on a structured time schedule, they came in as the second-highest rated policy option. Here, the main hurdle would be the high purchase cost of electric buses and/or converting buses from fossil fuel to electric usage. More research should be conducted to understand how policy can incentivize school districts and municipal decision-makers to invest in EV technology. Incentivization schemes like subsidization might be one way to do so, as might federal or state grants.

Municipal owned vehicles scored about average, due to their lack of widespread adoption. However, because these vehicles do not require coercion or incentive to participate in scheduling for peak-management, this option may have more potential as time progresses and the costs of EVs decrease. This is largely true for all of the options, but more so for electric military vehicles and garbage trucks. Once costs decrease in these areas, decision makers might want to start integrating into EV technology.

All five technology options present great promise for the future of V2G and VGI technologies. Now that research has narrowed down policy options, future research will want to focus on incentivizing individual car owners and municipal decision makers to adopt these technologies. Future research should also analyze summer time peak management scheduling for a variety of municipal and government owned EVs.

References

1. R. Wiser, A. Mills, J. Seel, T. Levin and A. Botterud, *Impacts of Variable Renewable Energy on Bulk Power System Assets, Pricing, and Costs* (Golden, CO: National Renewable Energy Laboratory with Argonne National Laboratory, 2017).
2. L. Wood, R. Hemphill, J. Howat, R. Cavanagh, S. Borenstein and L. Schwartz, *Recovery of Utility Fixed Costs: Utility, Consumer, Environmental and Economist Perspectives* (Berkeley, CA: Lawrence Berkeley National Laboratory, 2016).
3. California Energy Commission, "2017 integrated energy policy report", Sacramento, CA, February 2018. https://www.energy.ca.gov/2017_energypolicy/.
4. Carbon Emissions Reduction Taskforce, "Carbon emissions reduction taskforce, report to the Washington state governor's office", Report, Washington State Governor's Office, Olympia, WA, 14 November 2015. http://www.governor.wa.gov/sites/default/files/documents/CERT_Final_Report.pdf.
5. Oregon Department of Environmental Quality, "Oregon greenhouse gas emissions data, 2012 sector share of total emissions", December 2015. http://www.oregon.gov/DEQ/AQ/Pages/Greenhouse-Gas-Inventory-Report.aspx.
6. P. De Martini, L. Kristov and L. Schwartz, *Distribution Systems in a High Distributed Energy Resources Future: Planning, Market Design, Operation and Oversight* (Berkeley, CA: Lawrence Berkeley National Laboratory, 2015). https://emp.lbl.gov/sites/all/files/lbnl-1003797.pdf.
7. N. Swalnick, "Climate mayors and electrification coalition partner to rapidly transition municipal fleets to electric vehicles", *Act-News*, 13 September 2018. https://www.act-news.com/news/municipal-fleets-electric-vehicles/.
8. Nissan USA, "Range and charging", *Nissan USA*. https://www.nissanusa.com/vehicles/electric-cars/leaf/range-charging.html. Accessed: 28 November 2018.

9. W. Cox, "School buses: America's largest transit system", *Newgeography.com*, 19 December 2014. http://www.newgeography.com/content/004801-school-buses-americas-largest-transit-system.
10. B. Plumer, "The wheels on these buses go round and round with zero emissions", *The New York Times*, 12 November 2018. https://www.nytimes.com/2018/11/12/climate/electric-school-buses.html.
11. Lion Electric Company, "Power in progress", Brochure. https://thelionelectric.com/documents/en/BrochureLionCang.pdf.
12. WLNY CBS New York, "White plains unveils state's first all-electric school buses", 14 November 2018. https://newyork.cbslocal.com/2018/11/14/electric-school-buses-white-plains-national-express/.
13. B. Lillian, "Electric school bus batteries to support New York grid in summer", *NGT News*, 6 July 2018. https://ngtnews.com/electric-school-bus-batteries-to-support-new-york-grid-in-summer.
14. C. Ockedahl, "Tesla veteran helps Mack create an electric garbage truck", *Trucks.com*, 7 June 2018. https://www.trucks.com/2016/06/07/mack-trucks-shows-electric-garbage-truck/.
15. J. O'Dell, "From Tesla to trash: Wrightspeed's electric garbage truck journey", *Trucks.com*, 21 February 2017. https://www.trucks.com/2017/02/21/tesla-electric-garbage-trucks-wrightspeed/.
16. J. O'Dell, "Peterbilt unveils battery-electric garbage truck", *Trucks.com*, 9 May 2017. https://www.trucks.com/2017/05/09/peterbilt-battery-electric-garbage-truck/.
17. F. Lambert, "Volvo unveils new all-electric garbage truck with up to 200 km of range", *Electric.co.uk*, 9 May 2018. https://electrek.co/2018/05/09/volvo-all-electric-garbage-truck/.
18. W. Pentland, "Nissan pilots vehicle-to-grid technology In Denmark", *Forbes*, 8 December 2015. https://www.forbes.com/sites/williampentland/2015/12/08/nissan-pilots-vehicle-to-grid-technology-in-denmark/.
19. Nissan Motor Corporation Global Newsroom, "Enel Energia, Nissan Italia and IIT join forces for the development of electric mobility in Italy", Press Release, 5 February 2017. https://newsroom.nissan-global.com/releases/enel-energia-nissan-italia-and-iit-join-forces-for-the-development-of-electric-mobility-in-italy.

20. M. Kane, "Nissan LEAF Is Germany's First V2G-Approved Electric Car", *Inside EVs,* 23 October 2018. https://insideevs.com/nissan-leaf-germanys-first-v2g-electric-car/.
21. T. Casey, "US military will get thousands more EVs as SPIDERS web grows", *Clean Technica,* 2 November 2013. https://cleantechnica.com/2013/11/02/us-military-electric-vehicles/.
22. D. Unger, (2013, October 31). "US military warms to electric cars", *The Christian Science Monitor,* 31 October 2013. https://www.csmonitor.com/Environment/Energy-Voices/2013/1031/US-military-warms-to-electric-cars.
23. Naval Facilities Engineering Command, "Technology transition final public report: Smart Power Infrastructure Demonstration for Energy Reliability and Security (SPIDERS)", Report, Washington, DC, 31 December 2015. https://www.energy.gov/sites/prod/files/2016/03/f30/spiders_final_report.pdf.
24. D. F. Kocaoglu and T. U. Daim, *Hierarchical Decision Modeling: Essays in Honor of Dundar F. Kocaoglu* (Cham, Switzerland: Springer, 2015).
25. N. J. Sheikh, K. Kim and D. F. Kocaoglu, "Use of hierarchical decision modeling to select target markets for a new personal healthcare device", *Health Policy and Technology* 5, 2 (2016) 99–112. doi:10.1016/j.hlpt.2015.12.001.
26. M. Abbas, "Consistency analysis for judgment quantification in Hierarchical Decision Model", Dissertation, Portland State University, 2016.

Chapter 2

Technical Transformation: Cloud Services

Cody Miller*, Wendy Lally*, Liyan Xiao*, David Burchfield*, Shihab Hanayneh* and Tugrul Daim*,†,‡

*Portland State University, Portland, Oregon, USA
†Higher School of Economics, Moscow, Russia
‡Chaoyang University of Technology, Taiwan

Abstract

Cloud migration is a complex process, and many decisions must be made before the process can be completed. This paper identifies the key criteria of cloud migration and offers a model for determining one of the decisions: which cloud service strategy (Infrastructure as a Service (IaaS), Platform as a Service (PaaS), or Software as a Service (SaaS)) to use for any particular cloud migration project? This paper describes some of these types of services as well as presents a Hierarchical Decision Model (HDM) structure for choosing a service strategy.

The authors have created the HDM model, which may be used as a basis for cloud service strategy decision-making at a wide variety of companies. This implementation of the model has been designed for a fictional company, Best Men's Fashion (BMF) LLC, and the pairwise comparison judgments are based solely upon the priorities of that company.

Keywords: Technology assessment, infrastructure as a service, platform as a service, software as a service.

1. Introduction

The cloud service strategy Hierarchical Decision Model (HDM) model was developed using four levels of criteria. The first level—Mission Level—was crafted to "determine the model of cloud service strategy for the company". In the second level called the Objective Level, criteria were gathered from a literature review and the opinions of experts. The four objectives in this level are Technical, Security, Economic and Management. To limit the scope of this project as well as to keep the expert pairwise comparison data points manageable, the team focused on two criteria per objective in the third level. So, the criteria of the Security objective are Protection and Migration: Compliance, Scalability and Migration: Technical Complexity for the Technical objective, Service Charges and Migration: Costs for the Economic objective, and Support Capabilities and Migration: Business Complexity for the Management objective. The last level contains three cloud service strategies—Software as a Service (SaaS), Platform as a Service (PaaS) and Infrastructure as a Service (IaaS)—for the HDM to compare.

This model evaluation had two panels of experts. The first panel, which evaluated the priorities of the objectives and criteria in relationship to the mission, comprised five members of the Best Men's Fashion (BMF) LLC executive team who were project team members acting on behalf of the company. These experts had direct knowledge of BMF's strengths and weaknesses and could make decisions on what they thought would be best for the company. The second panel consisted of two external experts with significant cloud strategy experience. This panel was tasked with evaluating each of the strategies in relation to the third level criteria. Since they had no knowledge of BMF's internal climate and (fictional) situation with IT staffing, they did not participate in the upper tier evaluation.

There were two rounds of analysis in the HDM modeling tool. The first round was negated as it took in all seven of the experts' opinions into account for all tiers. This round was inconclusive in determining a cloud strategy, so their opinions were critiqued in class, revaluated by the team, and then redeveloped into a new strategy. The second round was crafted in the same way, with the experts divided into business and cloud expertise and who were only allowed to influence tiers on which they had extensive knowledge.

Scalability, Protection and Service Charges were top-ranked concerns among all the criteria in both the first and second rounds. As a result, the team is highly confident that these are the most important criteria a company should evaluate when choosing a cloud strategy.

The results of the second round were more conclusive when compared to the first round results with regard to the strategy choice. In the first round, there was little differentiation between the scores of the strategies, although IaaS was a slight leader. In the second and more focused round, IaaS ranked most successfully with a score of 0.38. SaaS was second with a score of 0.32 and PaaS came in last with 0.30. Clearly, IaaS would be a sound choice for BMF. The company could now move on to comparing just IaaS vendors.

This model can also be used for the decision processes of other companies in the same situation. These companies would need to complete a pairwise comparison round of "mission to objective" and "objective to criteria" with their business leaders. If changes in the market are not great, the cloud expert's judgment of the criteria to strategy could be reused to save time. However, if the cloud market has changed greatly, then the cloud expert tier should be re-evaluated before making conclusions.

Moving to the cloud is a daunting prospect for any company, large or small. In addition to choosing which Internet Service Provider (ISP) to align with, the company must first determine a strategy of where they should house their data, codebase and business processes during the migration. It must also determine which of their business systems should be targeted for the cloud. Concerns about data placement (on-premise or off-premise?) or a hybrid placement must be considered before engaging an ISP.

This paper will now describe the three broad types of internet service strategies: IaaS, PaaS and SaaS, as well as present an HDM structure for choosing a service strategy.

The goal of this paper was to create an HDM model that other companies could use as a basis for their decision-making. This model was designed, in particular, for a fictional company, BMF LLC, and the value judgments are based solely upon the priorities of a small company with limited staff, retail and supply chain websites, and the need to focus on its core business.

2. Company Profile

BMF LLC is a 3-year-old start-up headquartered in Portland, Oregon. In addition to its Portland office, the company also has locations in China and Mexico, which are primarily focused on manufacturing.

BMF is set up to be a luxury fashion online retailer that will style and groom male executives who desire to look great at work. This company will send a pre-selected and ready-to-wear outfit to the customer, who can then choose to send unwanted items back at no cost. BMF also has an online retail presence that is hosted on a premise with several servers dedicated to the supply chain, including custom applications for their stylists.

The company is currently funded by venture capital and has ramped up to 100 employees. Of those employees, there are two IT support staff, five developers and one IT manager. The IT manager and support staff handle both the server support for the website and supply chain software, as well as employee hardware and software issues. Collaboration cloud software has been used in this company since it was funded, and support for this is included in software support. The IT manager spends 50% of his time as a support staff and thus would like to have more time to focus on IT strategy instead.

The five developers consist of the following: a Python developer on internal tools that are custom to this company, two front-end developers for the website (and as needed on the tools' User Interface (UI)), and two java and python developers to develop server side code for the website and supply chain integration.

3. IT Strategy and Considerations

The IT staff has come to the consensus that the following systems will be targeted to move to an off-premise cloud solution:

- Website—bestmensfashion.com—java and python.
- Supply chain servers with custom java and python applications.

Based on these targeted technology systems, this HDM project will evaluate the three cloud strategies (IaaS, PaaS and SaaS) and determine which one will be the most successful for BMF. Further evaluation, following the service type decision, will be needed to choose the right ISP vendor.

3.1. *Benefits of cloud*

The IT and Business staff have evaluated that the benefits of moving to the cloud are large with regard to long-term and on-demand scalability, which is needed due to business projections of 20% year on year growth of website traffic as well as seasonal demand that will cause fluctuations to the number of users. It is expected that website uptime will be increased after moving to a cloud model, due to failover and redundancy mechanisms.

An additional benefit is that the core business of creating the best experience for the customer can be the focus on more of its employees. For example, IT staff can be less tasked with maintaining the server lab, and be more targeted on employee support and creating web applications needed for this unique product experience.

3.2. Challenges of cloud

One of the largest challenges for BMF's cloud strategy is that the migration process will be complex, both technical- and process-wise. Technically, the migration will involve moving all the codes, data and developers to the cloud model. One of our highest concerns is addressing privacy issues for both company secrets and customer data.

There are also substantial business migration concerns, for example, the BMF's supply chain processes are executed by non-technical business analysts and stylists. Training (for the new site) is needed for these non-technical users to prepare them for possible differences once the code is migrated. All processes will need to be tested and validated before the new site and supply chain are opened for business. It should be expected that there are difficulties in this process and possibly downtime if the migration process is not tightly managed.

Another concern is the high costs of a cloud solution. The management expects that there will be substantial upfront costs for the technical migration, training and time spent on training their employees. Reoccurring monthly costs will also need to be managed, but that management strategy will largely be dependent upon which ISP is ultimately chosen, as an ISP's monthly costs can vary due to support tiers, time of day and the amount of data.

Off-premise cloud solutions inherently create a dependence on the ISP. This dependence will be for mission critical systems. In the case of BMF, if their systems go offline, business can suffer. If their systems are hacked, the company's reputation and trust can be lost. Thus, in the event of such scenarios, businesses have a foundational reliance upon their ISPs. In spite of this, BMF's management still continues to desire the move due to the obvious benefits.

4. Literature Review

Hierarchical Decision Modeling (HDM) is a technical tool used in project selection, resource allocation and evaluation decision-making. Its objective is to assist the user, by a series of pairwise comparisons, to reach quantifiable judgmental value using ratio scales. The underlying assumption is that each decision has a number of perspectives and each perspective has number of criteria to consider [1].

Thus, combining the perspectives—quantifiable or non-quantifiable—and the supporting criteria will help in determining the strategy (decision). The HDM is a process using multi-level decisions and utilizing multiple criteria by separating the overall system into several hierarchical levels.

The HDM is also a process based on reaching out to an independent panel of selected experts, who responds to questions by dividing 100 points between two alternatives at a time. The allocation of the points represents each expert's independent judgment with respect to a specific criterion. The 100-point scale is from 1 to 99. The zero value is avoided to eliminate mathematical difficulties; however, if such a consideration is given, each expert selects 50 points. This means the judgment is neither important nor unimportant [1].

The HDM is based on a pairwise comparison analysis using linear algebra and matrix analysis. The goal is to find the eigenvalue and the eigenvector for each consideration in the matrix. In other words, pairwise comparison is a method used to determine how to evaluate alternatives by providing an easy and reliable means to rate and rank decision-making criteria. Weights are used and assigned to criteria and the results are normalized. The comparison is implemented in two stages:

1. Determining qualitatively which criteria is more important (i.e., establish a rank order of the criteria), and

2. Assigning a quantitative weight to each criterion, such that the qualitative rank order is satisfied.

The process is based on three steps that differ in their underlying scale. First, the measurement is based on a range from an ordinal perspective (i.e., weighting by ranking). The second step is about constructing an interval by weighted ranking, while the third step calculates the ratio scale—the pairwise comparison value. The three steps are summarized below, based on the document **HDM** by Dundar Kocaoglu.

- **Step 1—Completion of the pairwise comparison matrix:** Two considerations are evaluated at a time in terms of their relative importance. Index values from 1 to 99 are used. If criterion A is exactly as important as criterion B, this pair receives an index of 1. If A is much more important than B, the index is 99. All degrees are possible in between when comparing A to B. For a "less important" relationship, the fractions would be closer to 50 points. The values are entered row by row into a cross-matrix. The diagonal of the matrix contains only values of 1. The right upper half of the matrix is filled until each criterion has been compared to every other one [1].
- **Step 2—Calculating the criteria weights:** The weights of the individual criteria are calculated. First, a normalized comparison matrix is created: each value in the matrix is divided by the sum of its column. To get the weights of the individual criteria, the mean of each row of this second matrix is determined. These weights are already normalized; their sum is 1.
- **Step 3—Assessment of the consistency matrix:** A statistically reliable estimate of the consistency of the resulting weights is made.

4.1. *How objectives and criteria are determined*

Theoretically, each level of the hierarchy consists of multi-dimensional alternative choices or decision elements, as noted

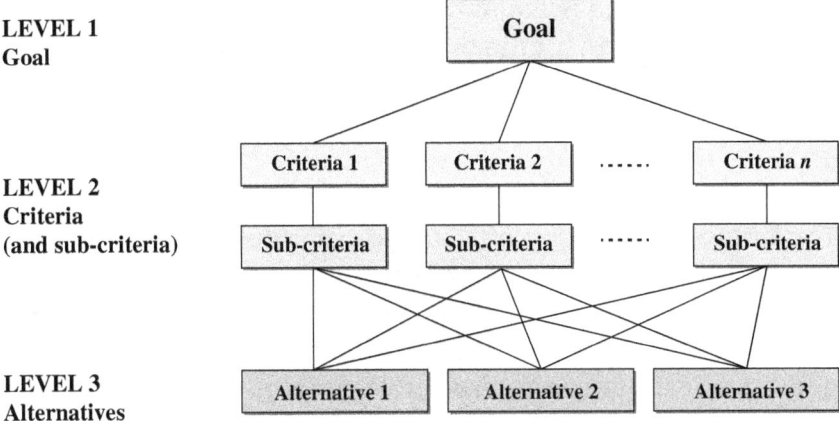

Figure 1. HDM conceptual framework.

in Figure 1 as Level 1. Multi-criteria objectives that lead to multiple subcriteria are shown in Level 2. At the bottom of the figure, multiple output results from multiple actions are shown in Level 3.

The decision element at a specific level has an impact on several elements at the next nod level in the connecting lines. Let's say, we are seeking to make an operational level decision to produce a cloud model of technology that contributes to several or maybe all subcriteria at the target level. Consequently, reaching our fulfillment level (i.e., the goal) that contributes to several or all the objectives. Figure 1 depicts how the goal, criteria and alternatives are related.

The process of evaluation between each internal relationship in such a hierarchy requires assigning a numerical value to each branch of the hierarchical network structure, as shown in Figure 1. The values are assigned to represent the relative contribution of one element to the next on different levels. As this process is completed, an evaluation model is developed to obtain the relative measure of effectiveness for each element at the bottom of the decision hierarchy, in terms of the elements at the top. In other words, each of the items (that makes Level 2) has a percentage value if the sum is equal to 1. Also, the sum value of each subcriteria is equal to the respective criteria at Level 2 (i.e., the upper limit for the number of relationships is defined by the product of the number of elements at the sublevels).

4.2. Use of experts and Delphi

The Delphi method is a structured communication technique (or process) developed to be systematic and interactive with an iterative component that relies on a panel of experts. The experts are preselected based on predefined criteria; each answers questionnaires or completes pairwise comparisons of an HDM model in two or more rounds. After each round, a facilitator provides an anonymized summary of the experts' judgments in a result table. Results from the previous round are provided as well, as are the reasons for the judgments [2].

Experts are encouraged to revise their earlier answers after considering the replies of the other members in the panel. The objective during this process is to decrease and converge towards the "most reasonable" answer or judgement. Finally, the process is stopped after a predefined stop criterion, e.g., number of rounds, achievement of consensus, or stability of results (i.e., reduced inconsistency level).

The Delphi method is based on the principle that decisions from a structured group of individuals are more accurate than those from unstructured groups. The technique can also be adapted for use in face-to-face meetings; it is called the mini-Delphi. The Delphi method has been widely used for business forecasting, which is commonly used among fund managers and stock picking analysts and has certain advantages over another structured forecasting approach.

There are four key characteristics to implement a successful Delphi technique:

(1) Anonymity of the participants;
(2) Structuring of information flow;

(3) Regular feedback;
(4) Role of the facilitator.

4.3. *Cloud computing models*

Cloud-computing providers offer services in three main different models: IaaS, PaaS and SaaS, which are often portrayed as being layers in a stack. However, such an understanding should not lead to the misconception that these platforms need to be implemented in coordination or in an order.

Thus, it is common to implement the SaaS without using the underlying PaaS or IaaS layers. It is also equally possible to run a program on IaaS and access it directly, without wrapping it as a SaaS [3].

The following definitions are based on The NIST Definition of Cloud Computing:

- "Software as a Service (SaaS). The capability provided to the consumer is to use the provider's applications running on a cloud infrastructure. The applications are accessible from various client devices through either a thin client interface, such as a web browser (e.g., web-based email), or a program interface. The consumer does not manage or control the underlying cloud infrastructure including network, servers, operating systems, storage, or even individual application capabilities, with the possible exception of limited user-specific application configuration settings.
- Platform as a Service (PaaS). The capability provided to the consumer is to deploy onto the cloud infrastructure consumer-created or acquired applications created using programming languages, libraries, services, and tools supported by the provider. The consumer does not manage or control the underlying cloud infrastructure including network, servers, operating systems, or storage, but has control over the deployed applications and possibly configuration settings for the application-hosting environment.

- Infrastructure as a Service (IaaS). The capability provided to the consumer is to provision processing, storage, networks, and other fundamental computing resources where the consumer is able to deploy and run arbitrary software, which can include operating systems and applications. The consumer does not manage or control the underlying cloud infrastructure but has control over operating systems, storage, and deployed applications; and possibly limited control of select networking components (e.g., host firewalls)." [3]

5. HDM Model

The HDM model (see Figure 2) was developed using four levels of criteria—Mission, Objectives, Criteria and Strategy. The explanation and rationale for choosing these criteria are as follows.

5.1. *Mission*

The "Mission" of this HDM model is to "determine the model of cloud service strategy for our company". In this specific

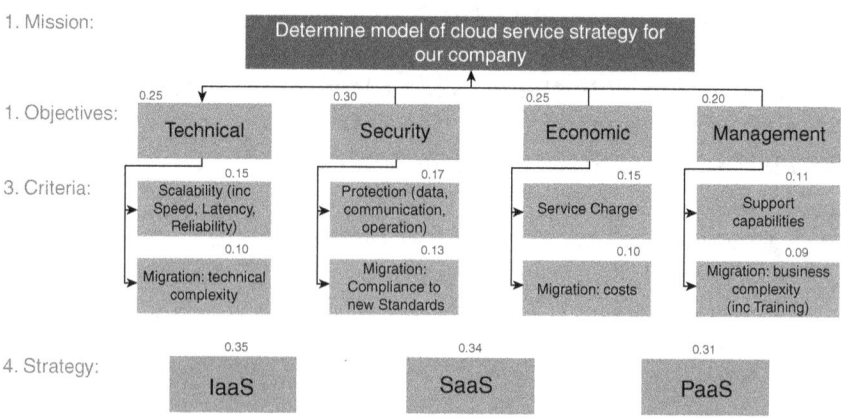

Figure 2. Team 1 HDM model.

scenario, a mid-size fashion company is looking to move its IT operations to the cloud. While this mission is specific to a fashion company, this model could be applied to any business where different experts would be required.

This mission is a common scenario many businesses face, where there are pros and cons to moving their IT infrastructure from in-house to the cloud. There are many cloud service options a company must evaluate and this model is designed to aid in choosing which cloud service model a business should move its operations to. This model does not recommend any specific cloud service provider, though it does recommend choosing one of the three cloud service options (SaaS, PaaS or IaaS). Additional models would be required to choose a specific provider.

5.2. *Objectives*

The HDM model consists of four criteria (see Figure 2): Technical, Security, Economic and Management. Objectives were chosen based on a thorough analysis of expert opinions and literature reviews. When a business is looking to move its services to the cloud, they are looking for specific benefits to the company. Examples include increased focus on business, faster time to market, increased business agility, reduced operational costs, and lower development costs. These four objectives cover all the concerns and objectives a company must consider when moving its IT services to the cloud. It is important to note that additional objectives were evaluated and considered by the team. These four objectives were deemed to be the most important objectives a company must evaluate. Due to time limitations on the scope of this project, no more than four objectives were added to the model. More objectives could be added to future models, for example, Political.

1. **Technical objective:** Evaluating technical considerations when deciding to move operations to a cloud service.

2. **Security objective:** Evaluating security considerations when deciding to move operations to a cloud service.
3. **Economic objective:** Evaluating financial considerations when deciding to move operations to a cloud service.
4. **Management objective:** Evaluating business/management considerations when deciding to move operations to a cloud service.

5.3. Criteria

The criteria in Figure 2 consist of two criteria per objective. Due to time limitations on the scope of this project, no more than two criteria per objective were added to this model. The original model consisted of four criteria per objective. To narrow this down to only two expert opinions, a literature review was used to reduce the number of criteria by 50%.

1. **Technical criteria:** Technical objectives (see Figure 2) consisted of two criteria: Scalability and Migration—Technical Complexity.

    i. *Scalability* includes speed, latency and reliability. This criterion would pertain to the company's current technical architecture and to which cloud service strategy would create the best scalability for the business. Scalability is a criterion all businesses must consider when deciding on a cloud service strategy.

    ii. *Migration—Technical Complexity* pertains to the company's current technical architecture and to which cloud service strategy would allow for the most efficient migration to the cloud. Migration can be a very costly endeavor with little Return on Investment (ROI) if it is not evaluated properly. Examples include trying to move existing services to a cloud service strategy that does not support current infrastructure.

2. **Security criteria:** Security objectives consisted of two criteria: *Protection* and *Migration—Compliance to New Standards*.
    i. *Protection* refers to security measures in regard to data center protection (e.g., building, fire, surveillance, etc.), communication protection (e.g., data encryption, secure cryptographic protocols, firewall, etc.) and operation protection (e.g., access control, role management, virus protection, etc.).
    ii. *Migration — Compliance to New Standards* is a method that will help companies to avoid being fined for compliance violations, and to manage risk factors as well as processes and decision rights. Examples include cloud encryption standards (FIPS 140-2), Payment Card Industry (PCI) data security standard and identity management that monitor application access and authorization.
3. **Economic criteria:** Economic objectives consisted of two criteria: *Service Charge* and *Migration — Costs*.
    i. *Service Charge* defines how the cloud service strategy is charged. Examples include volume-based, time-based and account-based. This criterion also considers the available booking concept, such as pay-per-use, subscription fee and market-based prices.
    ii. *Migration — Costs* refers to the costs to consider when moving existing infrastructure to the specified cloud service model.
4. **Management criteria:** Management objectives consisted of two criteria: *Support Capabilities* and *Migration — Business Complexity*.
    i. *Support Capabilities* refer to the support offered and the mechanism (e.g., phone, online, etc.) it is under. This includes information such as multilingual support, worldwide offices and local contact options.

ii. *Migration — Business Complexity* defines the business complexity in migrating the business from its current solution to the cloud service strategy. This includes all management functionality including training time and the ease of moving employees over to the new platform.

5.4. *Strategy*

The Strategy of this model refers to the cloud service strategy a company should use. To limit the scope of this project, cloud service strategies were narrowed down to the three most common cloud service strategies—IaaS, SaaS and PaaS—that companies move their services to.

It is important to note this model and strategy do not encompass all the different cloud service providers within a specific cloud service strategy. A different model would need to be created to evaluate a specific provider within a cloud strategy option. Choosing a cloud service strategy will greatly narrow the scope of decisions a company must consider when moving their services to the cloud.

6. Results and Discussions

Our expert panel consisted of seven experts—five team members acting as executives from BMF LLC and two external cloud service consultants. All seven members completed the pairwise comparisons on all three levels.

We conducted two rounds of analysis on the results from the HDM tool. In the first round, we took into account the inputs from the seven experts on all three levels of the pairwise comparisons. Since we have explanations on our Mission, Objectives and Criteria in the online HDM tool, we assumed that all the experts knew the company's needs, as well as the technical aspects of all the different types of cloud services.

The results of our first-round analysis show that Security is the company's top concern, among all of its objectives, when migrating its information technology to cloud servers. Among its criteria, Protection, Scalability and Service Charge were ranked as the top three. These are the criteria the company should pay close attention to when making decisions. Also, IaaS was the preferred cloud service with a score of 0.35. SaaS came in second with a close score of 0.34. Even though IaaS was the winning choice of cloud service, there is no major differentiation between IaaS and SaaS because of their close scores. This makes the decision inconclusive.

Considering the suggestions and recommendations during our class presentation, we conducted a second round of analysis on the model and pairwise comparison data. In this round, we separated the inputs of the team members and the external experts: inputs from the five team members were used for Level 1 and Level 2 comparisons, while the inputs from the two external experts were used for Level 3 comparisons.

The reason for the second round of analysis is that the executives understand the company's Mission, Objectives and Criteria under each objective. They are not technical experts and may not be able to make sound decisions on the third level of alternatives. As for the external cloud experts, although they are versed in the technical details of the three types of cloud service alternatives, they are not familiar with the inner workings of the company since they are not members. The results of the second round of analysis show that Security remains the top concern of the company when migrating to the cloud. Scalability, Protection and Service Charge again ranked as the top three among all the criteria, albeit with minor differences in actual weights. IaaS came out as the winner again with a score of 0.38, while SaaS came in as the second choice with a score of 0.32. The difference between the IaaS and SaaS strategies is more significant as compared to the first round. IaaS is the clear winner. We believe that the result from the second round of comparison is more convincing.

Detailed HDM model comparison results and our analysis are discussed in the following sections of this report.

6.1. First round of analysis

All seven experts were asked to make the pairwise comparison of all objectives, criteria and alternatives. Their pairwise comparison results were counted towards the final decision.

The Level 1 comparison results show that Security is the company's top concern among all objectives when migrating to the cloud (see Table 1). This is due to the nature of BMF's business sector. As the company is an online retail company, it has lots of confidential customer identification and finance information, as well as online transaction information. These information needs to be kept at the highest level of privacy. Any leakage could be fatal to the company's reputation and might be subject to fines if any online financial transaction related to the federal compliance code is violated.

Migration is ranked as the least significant objective. This shows that the company is confident in its management ability, especially within its IT department. Some of the IT management team members have previous experience in cloud migration and are familiar with the process. The company is least concerned over managing the migration.

The Level 2 comparison results show that Protection, Scalability and Service Charge are among the top three criteria the company would consider when choosing the right cloud

Table 1. Level 1 comparison results.

	EXP 1	EXP 2	EXP 3	EXP 4	EXP 5	EXP 6	EXP 7	AVG	Standard Deviation
Technical	0.22	0.49	0.18	0.14	0.27	0.16	0.30	0.25	0.12
Security	0.42	0.28	0.33	0.23	0.38	0.23	0.22	0.30	0.08
Economic	0.23	0.12	0.27	0.43	0.16	0.27	0.25	0.25	0.10
Management	0.13	0.11	0.22	0.20	0.18	0.34	0.23	0.20	0.08
Inconsistency	0.09	0.06	0.01	0.01	0.01	0.02	0.01		
							Total	1.00	

service type. Business migration and technical migration ranked the lowest among all the criteria (see Table 2).

Since security is the company's top concern when migrating to the cloud, it is no surprise that protection is the most important criterion to consider. After migrating to the cloud, the company needs to work closely with the cloud service provider to provide the satisfactory level of data protection, communication protection and operation protection.

Scalability is one of the major reasons the company will want to migrate to the cloud. The capability to meet the company's growth is essential. The company projected a year over year growth of 20% in the coming years. The cloud service choice needs to be able to handle this growth without significant successive migration efforts or additional charges.

Service Charge as a repeated cost is important to the financial health of a company. Whether this charge will increase significantly over time with the growth of the company needs to be carefully considered and counted into the total cost of production. Migrating to cloud computing is expected to be a cost-effective way of doing business. A good calculation and

Table 2. Level 2 comparison results.

		EXP 1	EXP 2	EXP 3	EXP 4	EXP 5	EXP 6	EXP 7	AVG	Weights
Technical	Scalability	0.75	0.70	0.40	0.70	0.55	0.41	0.63	0.59	0.15
	Technical migration	0.25	0.30	0.60	0.30	0.45	0.59	0.37	0.41	0.10
Security	Protection	0.70	0.75	0.40	0.40	0.75	0.27	0.61	0.55	0.17
	Migration compliance	0.30	0.25	0.60	0.60	0.25	0.73	0.39	0.45	0.13
Economic	Service charge	0.60	0.75	0.60	0.60	0.75	0.39	0.66	0.62	0.15
	Migration cost	0.40	0.25	0.40	0.40	0.25	0.61	0.34	0.38	0.09
Management	Support	0.60	0.75	0.60	0.50	0.60	0.36	0.56	0.57	0.11
	Business migration	0.40	0.25	0.40	0.50	0.40	0.64	0.44	0.43	0.09
									Total	1.00

estimate of cloud service charges will help improve the profit margin of the company.

The company has an experienced IT department. BMF's IT manager has previous experience in cloud migration; this helps to make the migration process easier for both business and technical areas. Also, the IT staff's knowledge in cloud server management adds confidence to the top management of this company over the technical aspect of this migration.

The Level 3 comparison results show that IaaS, with a score of 0.35, is the first choice for the cloud service that will suit the company's needs. SaaS, with a score of 0.34, is a close second choice (see Table 3). The difference between IaaS and SaaS is very small. Two out of seven experts scored IaaS as the first choice, while three out of seven experts scored SaaS as the first choice. IaaS get the highest average score partially due to one of the experts giving IaaS a very high score of 0.5. It could be

Table 3. Level 3 comparison results.

To Determine the Right Cloud Service Strategy	IaaS	SaaS	PaaS	Inconsistency
EXP 1	0.50	0.21	0.29	0.02
EXP 2	0.27	0.31	0.42	0.01
EXP 3	0.33	0.35	0.32	0
EXP 4	0.31	0.4	0.29	0.01
EXP 5	0.41	0.32	0.27	0
EXP 6	0.33	0.48	0.19	0.01
EXP 7	0.32	0.31	0.37	0.04
Mean	0.35	0.34	0.31	
Minimum	0.27	0.21	0.19	
Maximum	0.50	0.48	0.42	
Std. Deviation	0.07	0.08	0.07	
Disagreement				0.064

viewed as an outlier in this set of data; this makes the final decision in IaaS less convincing. By simply looking at the scores, it seems that both IaaS and SaaS could be the final decision from this model.

Based on our research and literature review, we are confident of the HDM model we set up for this problem. We believe that our HDM model covers all the criteria that need to be considered when making this decision. After evaluating the comments from the class presentation and discussion within our group, we found out that our evaluation method in this round of analysis was flawed. It had been an inaccurate assumption to include all the seven experts' pairwise comparison inputs in each of the three levels of comparison.

The company executives in the expert panel are not technical experts. They have very limited knowledge on how cloud computation and cloud migration works. Their pairwise comparisons in Level 3 on each alternative were not reliable, though their inputs in Level 1 and Level 2 comparisons were useful since they are most familiar with the company's current condition and future needs. In contrast, the two external experts' inputs in Level 3 comparisons were valuable because they are familiar with the technical issues and barriers a company could face during migration. However, since they are not familiar with the company's internal operation, their inputs on Level 1 and Level 2 comparisons were mostly based on their general knowledge of companies of similar scale, and were therefore less reliable.

To resolve this issue in the first round of analysis, we conducted a second round of analysis on the HDM model.

6.2. *Second round of analysis*

To test the HDM model and fix our problem in the expert panel use, we conducted a second round of analysis. During the

Table 4. Level 1 comparison results.

	EXP 1	EXP 2	EXP 4	EXP 6	EXP 7	AVG	Standard Deviation
Technical	0.22	0.49	0.14	0.16	0.30	0.26	0.14
Security	0.42	0.28	0.23	0.23	0.22	0.28	0.08
Economic	0.23	0.12	0.43	0.27	0.25	0.26	0.11
Management	0.13	0.11	0.20	0.34	0.23	0.20	0.09
Inconsistency	0.09	0.06	0.01	0.02	0.01		
					Total	1.00	

second round, we divided the seven experts into two panels. Expert Panel 1 consisted of the five team members acting as the executive team, who would make pairwise comparisons in Level 1 and Level 2 of this model, while Expert Panel 2 consisted of the two external cloud migration consultants, who would make pairwise comparisons in Level 3 only.

The results of the Level 1 comparison show that Security is still the company's top concern in migration to the cloud, with Management being of the least concern. These results correspond to those from the first round analysis. Thus, Security is indeed critical to BMF, as it is an online retail company. Security should be given the highest level of consideration when making the decision on cloud migration (see Table 4).

The Level 2 comparison results show that Scalability, Protection and Service Charge are still ranked as the top three decision criteria for this migration. These results are also consistent with those from the first round of analysis, with minor differences in weights only. Technical Migration and Business Migration are still ranked the lowest among all the criteria (see Table 5).

We could interpret the similarities in the Level 1 and Level 2 comparison results in the two rounds of analysis as that of the internal and external experts having a similar understanding of

Table 5. Level 2 comparison results.

		EXP 1	EXP 2	EXP 4	EXP 6	EXP 7	AVG	Weights	Standard Deviation
Technical	Scalability	0.75	0.70	0.70	0.41	0.63	0.64	0.17	0.13
	Technical migration	0.25	0.30	0.30	0.59	0.37	0.36	0.09	0.13
Security	Protection	0.70	0.75	0.40	0.27	0.61	0.55	0.15	0.20
	Migration compliance	0.30	0.25	0.60	0.73	0.39	0.45	0.13	0.20
Economic	Service charge	0.60	0.75	0.60	0.39	0.66	0.60	0.16	0.13
	Migration cost	0.40	0.25	0.40	0.61	0.34	0.40	0.10	0.13
Management	Support	0.60	0.75	0.50	0.36	0.56	0.55	0.11	0.14
	Business migration	0.40	0.25	0.50	0.64	0.44	0.45	0.09	0.14
						Total		1.00	

the objectives and criteria associated with cloud migration. BMF, as an online retail company, has similar concerns and issues as with any other companies from the same line of business.

Level 3 comparison results show that IaaS, with a score of 0.38, is the first choice for the cloud service that will suit the company. SaaS, with a score of 0.32, is the second choice. The differences between IaaS and SaaS are significant. IaaS is the clear winner in this round of analysis (see Table 6).

Results from both the experts are very similar to each other, where there was minimal disagreement. This makes the results of the second round analysis more convincing than those in the first round. IaaS is the definite choice of the cloud service that this company should take.

The similarities between the results from both rounds of analysis prove that our HDM model is well set up and stable. The use of expert panels in the second round of analysis is more appropriate.

Table 6. Level 3 comparison results.

EXP 3

Level-3	Scalability	Technical Migration	Protection	Migration Compliance	Service Charge	Migration Costs	Support	Business Migration	Results
IaaS	0.48	0.27	0.54	0.16	0.16	0.53	0.25	0.51	0.36
SaaS	0.21	0.16	0.16	0.6	0.57	0.15	0.43	0.19	0.32
PaaS	0.31	0.57	0.3	0.25	0.27	0.32	0.33	0.31	0.32
Inconsistency	0.00	0.00	0.00	0.00	0.00	0.00	0.00	0.00	

EXP 5

Level-3	Scalability	Technical Migration	Protection	Migration Compliance	Service Charge	Migration Costs	Support	Business Migration	Results
IaaS	0.45	0.54	0.4	0.29	0.27	0.4	0.45	0.39	0.39
SaaS	0.3	0.26	0.33	0.36	0.4	0.27	0.3	0.32	0.32
PaaS	0.25	0.2	0.27	0.35	0.33	0.33	0.25	0.29	0.29
Inconsistency	0.00	0.00	0.00	0.00	0.00	0.00	0.00	0.00	

Final Result

Level-3	Scalability	Technical Migration	Protection	Migration Compliance	Service Charge	Migration Costs	Support	Business Migration	Results
IaaS	0.465	0.405	0.47	0.225	0.215	0.465	0.35	0.45	0.38
SaaS	0.255	0.21	0.245	0.48	0.485	0.21	0.365	0.255	0.32
PaaS	0.28	0.385	0.285	0.3	0.3	0.325	0.29	0.3	0.30

7. Further Analysis of the Results

7.1. *First round of analysis*

7.1.1. *The top three objectives*

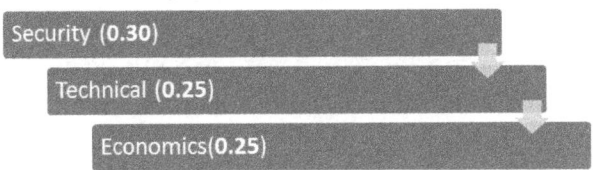

The top three objectives were Security (0.30), Technical (0.25) and Economics (0.25). Management received the lowest score of 0.20. Our first round of analysis determined that Security had the most influence in our Level 1 comparisons, while Management of the new software structure was assumed to be the least demanding of all four levels of our comparisons. The tie between Technical and Economics created questions about the first round of analysis. Since Technical and Economics were part of the company's top four objectives, it was important to understand why they were considered equally important; the Technical and Economic objectives are completely different aspects of the business. Understanding the similarities and differences leads us to the eight criteria that are a level below the objectives.

7.1.2. *The top three criteria*

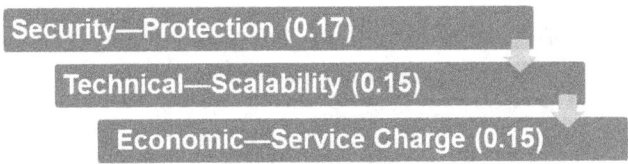

Protection is a criterion of Security and has the highest score of 0.17. Scalability is a criterion of the Technical objective and has the score of 0.15. Service Charge is a criterion of the Economic objective and also has a score of 0.15. Migration Cost was also a criterion of the Economic objective and has a score of 0.09. The tie between Scalability and Service Charge also led us to believe something was skewing our results. We began to ask ourselves whether we were having the right people to do the correct pairwise comparisons. At this point, we considered changing the individuals who were to perform pairwise comparison for each level of the model. The strategy ranking will shed more light to our decision.

7.1.3. *The strategy rankings*

IaaS scored 0.35, SaaS scored 0.34 and PaaS scored 0.31. Since we had a tie in the objectives between Technical and Economic, and a tie in the criterion between Scalability and Service Charge, it resulted in no true winner of the strategy. We knew then that we had to make some adjustments to our analysis of the model. This situation solidified our decision to create a second round of analysis.

7.2. *Second round of analysis*

7.2.1. *The top three objectives*

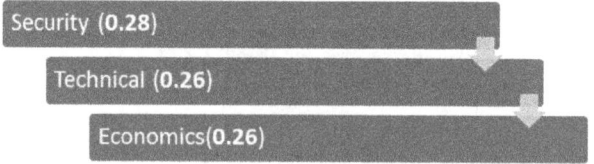

Removing the cloud expert opinions from the comparisons that should have been made by the company did not affect the top three objectives rankings. Since Security decreased by two points, it showed that the cloud experts valued Security more than BMF's upper management. This was good information to know, in that BMF should research more into the importance of security to ensure that all staff members buy into the company's objectives, which is very important.

7.2.2. The top three criteria

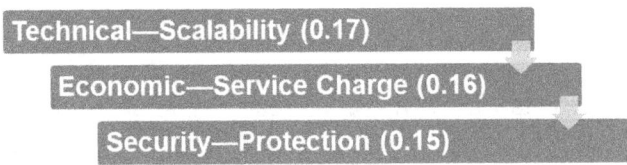

Technical—Scalability (0.17)
Economic—Service Charge (0.16)
Security—Protection (0.15)

Removing the cloud expert opinions from the comparisons reordered the criterion ranking and showed that Scalability was the most important objective to BMF, as it wanted to be able to grow quickly, with the lowest variable cost. Although Security is still important to BMF because it was one of its top three objectives, it is worth asking why Security should be the first objective when Protection is the third criterion? Migration compliance would be the answer to that question since it scored 0.13, which ranked this criterion in fourth place, with part of it beneath Security.

7.2.3. The strategy rankings

IaaS (0.36)
SaaS (0.32)
PaaS (0.32)

The strategy rankings did not change order, though SaaS lost two points and IaaS gained one, thus making IaaS a more definitive leader. With the second round of analysis and the new scoring of the model, IaaS is now the best choice because it scored significantly higher than SaaS and PaaS on Scalability and Protection. IaaS did not score the highest on Service Charge, meaning that it was not the cheapest. However, results show that it has the highest security rating. Therefore, BMF must spend more money on Service Charge to increase their customer's data security.

8. Conclusion

BMF LLC is a small online fashion retailer that is in the process of scaling up its operations. To scale efficiently and maintain security, BMF must make the choice of which cloud-based strategy will work best; this decision is very complex. An HDM model was constructed and used as an aid to make this complex decision. It was built with four levels—Mission, Criteria, Subcriteria and Strategy. At first, through experts' evaluation the HDM model evaluated all four levels. When the first model produced inconclusive results, the team re-evaluated and delivered a new HDM model delivery strategy. With this new delivery strategy, professionals from BMF evaluated the first three levels of the HDM model (Mission, Criteria and Subcriteria), while the cloud professionals evaluated the last level (Strategy). The new HDM model delivery strategy rectified the inconclusive results of the first model delivery, and identified IaaS as clearly the best choice for BMF.

9. Limitations and Future Work

The cloud service provider model can be used for any company planning on migrating their software services to the cloud. The senior management in such a company should perform pairwise comparisons for both the objectives and criteria and leave the experts to choose the best strategy. Experts' opinions are also relevant for any other uses of the model. Here, the weights for the objective and criteria are relevant only to BMF because they are unique to its needs. The strategy weights are global and can be used for any other applications.

Future work should be conducted to find out which IaaS platform should be purchased. If an HDM model was created for this purpose, the pairwise comparisons should be conducted by the senior management of BMF or subcontracted to a consulting firm. Vendors selling the IaaS service should not be doing the pairwise comparisons.

References

1. D. Kocaoglu, "Hierarchical Decision Modeling", Engineering and Technology Management Department, Portland State University, Portland, 1987.
2. A. K. Sadhu, "Delphi technique", *Managementversity*, 29 November 2014. http://managementversity.com/delphi-technique/. Accessed: 10 March 2017.
3. P. Mell and T. Grance, "The NIST definition of cloud computing", Technical report, National Institute of Standards and Technology, US Department of Commerce, September 2011. http://nvlpubs.nist.gov/nistpubs/Legacy/SP/nistspecialpublication800-145.pdf. Accessed: 16 March 2017.

Chapter 3

Technical Transformation: Cloud Computing

Amit Pingle* and Tugrul Daim*,†,‡

*Portland State University, Portland, Oregon, USA
†Higher School of Economics, Moscow, Russia
‡Chaoyang University of Technology, Taiwan

Abstract

This project used a Hierarchical Decision Model (HDM) approach by dividing the model hierarchies into Mission, Objectives, Goals, Strategies and Actions, also called MOGSA. The fundamental criteria used for assessment are: Innovation Factor, Technological Factor, Usability Factor and Economic Factor. These four primary evaluation criteria were further divided into seven subcriteria: Complexity, Compatibility, Security, Architecture, Usefulness, Ease of Use and Cost. These criteria were evaluated using a pairwise comparison method. Four cloud computing platforms were considered for the project: Amazon Web Services, Google Cloud Platform, IBM Bluemix and Microsoft Azure. A group of experts was used to measure and compare results of the HDM. This group includes application developers working in different domains and had been using cloud computing platforms. The range of inconsistency recorded was between 0.02 and 0.04, whereas the disagreement between the judgments was 0.055. Despite individual responses by some of the evaluators, Amazon Web Services was the preferred cloud computing platform, thus making the HDM a better methodology to quantify and counterbalance all individual preferences while making complex decisions.

Keywords: Technology assessment, cloud computing, web services.

1. Introduction

Cloud service is an integral part of today's business. With rapidly increasing amounts of data, Internet of Things (IoT) and applications, their presence in our lives today demand high storage and computing power. Cloud computing makes it easier for businesses by providing them high computing power as an alternative to investing in costly infrastructure. Using cloud computing, people and enterprises can operate any application on a plug and play basis without really investing on hardware. Organizations not only get save a lot of money, maintenance also becomes easier since the platform provider takes care of its speed and technical abilities.

Hence, it becomes very important for any business to carefully choose the appropriate cloud service provider that can provide the desired speed and computing power required for the business.

An application developer wanted to choose the best cloud computing platform for one of the application he had developed. He was undecided which cloud service provider he should choose. Many possible decisions he had to make were discussed, such as choosing a cloud computing platform and the type of hardware that would be compatible with it. After much discussion without any results, he decided to use the HDM because it would help solve the problem in a better way. While going through the process it was observed that many application developers face the same problem in choosing a cloud service provider. Instead of using the HDM for just one developer, he decided to use it to help other developers to choose between Amazon Web Services, Microsoft Azure, Google Cloud Platform and IBM Bluemix.

To make the HDM and research more vigorous, the panel of experts would be expanded to 13 application developers from different domains, where each would give their valued assessments.

2. Methodology

The methodology that has been used in this project was the Hierarchical Decision Model (HDM), which was developed by David Cleland and Dundar Kocaoglu. For any HDM, the basic structure of the hierarchy is presented in the MOGSA form [1]. This model consists of five levels—Mission, Objectives, Goals, Strategies and Actions. Each of these levels has a specific function for the model [2]. Nevertheless, it is not essential to have all five levels in a model, though it needs to have at least three levels, which are Mission, Objectives (criteria) and Actions (decisions).

3. Hierarchical Decision Model

The HDM—a multilayered method for studying complex decisions—was developed in 1979 using a similar concept as the Analytical Hierarchy Process (AHP) methodology, but with a different pairwise comparison scale and judgmental quantification technique [3]. Depending on how simple or complex the decision-making problem is, the number of hierarchical levels is determined.

HDM is a methodology that breaks down a problem into different hierarchies or sublevels. The approach an HDM takes is that it considers any problem as an association of subproblems, which can be broken down into hierarchies or levels. The most common approach in a HDM consists of three important decision hierarchies: Impact or Mission level, Target or Objective level and Operational or Action level [4]. Each level comprises of multidimensional components [4].

The top level which is the objective, leads to benefits. The bottom level, which is the alternative, results from multiple actions. Each decision element at every level has an impact on different elements at the next higher level. A hierarchy can be determined as a completed hierarchy if each element of the given hierarchy is evaluated with respect to each element in the next hierarchy [2]. Any complex decision problem can be expressed as an analytical hierarchical decision.

3.1. *Pairwise comparison*

Decision elements at every level are compared with each other. The expert panel assigns weights to each element, which contributes to the decision element in the next level. A total of 100 points is allocated between two decision elements. The formula for the pairwise comparison is given by [3]

$$N = (n - 1)/2,$$

where N is the number of pairwise comparisons, and n is the decision elements at every level.

3.2. *Inconsistency*

Inconsistency occurs when there is an intentional or unintentional error while performing a pairwise comparison by an expert. There are two types of inconsistency: ordinal and cardinal [6]. In ordinal inconsistency, the ranking order of elements should be upheld. For example, if someone likes apples more than oranges, and oranges more than grapes, then that person should like apples more than grapes. However, if that person prefers grapes over apples, then that is accounted as ordinal inconsistency. Cardinal inconsistency occurs when the element's proportion is not upheld. For example, if someone regards apples as being two times more valuable than oranges and oranges as being three times more valuable than grapes, then that person should regard apples as being six times more valuable than grapes, or else cardinal inconsistency could be observed. It is observed that an inconsistency of 10% (or 0.10) is considered as an acceptable inconsistency [3].

3.3. *Disagreement*

Unlike inconsistency, disagreement is calculated based on the differences between the opinions or evaluations of the expert panel. If the disagreement among the expert panel is beyond a certain range (which is considered to be 10%), then a brief session should be conducted with the experts to try to convince that particular expert whose judgment was different from the rest. There are additional tests such as an *F*-test that can be performed to determine whether the disagreement among experts in a panel is statistically significant or not [6]. "Understanding and resolving the disagreement is an important aspect of the research and for building the decision model" [6].

3.4. *Decision model*

The objective of the model in this paper is to choose the best cloud computing service provider for an application developer.

Figure 1 illustrates the process of decision-making for this paper using a HDM.

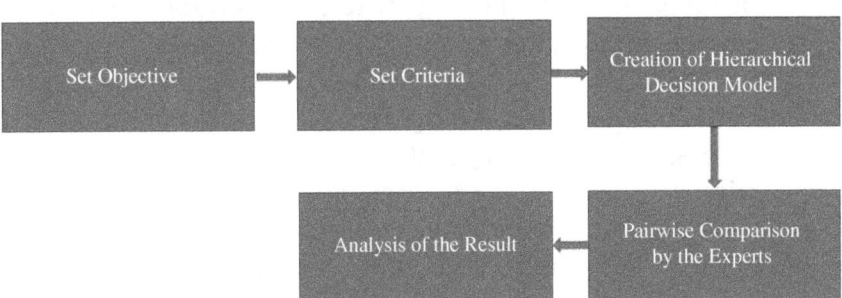

Figure 1. A HDM's process of decision-making.

3.5. Expert panel

An expert panel consists of application developers (across different industries) who use different cloud computing platforms. All the experts are significantly related to the model's objective and decision elements. The experts are from different backgrounds, age and sex.

3.6. Decision elements and model levels

Each of the 13 experts accessed the HDM software and gave their professional judgments and quantification with regard to each of the following criteria:

1. **Innovation:** Is it easy to scale up the cloud if the need increases or is it compatible with the other applications?
2. **Technological:** What is the architecture of the platform or how secure is the cloud?
3. **Usability:** What is the platform's ease of use or is the chosen platform useful for a certain application?
4. **Economical:** What is the total monetary expense of the proposed cloud computing platform?
    - **Level 1:** Mission (selecting a cloud computing platform for an application).
    - **Level 2:** Four Primary Evaluation Criteria using a pairwise comparison method. A total of a hundred points was divided between the two criteria in proportion to their relative importance to the problem objective.
    - **Level 3:** Seven Secondary Evaluation Criteria with respect to the Level 1 criteria using a pairwise comparison method. A total of a hundred points was divided between the two criteria in proportion to their relative importance to the problem objective.
    - **Level 4:** Four cloud computing service providers using a pairwise comparison method. A total of a hundred points was divided between the two criteria in proportion to their relative importance to the problem objective.

3.7. Cloud computing complexity

A cloud system is used to sustain effective teleworking and tools establishment supporting virtual teams of employees situated around the globe. Huczynski and Buchanan [7] define a virtual team as a team "that relies on technology mediated communication, while crossing boundaries of geography, time, culture and organization to accomplish an interdependent task" [7]. It can be said that every team functions in its own way. Additionally, physical separation of the team members across the boundaries will be different for each team, not only making it more unique than other teams but also creating inconsistencies within as teams grow and evolve over time. Shin [8] argues that there is a range or scope of "virtualness"— the larger the dispersion in a team, the more amplified is this virtualness [8]. Hence, it is very important that there should be an effective way of interaction, communication and level of engagement between team members, in order to implement successful virtual teams. This is the most vital yet challenging part of team management for such teams. According to Duart and Snyder [9], critical success factors in managing virtual teams and facilitating effective interaction between members consist of the right selection of IT tools, team members' competencies, the leadership and culture of the team, the process standardization and training of team members towards these goals [9].

3.8. Cloud computing compatibility

The existing competition between the big service providers of cloud computing has made the service incompatible. Current provided solutions by these vendors are not really compatible with each other [10], as they tend to lock in existing customers into their provided services or infrastructure, and prevent data or software portability [11]. In addition, dominant vendors are not willing to accept the common standards, which ultimately results in incompatible platforms [12], which again increases the lock-in effect. This greatly prevents many small and medium businesses from entering the cloud market. The European Network and Information Security Agency (ENISA) and European Commission states that the vendor lock-in problem is a high risk that cloud infrastructures are facing [13]. It can be said that cloud compatibility is the solution for this problem, which will not only improve this situation but also benefit both consumers and providers.

Cloud compatibility can provide customers with freedom to select the appropriate service for their businesses. It allows them to compare existing offerings and evaluate their functionalities and features. It will also allow businesses to easily switch between cloud providers without unnecessary settings and formalities, as well as risking applications on the existing cloud. Moreover, it will provide opportunities for SMEs by opening an exchangeable cloud market, though this may lead to the problem of incompatible solutions due to the varying standards and frameworks that cloud services might be operating in. To avoid this issue, researchers and scientists need to come together and propose a set of principles that all providers and solutions can follow [14].

3.9. *Cloud computing architecture*

Cloud computing is basically separated into two parts: front end and back end [15]. The internet connects both these parts together to form a cloud computing system. The front end deals with the customer's interaction with the system and it consists of the client's computer and application needed to operate the cloud, whereas the back end is nothing but the cloud getting accessed in the system, which comprises of computers, servers and storage devices. The central server of the system is used to monitor traffic, manage the system and direct client demands. It is bound to certain rules and protocols and makes use of "Middleware", a software that facilitates communication between various networked computers [15].

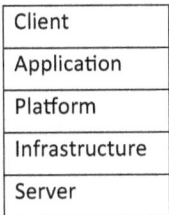

Figure 2. Tiers of a cloud computing design.

Figure 2 shows the different tiers of cloud computing design.

A client using a cloud system possesses computer hardware and/or software to access the specific application on the cloud, the combination of which can be designed to deliver the application instantly using cloud services [16].

A cloud application provides a "Software as a Service" (SaaS) service through the internet without actually having to install and run the application on a client's system [16]. Some of the important characteristics of a SaaS are: it has network-based access and the software is managed over the internet, enabling operations from a central location via the internet [16].

Another important part of cloud computing is the "Platform as a Service" (PaaS). It provides a computing platform that consists of all the applications a client wants to deploy with any given cloud infrastructure. Using the PaaS, developers can control and manage all the systems and settings required to design, test and deploy any web application [17].

Lastly, the required infrastructure is provided using the "Infrastructure as a Service" (IaaS). This service is scalable and the client only has to pay for the amount he uses. Thus, cloud computing makes its customer achieve faster delivery in less cost [17].

3.10. *Cloud computing security*

It is important for any business choosing a cloud service provider to trust it and its system with their data. The decision of moving to a cloud system comes with its own risks: data security, VM security and other compliance issues. Hence, these three elements drive the Confidentiality, Integrity and Availability (CIA) of any cloud system. Hence, the CIA is widely treated as the convention to determine the reliability of the system under these three factors.

3.10.1. *Confidentiality*

Confidentiality represents the ability of a system to protect important assets and information of a business from getting disclosed to unsanctioned users. System confidentiality should be able to minimize the threat of losing a client's data to hackers or unauthorized users. In some cases, service providers may even have fraudulent members who gain access to or want to use/corrupt a client's data. A virtual machine network also possesses unavoidable confidentiality requirements to protect its data [18].

3.10.2. *Integrity*

The integrity of a system simply means that a client's assets or information has not been altered by any unsanctioned user. It maintains the security and accuracy of the asset. To ensure such security, a system needs to be protected against web-based attacks, which are common in the cloud environment. Such attacks can easily access, edit or delete user data, virtual machine metadata and WSDL files [19].

3.10.3. *Availability*

Availability in the cloud service system represents the quick and rapid delivery of applications to the user. Businesses and enterprises need to maintain quick delivery of applications to their users, as downtimes (no matter how short) can cause an irrecoverable loss to an enterprise.

A general service-level agreement between a provider and a business will specify the rate or downtimes of the applications or service. It can be 99.999% or 80%. More resources can be provided to maintain this agreement [20].

3.10.4. *Cloud computing usability*

In order for reliable usage of the cloud, the cloud computing environment needs to consist of all the foundations and supporting devices of IT. These include networks, which can enable the smooth processing of applications and services. The Cloud Computing Environment (CCE) delivers a single point of access to the user, which can be used to access computing resources, infrastructure, services and applications. It also enables the business to respond rapidly to changes in the market or technologies [21].

The public cloud computing system includes a restricted class of servers, storage devices and networks. Amazon's EC2 offers small, large and extra-large classes of servers, although the extra-large server seems to be of no match to any of the enterprise class servers in data centers.

To achieve powerful computational resource, one has to upgrade the system using powerful servers and CPUs to increase speed. However, this can require additional servers, which require more storage and network bandwidth. To exploit this, the applications may need to partition workloads to run on multiple instances sharing the same computer architecture. But many class applications are not suitable for such an environment and still depend on traditional methods of creating spare space to meet additional workloads. This contradicts the gains from clouds and affects its flexibility [22].

From the user's point, cloud computing usability can be complicated as the usage of computing resources might be dependent on multiple businesses, as they own these resources which include public and private clouds and virtual resources with no enterprise control [23].

3.10.5. *Cloud computing cost factor*

There are several aspects of cloud projects, such as data privacy and protection, business stability of the provider, pricing structure, legal background, interoperability of the system and solutions, etc., out of which the cost-benefit analysis of the system is the most fundamental one that determines whether a business chooses a commercial cloud or builds an on-premise system. However, research shows that the ultimate decision results from an impact of several other variables, such as the data transfer ratio, storage capacity and requirements, workload factors and licensing policies. Hence, the cost-benefit analysis of any IT project/system is complex and depends on future business processes. Financial calculations always consider the fact that the value of invested money changes over time and that every payment or income is worth more if received immediately rather then after some time in the future. This is presented in the concept of the Net Present Value, which is given as

$$PV = FV (1 + k) T.$$

In this expression FV refers to the future value of capital, PV is the present value of capital, T stands for the considered years of investment and k is the price of capital (or the annual interest rate). If certain periodic revenue is expected over several years, T ($T = 1, 2, Y$), then we can calculate NPV according to

$$NPV = \sum_{\gamma=0}^{\gamma-1} \frac{C_\gamma}{(1+k)\gamma}.$$

Unlike classical financial models when analyzing IT related investments we have to consider the impact of Moore's Law [21], which was elaborated upon in detail [16] and is briefly presented in the rest of the chapter. Walker's model recognizes the fact that a CPU bought today will be two times worse (in terms of performance) than one bought for the same amount of money in two years.

3.11. Analysis and key findings

The model of the project has three levels. The first and second levels show the primary and secondary evaluation criteria, respectively, while the last level shows the alternatives for the project. Therefore, it has been decided to quantify the model by sending it to 13 experts and getting their responses that can be helpful to choose one of the choices. The responses that came from the 13 experts showed that the Technological factor was the most important primary criteria to be considered—the mean for the Technological factor calculated by all the experts was 0.369, which can be seen in Appendix B. For the secondary selection criteria, Usefulness was the most important criteria that should be considered and the mean for this factor calculated by all the experts was 0.604. The responses from the expert panel also show that the Economical factor was the second-most important criteria that should be considered—the mean for the Economical factor was 0.268. For the secondary selection criteria the responses from the expert panel showed that Security should be considered as the second-most important criteria, where the calculated mean was 0.581. The third-most important criteria was the Innovation factor with a mean of 0.191. After the final results had been collected from the 13 experts, the final calculation results using the given criteria showed that Amazon Web Services should be chosen as the cloud computing platform to develop the application, as it received the highest mean value of 0.36 (see Appendix A). However, Amazon Web Services also has a biggest standard deviation of 0.07, as compared to other alternatives. It is noteworthy that the data obtained from our panel of experts did not show a high disagreement value, which at 0.055 gives us a good indication that the experts' opinions about the decision were very close. The advantage suggested by this low disagreement value is that there is little value in investing any further efforts to further lower the disagreement value. From the results, the second-best option for a cloud computing platform was Microsoft Azure.

3.12. *Future research and limitations*

This model tries to cover the important factors an application developer should consider when choosing a cloud computing platform. There are other cloud service providers to choose from, but we shall limit this model to four alternatives only. This model does not differentiate between mobile cloud computing or the different applications used in different industries.

In future research, more primary and secondary selection criteria such as SaaS, PaaS, IaaS, Backup & Recovery, Upgrade, Training & Support, etc., should be considered. A pairwise comparison for Small Scale, Medium Scale and Large Scale businesses would be worth doing. It would be interesting to see the results from the same experts for different primary and secondary criteria selection models.

3.13. *Conclusion and recommendation*

In conclusion, the proposition of choosing Amazon Web Services as the best cloud computing platform to develop applications was proven correct by using pairwise comparisons and the HDM. It is interesting to note that while not all individual values rated Amazon Web Services as the highest, it was the clear front-runner once all the calculations were done. The HDM is a useful tool to help decision-making or as a classification among alternatives with many different criteria to consider. In this model, multiple criteria were used in a decision regarding the best choice for a cloud platform. Even though each of the experts already had a mode of platform in mind, their answers were different once all the comparisons were made. The HDM had taken the biases out of their decisions and left them with an alternative that fit what was felt to be the most important. The use of a HDM should be made a means of decision-making to any group of individuals with a multifaceted decision to make.

References

1. H. A. Alanazi, T. U. Daim and D. F. Kocaoglu, "Identify the best alternatives to help the diffusion of teleconsultation by using the Hierarchical Decision Model (HDM)", in *Portland International Conference on Management of Engineering and Technology (PICMET)*, 2015.
2. M. Adnan. "Title of dissertation", University of Minnesota, December 2011.
3. M. M. Lingga, "Developing a Hierarchical Decision Model to evaluate nuclear power plant alternative siting technologies", PhD dissertation, Portland State University, 2016.
4. D. I. Cleland and D. F. Kocaoglu, "Hierarchical Decision Model", *Engineering Management* (New York: McGraw-Hill, 1981), pp. 449–463.
5. N. J. Sheikh, K. Kim and D. F. Kocaoglu, "Use of hierarchical decision modeling to select target markets for a new personal healthcare device", *Health Policy and Technology*, **5**, 2 (2016) 99–112.
6. M. Abbas, "Analysis of decision inconsistencies in judgment quantification", in *Proceedings of Technology Management in the IT-Driven Services (PICMET)*, 2013.
7. A. A. Huczynski and D. A. Buchanan, *Organisational Behaviour*, 6th edition (England: Prentice Hall, 2007).
8. Y. Shin, "Conflict resolution in virtual teams", *Organisational Dynamics*, **34**, 4 (2005) 331–345.
9. D. L. Duart and N. T. Snyder, *Mastering Virtual Teams: Strategies, Tools and Techniques That Succeed*, 2nd edition (San Francisco: Jossey-Bass, 2001).
10. A. Sheth and A. Ranabahu, "Semantic modeling for cloud computing, part I & II", *IEEE Internet Computing Magazine*, **14** (2010) 81–83.
11. J. McKendrick, "Does platform as a service have interoperability issues?" *Zdnet.com*, 2010. http://www.zdnet.com/blog/service-oriented/doesplatform-as-a-service-have-interoperability-issues/4890.
12. The Economist, "Battle of the clouds", *The Economist*, 2009. https://www.economist.com/leaders/2009/10/15/battle-of-the-clouds.
13. D. Catteddu and G. Hogben, "Cloud computing—benefits, risks and recommendations for information security", European Network

and Information Security Agency, 2009. http://www.enisa.europa.eu/act/rm/files/deliverables/cloud-computingrisk-assessment/.
14. N. Loutas, E. Kamateri, F. Bosi and K. Tarabanis, "Cloud computing interoperability: The state of play", in *3rd IEEE International Conference on Cloud Computing Technology and Science*, 2011.
15. Y. Jadega and K. Modi "Cloud computing—concepts, architecture and challenges", in *International Conference on Computing, Electronics and Electrical Technologies (ICCEET)*, 2012.
16. P. Mathur and N. Nishchal, "Cloud computing: New challenge to the entire computer industry", in *1st International Conference on Parallel, Distributed and Grid Computing (PDGC)*, 2010.
17. B. P. Rimal and E. Choi, "A taxonomy and survey of cloud computing systems", in *5th International Joint Conference on INC, IMS and IDC*, 2009.
18. Q. Wang, C. Wang, J. Li, K. Ren and W. Lou, "Enabling public verifiability and data dynamics for storage security in cloud computing", in *Proceedings of the 14th European Conference on Research in Computer Security (ESORICS)*, 2009, pp. 355–370.
19. D. Chen and H. Zhao, "Data security and privacy protection issues in cloud computing", in *International Conference on Computer Science and Electronics Engineering*, Hangzhou, 2012, pp. 647–651.
20. J. Sen, "Security and privacy issues in cloud computing", *Architectures and Protocols for Secure Information Technology Infrastructures*, 2013, pp. 1–45.
21. P. Goyal, "Cloud computing environments and enterprise security", manuscript, 2010.
22. E. Overby, A. Bharadwaj and V. Sambamurthy, "Enterprise agility and the enabling role of information technology", *European Journal of Information Systems*, **15** (2006) 120–131.
23. P. Goyal, "Enterprise usability of cloud computing environment: Issues and challenges", in *Workshops on Enabling Technologies: Infrastructure for Collaborative Enterprises*, 2010.

Appendix A
Final Results

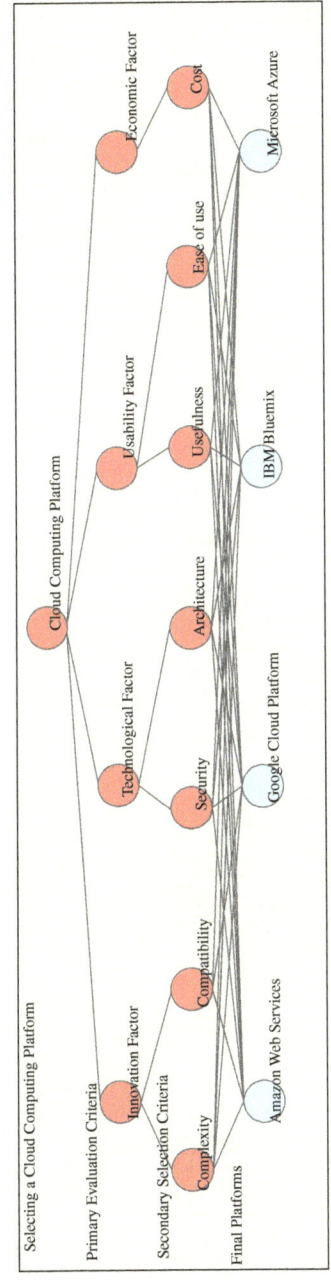

Cloud Computing Platform	Amazon Web Services	Google Cloud Platform	IBM Bluemix	Microsoft Azure	Inconsistency
Expert 2	0.25	0.26	0.25	0.24	0.04
Expert 10	0.43	0.27	0.12	0.18	0.02
Expert 11	0.45	0.19	0.12	0.24	0.02
Expert 12	0.37	0.21	0.17	0.25	0.02
Expert 13	0.36	0.26	0.18	0.19	0.02
Expert 1	0.3	0.19	0.19	0.32	0.02
Expert 3	0.37	0.24	0.16	0.23	0.02
Expert 4	0.24	0.24	0.27	0.24	0.04
Expert 5	0.29	0.16	0.16	0.39	0.03
Expert 6	0.4	0.12	0.17	0.3	0.04
Expert 7	0.4	0.25	0.16	0.19	0.02
Expert 8	0.43	0.12	0.21	0.24	0.03
Expert 9	0.45	0.13	0.28	0.15	0.03
Mean	0.36	0.2	0.19	0.24	
Minimum	0.24	0.12	0.12	0.15	
Maximum	0.45	0.27	0.28	0.39	
Std. Deviation	0.07	0.05	0.05	0.06	
Disagreement					0.055

Statistical *F*-test

The statistical F-test for evaluating the null hypothesis (Ho: ric = 0) is obtained by dividing between-subjects variability with residual variability:

Source of Variation	Sum of Square	Deg. of freedom	Mean Square	F-test value
Between Subjects:	0.25	3	.084	16.52
Between Conditions:	0.00	12	0.000	
Residual:	0.18	36	0.005	
Total:	0.43	51		
Critical F-value with degrees of freedom 3 & 36 at 0.01 level:				4.38
Critical F-value with degrees of freedom 3 & 36 at 0.025 level:				3.5
Critical F-value with degrees of freedom 3 & 36 at 0.05 level:				2.87
Critical F-value with degrees of freedom 3 & 36 at 0.1 level:				2.24

Appendix B

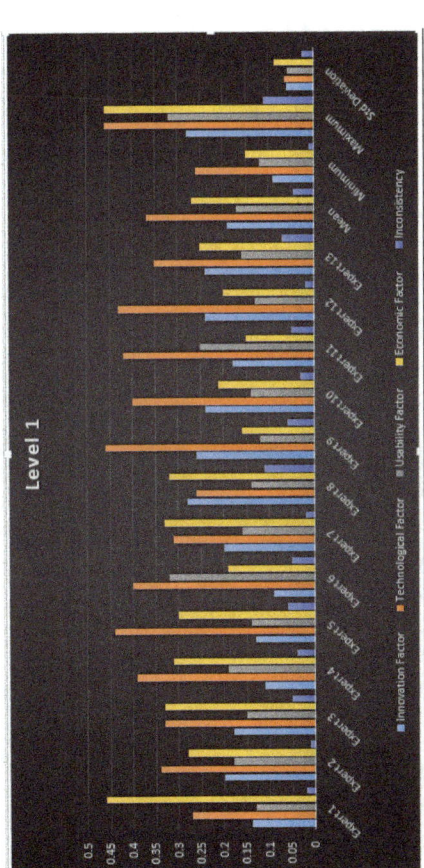

Level 1	Expert 1	Expert 2	Expert 3	Expert 4	Expert 5	Expert 6	Expert 7	Expert 8	Expert 9	Expert 10	Expert 11	Expert 12	Expert 13	Mean	Minimum	Maximum	Std Deviation
Innovation Factor	0.14	0.2	0.18	0.11	0.13	0.09	0.2	0.28	0.26	0.24	0.18	0.24	0.24	0.191538	0.09	0.28	0.060117406
Technological Factor	0.27	0.34	0.33	0.39	0.44	0.4	0.31	0.26	0.46	0.4	0.42	0.43	0.35	0.369231	0.26	0.46	0.06448017
Usability Factor	0.13	0.18	0.15	0.19	0.14	0.32	0.16	0.14	0.12	0.14	0.25	0.13	0.16	0.17	0.12	0.32	0.056568542
Economic Factor	0.46	0.28	0.33	0.31	0.3	0.19	0.33	0.32	0.16	0.21	0.15	0.2	0.25	0.268462	0.15	0.46	0.086683923
Inconsistency	0.02	0.01	0.05	0.04	0.06	0.05	0.02	0.11	0.06	0.03	0.05	0.02	0.07	0.045385	0.01	0.11	0.026961511

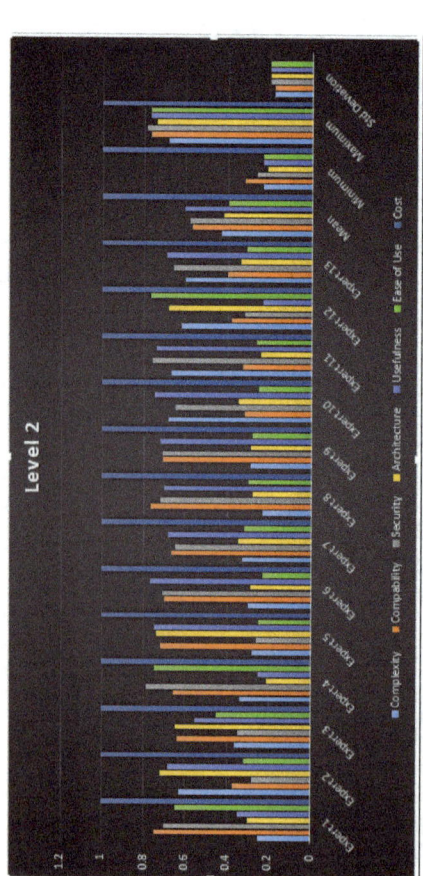

Level 2	Expert 1	Expert 2	Expert 3	Expert 4	Expert 5	Expert 6	Expert 7	Expert 8	Expert 9	Expert 10	Expert 11	Expert 12	Expert 13	Mean	Minimum	Maximum	Std Deviation
Complexity	0.25	0.63	0.36	0.34	0.28	0.3	0.33	0.23	0.29	0.68	0.67	0.62	0.6	0.429231	0.23	0.68	0.17783239
Compability	0.75	0.37	0.64	0.66	0.72	0.7	0.67	0.77	0.71	0.32	0.33	0.38	0.4	0.570769	0.32	0.77	0.17783239
Security	0.7	0.28	0.35	0.79	0.26	0.71	0.65	0.72	0.71	0.65	0.76	0.32	0.66	0.581538	0.26	0.79	0.198613786
Architecture	0.3	0.72	0.65	0.21	0.74	0.29	0.35	0.28	0.29	0.35	0.24	0.68	0.34	0.418462	0.21	0.74	0.198613786
Usefulness	0.35	0.68	0.55	0.25	0.75	0.77	0.68	0.7	0.72	0.75	0.74	0.23	0.69	0.604615	0.23	0.77	0.196410089
Ease of Use	0.65	0.32	0.45	0.75	0.25	0.23	0.32	0.3	0.28	0.25	0.26	0.77	0.31	0.395385	0.23	0.77	0.196410089
Cost	1	1	1	1	1	1	1	1	1	1	1	1	1	1	1	1	0

Technical Transformation: Cloud Computing

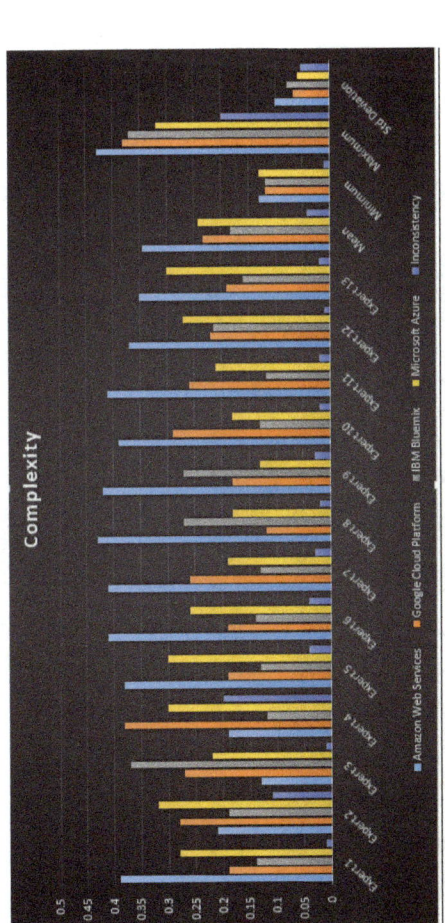

Complexity Level 3	Expert 1	Expert 2	Expert 3	Expert 4	Expert 5	Expert 6	Expert 7	Expert 8	Expert 9	Expert 10	Expert 11	Expert 12	Expert 13	Mean	Minimum	Maximum	Std Deviation
Amazon Web Services	0.39	0.21	0.13	0.19	0.38	0.41	0.41	0.43	0.42	0.39	0.41	0.37	0.35	0.345385	0.13	0.43	0.099967944
Google Cloud Platform	0.19	0.28	0.27	0.38	0.19	0.19	0.26	0.12	0.18	0.29	0.26	0.22	0.19	0.232308	0.12	0.38	0.066477295
IBM Bluemix	0.14	0.19	0.37	0.12	0.13	0.14	0.13	0.27	0.27	0.13	0.12	0.214	0.16	0.183385	0.12	0.37	0.077301939
Microsoft Azure	0.28	0.32	0.22	0.3	0.3	0.26	0.19	0.18	0.13	0.18	0.21	0.27	0.3	0.241538	0.13	0.32	0.059978629
Inconsistency	0.01	0.11	0.01	0.2	0.04	0.04	0.03	0.02	0.03	0.02	0.02	0.01	0.02	0.043077	0.01	0.2	0.053911133

76 Digital Transformation

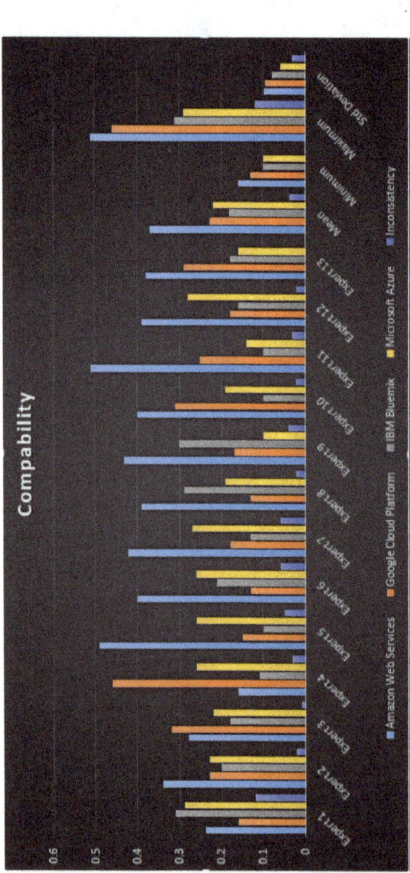

Compability Level 3	Expert 1	Expert 2	Expert 3	Expert 4	Expert 5	Expert 6	Expert 7	Expert 8	Expert 9	Expert 10	Expert 11	Expert 12	Expert 13	Mean	Minimum	Maximum	Std Deviation
Amazon Web Services	0.24	0.34	0.28	0.16	0.49	0.4	0.42	0.39	0.43	0.4	0.51	0.39	0.38	0.371538	0.16	0.51	0.096940373
Google Cloud Platform	0.16	0.23	0.32	0.46	0.15	0.13	0.18	0.13	0.17	0.31	0.25	0.18	0.29	0.227692	0.13	0.46	0.096276844
IBM Bluemix	0.31	0.2	0.18	0.11	0.1	0.21	0.13	0.29	0.3	0.1	0.1	0.16	0.18	0.182308	0.1	0.31	0.077260797
Microsoft Azure	0.29	0.23	0.22	0.26	0.26	0.26	0.27	0.19	0.1	0.19	0.14	0.28	0.16	0.219231	0.1	0.29	0.059225774
Inconsistency	0.12	0.02	0.01	0.03	0.05	0.06	0.06	0.02	0.04	0.02	0.03	0.02	0	0.036923	0	0.12	0.030925883

Technical Transformation: Cloud Computing

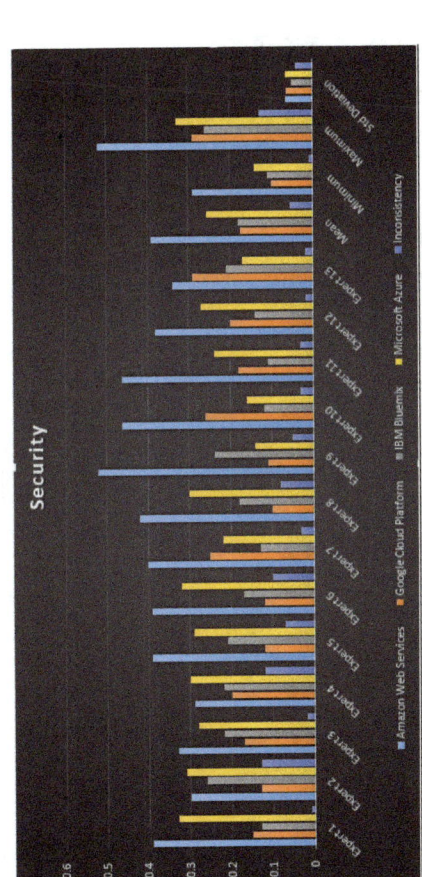

Security Level 3	Expert 1	Expert 2	Expert 3	Expert 4	Expert 5	Expert 6	Expert 7	Expert 8	Expert 9	Expert 10	Expert 11	Expert 12	Expert 13	Mean	Minimum	Maximum	Std Deviation
Amazon Web Services	0.39	0.3	0.33	0.29	0.39	0.39	0.4	0.42	0.52	0.46	0.38	0.34	0.39	0.175385	0.29	0.52	0.065828059
Google Cloud Platform	0.15	0.13	0.17	0.2	0.12	0.12	0.25	0.1	0.11	0.26	0.18	0.2	0.29	0.18	0.1	0.29	0.06186213
IBM Bluemix	0.13	0.26	0.22	0.22	0.21	0.17	0.13	0.18	0.24	0.12	0.11	0.14	0.21	0.256154	0.11	0.26	0.05016639
Microsoft Azure	0.33	0.31	0.28	0.3	0.29	0.32	0.22	0.3	0.14	0.16	0.24	0.27	0.17	0.054615	0.14	0.33	0.064490111
Inconsistency	0.01	0.13	0.02	0.12	0.07	0.1	0.03	0.08	0.05	0.03	0.03	0.02	0.02		0.01	0.13	0.041153247

78 Digital Transformation

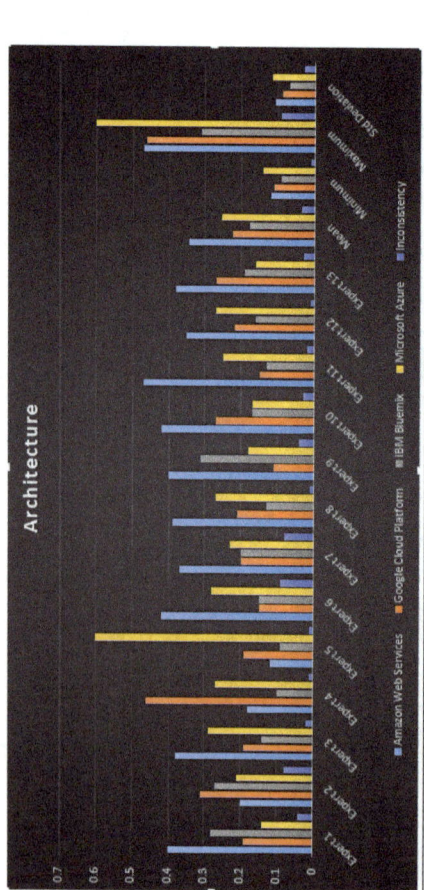

Architecture Level 3	Expert 1	Expert 2	Expert 3	Expert 4	Expert 5	Expert 6	Expert 7	Expert 8	Expert 9	Expert 10	Expert 11	Expert 12	Expert 13	Mean	Minimum	Maximum	Std Deviation
Amazon Web Services	0.4	0.2	0.38	0.18	0.12	0.42	0.37	0.39	0.4	0.42	0.47	0.35	0.38	0.344615	0.12	0.47	0.106818802
Google Cloud Platform	0.19	0.31	0.19	0.46	0.19	0.15	0.2	0.21	0.11	0.27	0.15	0.22	0.27	0.224615	0.11	0.46	0.089220269
IBM Bluemix	0.28	0.27	0.14	0.1	0.09	0.15	0.2	0.13	0.31	0.17	0.13	0.16	0.19	0.178462	0.09	0.31	0.069503735
Microsoft Azure	0.14	0.21	0.29	0.27	0.6	0.28	0.23	0.27	0.18	0.17	0.25	0.27	0.16	0.255385	0.14	0.6	0.115225531
Inconsistency	0.04	0.08	0.02	0.01	0.01	0.09	0.08	0.01	0.04	0.03	0.02	0.01	0.03	0.036154	0.01	0.09	0.029022538

Technical Transformation: Cloud Computing

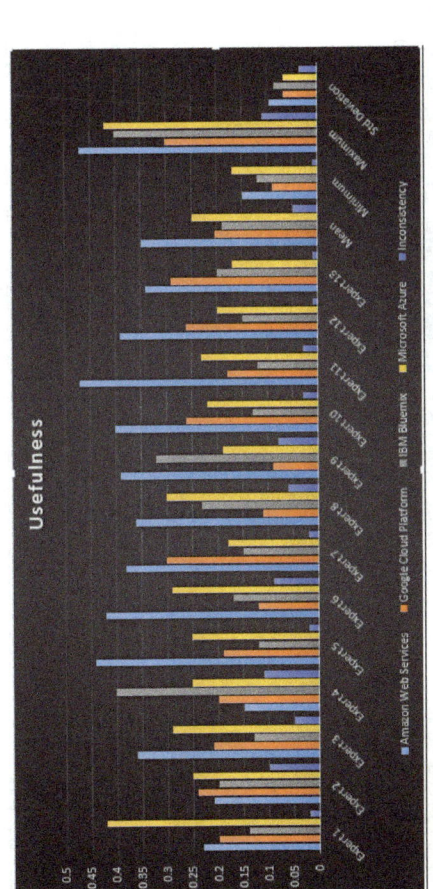

Usefulness Level 3	Expert 1	Expert 2	Expert 3	Expert 4	Expert 5	Expert 6	Expert 7	Expert 8	Expert 9	Expert 10	Expert 11	Expert 12	Expert 13	Mean	Minimum	Maximum	Std Deviation
Amazon Web Services	0.23	0.21	0.36	0.15	0.44	0.42	0.38	0.36	0.39	0.4	0.47	0.39	0.34	0.349231	0.15	0.47	0.094996626
Google Cloud Platform	0.2	0.24	0.21	0.2	0.19	0.12	0.3	0.11	0.09	0.26	0.18	0.26	0.29	0.203846	0.09	0.3	0.06702468
IBM Bluemix	0.14	0.2	0.13	0.4	0.12	0.17	0.15	0.23	0.32	0.13	0.12	0.15	0.2	0.189231	0.12	0.4	0.084603146
Microsoft Azure	0.42	0.25	0.29	0.25	0.25	0.29	0.18	0.3	0.19	0.22	0.23	0.2	0.17	0.249231	0.17	0.42	0.066641021
Inconsistency	0.02	0.1	0.05	0.11	0.02	0.09	0.02	0.06	0.08	0.03	0.03	0.01	0.01	0.048462	0.01	0.11	0.035787836

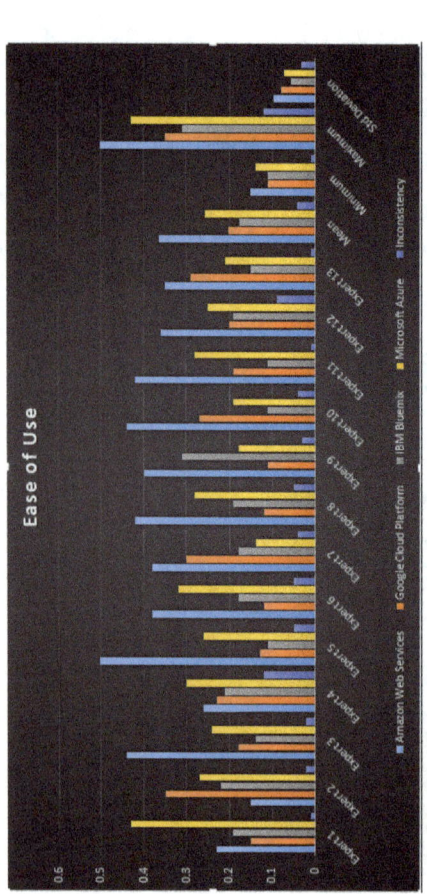

Ease of Use																		
Level 3	Expert 1	Expert 2	Expert 3	Expert 4	Expert 5	Expert 6	Expert 7	Expert 8	Expert 9	Expert 10	Expert 11	Expert 12	Expert 13	Mean	Minimum	Maximum	Std Deviation
Amazon Web Services	0.23	0.15	0.44	0.26	0.5	0.38	0.38	0.42	0.4	0.44	0.42	0.36	0.35	0.363846	0.15	0.5	0.096999868
Google Cloud Platform	0.15	0.35	0.18	0.23	0.13	0.12	0.3	0.12	0.11	0.27	0.19	0.2	0.29	0.203077	0.11	0.35	0.079097473
IBM Bluemix	0.19	0.22	0.14	0.21	0.11	0.18	0.18	0.19	0.31	0.11	0.11	0.19	0.15	0.176154	0.11	0.31	0.055307995
Microsoft Azure	0.43	0.27	0.24	0.3	0.26	0.32	0.14	0.28	0.18	0.19	0.28	0.25	0.21	0.257692	0.14	0.43	0.072818707
Inconsistency	0.01	0.02	0.02	0.12	0.05	0.05	0.04	0.05	0.03	0.04	0.01	0.09	0.01	0.041538	0.01	0.12	0.032620585

Technical Transformation: Cloud Computing 81

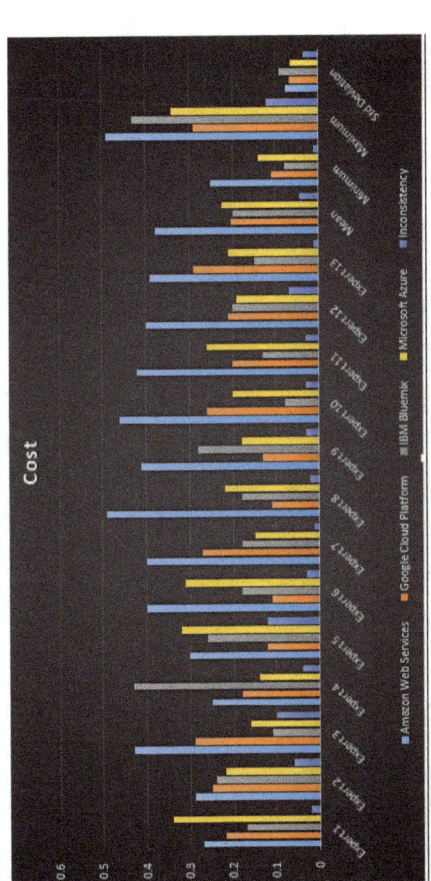

Cost Level 3	Expert 1	Expert 2	Expert 3	Expert 4	Expert 5	Expert 6	Expert 7	Expert 8	Expert 9	Expert 10	Expert 11	Expert 12	Expert 13	Mean	Minimum	Maximum	Std Deviation
Amazon Web Services	0.27	0.29	0.43	0.25	0.3	0.4	0.4	0.49	0.41	0.46	0.42	0.4	0.39	0.377692	0.25	0.49	0.07540489
Google Cloud Platform	0.22	0.25	0.29	0.18	0.12	0.11	0.27	0.11	0.13	0.26	0.2	0.21	0.29	0.203077	0.11	0.29	0.067993212
IBM Bluemix	0.17	0.24	0.11	0.43	0.26	0.18	0.18	0.18	0.28	0.08	0.13	0.2	0.15	0.199231	0.08	0.43	0.089671766
Microsoft Azure	0.34	0.22	0.16	0.14	0.32	0.31	0.15	0.22	0.18	0.2	0.26	0.19	0.21	0.223077	0.14	0.34	0.06575011
Inconsistency	0.02	0.06	0.1	0.04	0.12	0.03	0.01	0.02	0.03	0.03	0.03	0.07	0.01	0.043846	0.01	0.12	0.034287622

82 Digital Transformation

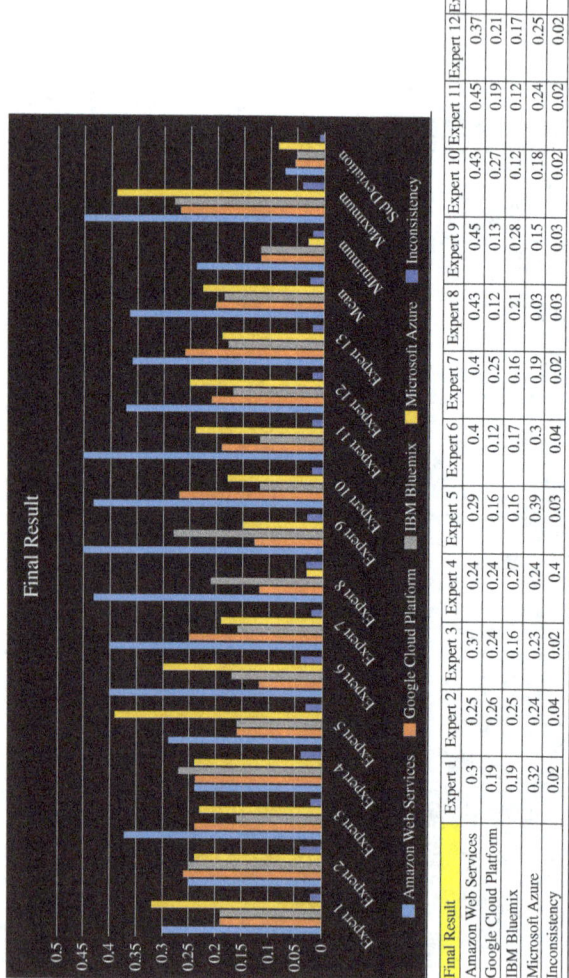

Final Result	Expert 1	Expert 2	Expert 3	Expert 4	Expert 5	Expert 6	Expert 7	Expert 8	Expert 9	Expert 10	Expert 11	Expert 12	Expert 13	Mean	Minimum	Maximum	Std Deviation
Amazon Web Services	0.3	0.25	0.37	0.24	0.29	0.4	0.4	0.43	0.45	0.43	0.45	0.37	0.36	0.364615	0.24	0.45	0.073099861
Google Cloud Platform	0.19	0.26	0.24	0.24	0.16	0.12	0.25	0.12	0.13	0.27	0.19	0.21	0.26	0.203077	0.12	0.27	0.05573518
IBM Bluemix	0.19	0.25	0.16	0.27	0.16	0.17	0.16	0.21	0.28	0.12	0.12	0.17	0.18	0.187692	0.12	0.28	0.051503049
Microsoft Azure	0.32	0.24	0.23	0.24	0.39	0.3	0.19	0.03	0.15	0.18	0.24	0.25	0.19	0.226923	0.03	0.39	0.087214736
Inconsistency	0.02	0.04	0.02	0.4	0.03	0.04	0.02	0.03	0.03	0.02	0.02	0.02	0.02	0.026923	0.02	0.04	0.008548504

Chapter 4

Technical Transformation: Internet of Things

Surekha Rani Chanamolu* and Tugrul Daim*,†,‡

*Portland State University, Portland, Oregon, USA
†Higher School of Economics, Moscow, Russia
‡Chaoyang University of Technology, Taiwan

Abstract

We are entering a new era of computing technology: the Internet of Things (IoT). Its foundation is the intelligence that embedded processing provides. Integrated microcontroller devices, which can provide the real-time embedded processing, are a key requirement for most IoT applications. However, the task of selecting the appropriate microcontroller for an IoT application is more difficult than it seems. Traditional microcontroller selection and management practices are inadequate.

This paper proposes a new methodology of selecting a microcontroller for an IoT application. A Hierarchical Decision Model (HDM) is used for the decision-making process and qualified experts' opinions are used as measurements. There are four levels in the hierarchy: Objective, Criteria, Subcriteria and alternatives. The last three of these are evaluated by a panel of experts and prioritized by their importance to the objective. The results are validated using inconsistency measures for the reliability of the experts' and group's agreement.

This model will result in a better selection methodology and help embedded designers select microcontrollers most suitable for their designs. It can also be improved by adding more criteria, such as flexibility, scalability, alternatives and expert panels.

Keywords: Technology assessment, microcontrollers, smart watermeter.

1. Introduction

The Internet of Things (IoT) market has seen exponential growth recently, with a large variety of different connected devices flooding the market. More than 8.4 billion connected devices were forecasted to be in use worldwide in 2017, up 31% from 2016 [1], with the number expected to rise to 26 billion by 2020 [2]. The IoT, as the name implies, creates an ecosystem of interconnected devices that interact with each other and the internet. These devices contain sensing and actuating elements as well as hardware and software that perform data aggregation, network connectivity and security. They are designed to perform tasks using the data they gather, or with information transmitted across the network by other IoT devices. IoT technology has already been successfully implemented in industries like healthcare, manufacturing, transportation, etc. As a result of the prices of important components such as sensors and processors going down, many IoT devices are being made "smart"—they are capable of performing tasks without human intervention. Competition in the market revolves around power consumption, wireless range and processing power [3].

IoT systems are centered on a Microcontroller Unit (MCU) that processes data and runs wireless network stacks. Selecting the right one is critical to the success of a company. As MCUs are becoming increasingly complex because of growing on-chip integration to reduce system costs, this decision will become more difficult [3].

Selecting the right microcontroller involves a variety of factors, both qualitative and quantitative. These include system requirements, availability, performance, size, reliability, maintainability, environmental constraints, support, correctness, safety, cost, manufacturer's history and track records [4].

Embedded designers face challenges to compare quantitatively and qualitatively in the absence of any mathematical model. There are a variety of steps that often-embedded designers follow to make the right decisions to finally select the most appropriate microcontroller. This paper proposes a mathematical model called the Hierarchical Decision Model (HDM), which is a selection process that will help designers identify the right components for the new or existing designs.

In this research the HDM was built in the decision-making process and utilized to select the best microcontroller for a Smart Water Meter IoT.

2. Methodology

The methodology that has been used in this research is "HDM", which had been introduced by David Cleland and Dundar Kocaoglu in 1987 [5]. The model prioritizes problems based on various criteria and arranges them in a hierarchy, which the decision maker is then guided through. The alternatives are presented in pairs, which the respondent divides 100 points between to reflect his judgment of each alternative relative to the other. For example, if the elements of a pair are given 50 points each, it means that both have the same importance in the respondent's judgment, but does not indicate whether they both are highly important or highly unimportant. Zero is not used in measurements to avoid computational errors. Therefore if one element of the pair is totally insignificant compared to the other, the respondent assigns values of 1 and 99 instead of 0 and 100. Other assignments are made similarly, for example, if one element is three times as important as the other one, they are given 75 and 25 points, respectively [5].

2.1. *Hierarchical decision model*

Based on literature review and consultations with experts, an HDM model was constructed (Figure 1). The model is created using the PSU ETM HDM software tool and is utilized in the

86 Digital Transformation

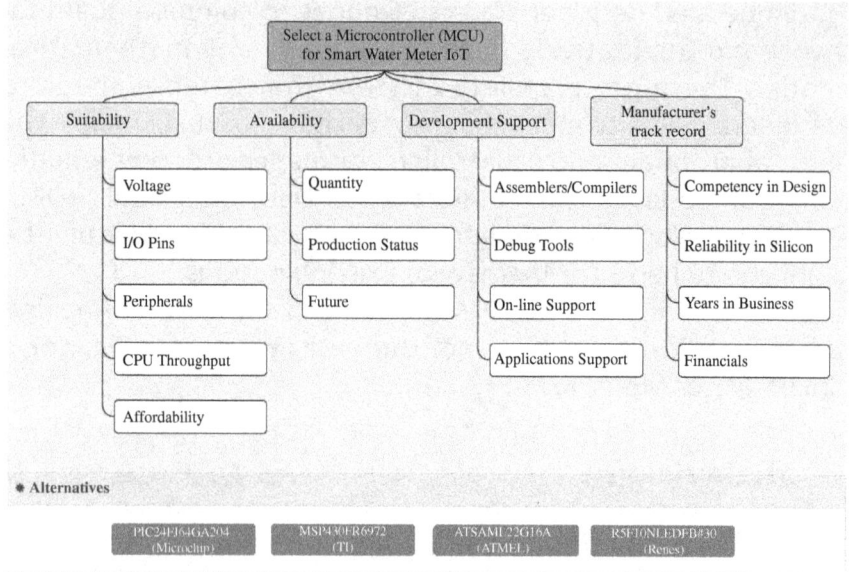

Figure 1. HDM model to select a microcontroller for the Smart Water Meter IoT.

decision-making activity, where qualified expert opinions are used as measurements.

There are four levels in the hierarchy: Objective, Criteria, Subcriteria and Alternatives. The last three are evaluated by a panel of experts and prioritized by their importance to the Objective. The results are validated using inconsistency measures for the reliability of the experts' and group's agreement.

2.1.1. *Level 1: Objective—Select a microcontroller for the Smart Water IoT*

The selection of a microcontroller for an application can be a challenging process. The online catalogue for MCUs provides a massive collection of thousands of various MCUs. There are so many good choices that it will be difficult to make a decision. The key to microprocessor selection is identifying the necessities of the design. The following criteria and subcriteria are a good basis to identify what to look for in terms of a microcontroller's form and function.

2.1.2. *Levels 2 and 3: Criteria and subcriteria [3, 4, 6−11]*

2.1.2.1. Suitability criteria

The suitability of the microcontroller depends on factors such as CPU throughput, peripherals, I/O pins, low power and the cost of the Smart Water Meter IoT system. Its datasheet, which contains information regarding device capabilities and functions, is essential to the selection process.

1. **Voltage:** Within the IoT ecosystem, sensors that measure parameters such as temperature and humidity or gas and water consumption are especially technically challenging. They operate on battery power yet must have a long operational lifetime to lower maintenance costs. So, the MCU must be energy efficient. This will result in a longer runtime [7, 11]. As can be seen in Figure 2, the current consumption

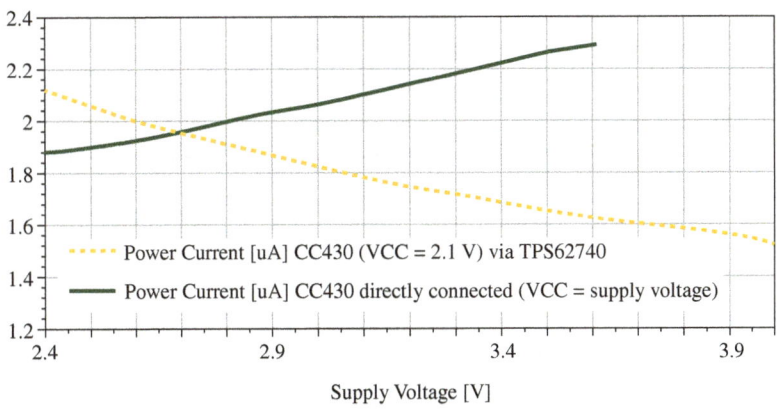

Figure 2. Current consumption depends on the supply voltage (green trace) [8].

depends on the supply voltage (green trace). So, low operating voltage is a key requirement for a low power-consuming MCU [8].
2. **I/O pins:** The devices that will be connected to a microcontroller and the way they will interface with it are major factors in the selection process. In many microcontrollers, there are pins that are configurable to either accept input signals or produce outputs. If an MCU with a certain number of I/O pins is required to develop a system, selecting one with more will increase the size of the MCU and the system as a whole. Choosing an MCU with the right number of I/O pins will help make the whole system compact [4, 6].
3. **Peripherals:** One of the most important criteria that drive MCU selection are the peripherals that it will be connected to. As shown in Figure 3, a typical IoT system includes a sensor unit, a controller, an interface to the communication block, a radio block for data communications and a power management system providing the energy required to power up the node [9].

When selecting an MCU, check for the availability of timers, serial interfaces, ROM, RAM, A/D converter, CRYPTO and the number of I/O ports; of the latter, too many will inflate cost while too few cannot do the job [4, 6, 10].

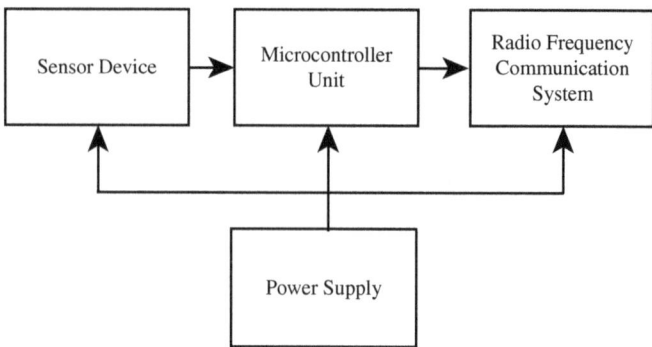

Figure 3. Typical sensor node architecture [9].

In addition, one needs to verify that the MCU has hardware features such as low power consumption, low wake time and autonomous peripherals (i.e., peripherals that do not require CPU intervention) that operate energy efficiently to help squeeze the most nano-amps out of the system, thus extending the battery life of the IoT device. Additional components such as a timer can handle A/D conversions, while a GPIO pin can handle UART transmissions without CPU intervention [11].

- **Communication interface:** The Smart Water Meter H/W subsystem sensor node needs to transmit/receive the data from/to the IoT host through a RF wireless subsystem. These subsystems are generally connected through a standard serial interface. The driving factor for choosing a serial interface over a parallel interface is the number of I/O pins and low power requirements. The most commonly used serial interfaces are SPI/I2C/UART.
- **Digital sensor interface (flow/level/temp):** The MCU will read data from digital sensors through GPIO ports and write it to on-board SRAM/SDRAM using a DMA controller.
- **Flash RAM/SDRAM for boot code, drivers and for applications:** The HW subsystem at the sensor node should have a memory interface for application drivers and data flow from the sensor interface to the internet cloud. The H/W subsystem at the sensor node will employ the on-chip DMA controller to transfer the data to/from the cloud.
- **Analog to digital converter (ADC):** The Smart Water Meter H/W subsystem at the sensor node has individual DC power requirements and is powered by a DC battery. Therefore, an ADC connected to the MCU through GPIO pins can be used to check the health of the battery.
- **LCD controller:** An LCD controller can be used to display the water flow/level and water utilization data at the sensor node (home monitors).
- **Captive buttons:** This interface is required to control and tune the meters to counter any errors in the analog low power sensor data.

- **Valve controller (relay drivers):** This interface is optional and can be used to control water flow remotely. It controls valve relays to control the physical valves that in turn control water flow.
- **Watchdog timer:** The processor on an MCU periodically sends a pulse to the watchdog timer to indicate that the system software is operating smoothly. If the watchdog timer does not receive this pulse within an allotted time frame (called the watchdog timeout), the watchdog timer asserts a reset output. This reset output is used to notify the system that the MCU has experienced an error, or reset the processor itself.

4. **CPU throughput:** The CPU core needs to have the computational power to handle tasks for the system's lifetime in the chosen implementation language. Once again, too much power inflates costs while too little does not meet minimum requirements. For a water meter, however, the main selection criterion is not computational power, but energy efficiency [7].
5. **Affordability:** Affordability is among the most important selection criterion. To implement the system within the budget, the cost of each component (MCU along with supporting ICs) needs to be minimized. Implementing on-chip features will compensate for inventory and assembly costs, and also reduce development time and effort by providing a ready integrated solution [6].

2.1.2.2. Availability criteria

Availability is an important consideration for commercial designs. Before implementing the system the device's availability should be checked. The guaranteed availability of microcontrollers in the future is an important aspect to consider. "The roadmap of an MCU family ensures that the time and investment made in choosing the family are future-proofing the designs. It simplifies migration and helps in optimization of the forthcoming product releases. Pin and package compatibility would ensure that the design could scale up or down easily by just replacing the processor with minimal or no board change. Software compatibility ensures that the software written for one device would be reusable on another device [12]".

2.1.2.3. Subcriteria

1. **Quantity:** One of the most important aspects of choosing a microcontroller is its ready availability in needed quantities, both now and in the future. A simple way of doing this is checking the stock on a vendor's website (Jameco, Digi-Key, Mouser, Newark/E14, etc.), as well as looking for pricing at different quantity levels. It is important to check lead times for the part to make sure that wait times are not too long to fill the order. Also ensure that the chosen MCU vendor caters to differing requirements for different stages of development, for instance, smaller quantities for engineering samples, and larger ones during production [12].
2. **Production status:** This identifies whether the MCU is in production or still under development. The availability of the microcontrollers needs to be checked before moving on to implement the system.
3. **Future:** Future availability of the MCU is important, especially to ensure that the End-of-Life (EoL) of the device will not be in the near future. If you want your product to be available in the market for at least a couple of years, you have to consider the number of years that the selected microcontroller has been in the market. "Many a times, MCU families come up with specific programming methods, programmers, IDEs, tools, etc. This means changing from one family to another not only changes the cost and other parameters of the MCU, but also affects the additional maintenance and overhead costs, and in some cases the total ownership cost as well [12]."

2.1.2.4. Development support criteria

Although system designers are primarily concerned with chip architectures and performance specifications, software and hardware development tools are also important [13]. They are critical for today's advanced embedded systems designs, specifically for the IoT market [14].

As software design continues to become more complex in today's embedded hardware design, complete solutions provide designers with all the necessary tools and embedded software. They are also fully integrated and seamless [15].

2.1.2.4.1. Subcriteria

1. **Assemblers/compilers:** 16- and 32-bit processors benefit from the advantages of high-level programming languages such as C. Design teams building applications for these processors will require a cross-assembler, a linker, a librarian and a debugger [13]. C/C++ compilers reduce overall development time and help bring products to the market faster.
2. **Debug tools:** The tools available to the design team are an important factor for the final product's success. Selecting the correct tools for an application can reduce development time while increasing the system's reliability and ease of maintenance [13]. Hardware development systems evaluation modules (EVMs), in-circuit emulators, logic analyzer pods, debug monitors, source-level debug monitors and a reference design for the specific application significantly increase developers' productivity.
3. **On-line support:** The MCU suppliers should have some facilities like a helpline, toll-free number, fax number, after-sales support, real-time executives, application examples, bug reports, utility software, including "free" assemblers, sample source codes, datasheets, user guides, etc. [16].
4. **Application support:** To help engineers bring their IoT devices to market faster, MCU suppliers must offer a diverse range of advanced design tools, production-ready sample applications, firmware development tools, software stacks and application demos. Also, whenever engineers face any kind of problems during product development, they should

have access to marketing/sales, field application engineers and dedicated applications support groups [16].

2.1.2.5. Manufacturer's track record criteria

Points like design challenges, on time delivery, performance, years in business and financial reports should be regarded as the track record of the manufacturers.

2.1.2.5.1. Subcriteria

1. **Competency in design:** Maturity in design and best practices.
2. **Reliability in silicon:** Product life cycles in the industrial market are at least 10–15 years. Hence, quality, reliability and longevity requirements are key to the success of the IoT.
3. **Years in business:** Numbers of years in the business.
4. **Finances:** Financial health of the company by checking financial reports and stock performance.

2.1.3. *Level 4: Alternatives*

Security and energy efficiency are increasingly becoming critical features because the IoT is introducing inter-connectivity to battery-powered devices.

2.1.3.1. PIC24FJ128GA204 from Microchip Technologies [17]

PIC24F is one of the 16-bit low-power microcontroller suitable for the Smart Water Meter IoT solutions (Figure 4). The following are the key features supported by this microcontroller.

- On-chip Security H/W Block with AES/DES/Triple DES Engine, Random Number Generator and OTP Key Storage, etc.
- Supports multiple low-power modes that include idle, sleep, doze, deep sleep and alternate clock modes using programmable power management block.

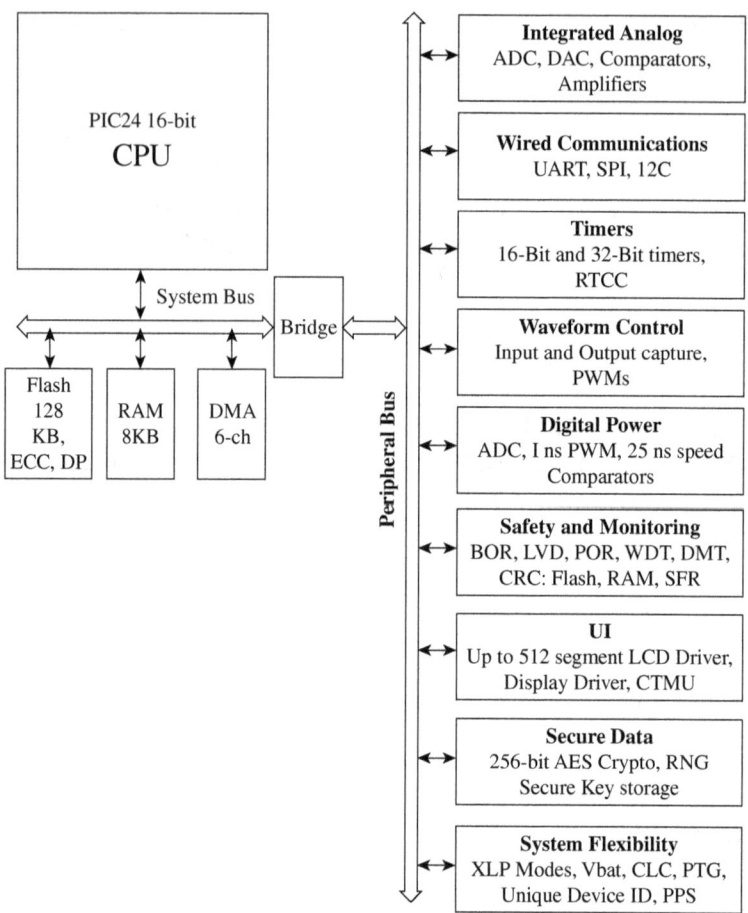

Figure 4. PIC24FJ128GA204 family general block diagram [17].

- On-chip analog controller supports ADCs, analog comparators and the time measurement unit for capacitive touch sensing.
- High CPU performance of up to 16 MIPS at a 32-MHz frequency with an optimized instruction set. An on-chip oscillator supports high-speed run-time and self-adjusts for accuracy and faster startup.
- Supported peripherals:
    - DMA Controller for bulk data transfers, Peripheral Pin Selection for Independent I/O Mapping, 16-Bit Timers/

Counters/CRC/PWM modules/Hardware Real-Time Clock (RTC)/Watchdog and Serial Peripheral Interfaces SPI/I2C/UART/IrDA.

2.1.3.1.1. Development environment

Microchip provides reputable development tools along with a support team. Their software offerings, provided via MPLAB, are integrated with their programmer, debugger and emulator tools, and are compatible with Windows, Linux and Mac. The tools are offered in three different tiers to fit a range of performance requirements and budgets. In addition, a variety of development boards and kits suitable for numerous different projects are available and complete with samples.

- **PIC24F Curiosity Development Board**—The PIC24F Curiosity Development Board is targeted at both first-time users and makers. It is a 16-bit development platform, both cost effective and fully integrated, and provides a feature-rich rapid prototyping board. The board is optimized for Microchip's MPLAB® X IDE and MPLAB Xpress Cloud-based IDE. It includes an integrated programmer/debugger and requires no additional hardware [18].

2.1.3.1.2. Availability of parts

Microchip has their own website (https://www.microchipdirect.com) to order parts. PIC24FJ128GA204 has 1,600 units available immediately. For orders of more than 1,600 units, there is a lead time of ~16 weeks.

2.1.3.1.3. Microchip Technologies (Atmel) company profile

Microchip Technologies was incorporated on 14 February 1989. The company develops, manufactures and sells specialized semiconductor products for embedded control applications. The Company has two segments: semiconductor products and technology licensing.

Microchip offers both general purpose and specialized 8-bit, 16-bit and 32-bit microcontrollers, a spectrum of linear, mixed-signal, power management, thermal management, Radio Frequency (RF), timing, safety, security, wired connectivity and wireless connectivity devices, as well as serial Electrically Erasable Programmable Read-Only Memories (EEPROMs), serial flash memories, parallel flash memories and Serial Static Random-Access Memory (SRAM) memories. It also licenses Flash-IP solutions that are used in a range of products. Microchip employs 14,234 employees worldwide as of 31 March 2018.

- **Financial performance**
  - Microchip followed a 3% revenue increase in 2016 with a 12% jump in 2017 to $3.8 billion. Profit in 2017 rose to $246 million as compared to $101 million in 2016 (Table 1 and Figure 5). Cash flow from operations was $1.38 billion in 2017, about $444 million higher than 2016 [19].
  - **Historical debt**—Microchip's level of debt (93.6%) compared to net worth is high (greater than 40%). This level

Table 1. Microchip past financial data [19].

Year Ended March 31	Revenues ($)	Operating Expenses ($)	Net Operating Cash Flows ($)	Net Income ($)
2018	3,981	1,484	1,420	255
2017	3,408	1,481	1,060	165
2016	2,173	853	744	324
2015	2,147	804	721	365
2014	1,931	670	677	395
2013	1,581	660	459	127
2012	1,383	403	412	337

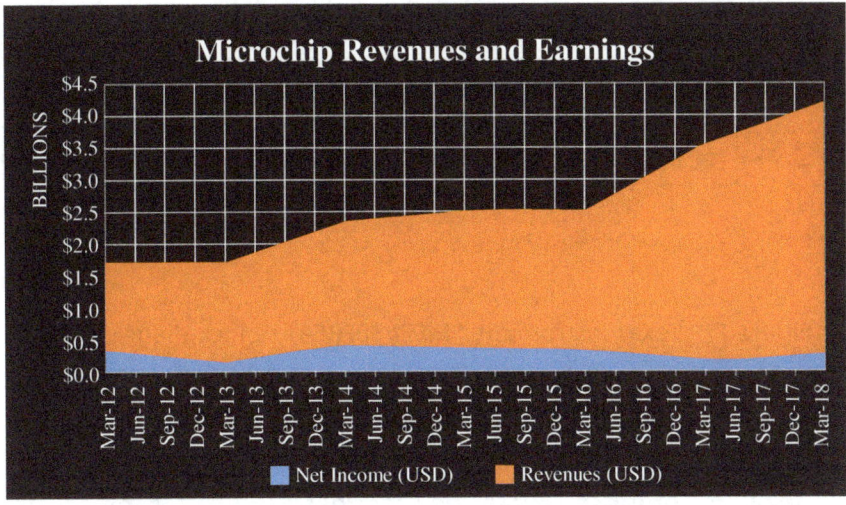

Figure 5. Microchip earnings and revenue history [19].

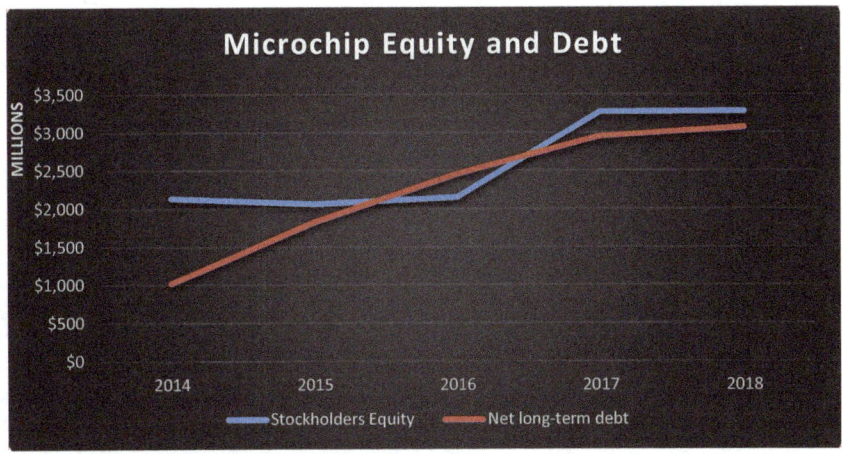

Figure 6. Microchip's historical debt as of 25 May 2018 [19].

has increased from 50.9% to 93.6% over the last 5 years (Figure 6). This debt is covered by the company's operating cash flow, while interest payments are well covered by its earnings [19].

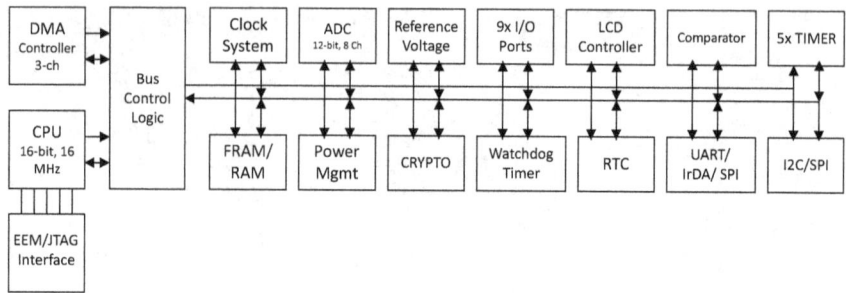

Figure 7. Texas Instruments' MSP430FR6972 block diagram [20].

2.1.3.2. MSP430FR6972 from Texas Instruments [20]

Texas Instruments' (TI) MSP430FR697x energy efficient microcontrollers comprise several devices featuring different peripherals (Figure 7). Their architecture and seven low-power modes are optimized for extended battery life. The devices contain a powerful 16-bit RISC CPU, 16-bit registers and constant generators that facilitate maximum code efficiency. The devices are microcontroller configurations with up to five 16-bit timers, a comparator, eUSCIs that support UART, SPI and I2C, a hardware multiplier, an AES accelerator, DMA, an RTC module with alarm capabilities, up to 52 I/O pins, and a high-performance 12-bit ADC. It also includes an LCD module with contrast control for displays with up to 116 segments [20].

2.1.3.2.1. Development tools and software

TI supports tools for the MSP microcontroller series. This includes development kits, plug-in boards and experimenter boards, among other support services. The MSP MCU development kits range from simple LaunchPad kits that are easy-to-use and facilitate MCU application development, to more feature-rich boards. The MSP430 and MSP430FR6989 are examples of the former, with the MSP430FR6989 LaunchPad kit facilitating extreme low-power MCU applications and featuring on-board programming, debugging and measuring tools.

Additionally, plug-in boards are available to expand the LaunchPad kits' functionality. BoosterPack plug-ins, for example, provide support for applications that involve capacitive touch, wireless communications, temperature measurements, etc. Development tools such as the eZ430, which is available in a simple USB drive form factor, provide the tools required for a full MSP microcontroller project. Additional hardware parts and MSPs are available through TI's feature-rich experimenter boards, which also come with examples to help developers get started.

2.1.3.2.2. Ultrasonic Water Flow Measurement reference design [22]

The Ultrasonic Flow Meter's reference schematic contains a 12-bit A/D conversion and LCD controller on a TI MSP430FR6972™ microcontroller, along with discrete AFE components and transducers. An AAA battery holder and a 14-pin JTAG module are present as well.

2.1.3.2.3. MSP MCU programmer and debugger

The MSP-FET: MSP430 Flash Emulation Tool provides a USB debugging interface to connect any MSP430 MCU to a computer for real-time, in-system debugging with multiple hardware breakpoints, trace capability, clock control and other advanced debugging features.

TI provides various tools to facilitate embedded software development for applications using the MSP microcontroller. These include IDEs such as Code Composer Studio and its Cloud version, which are Eclipse-based and support all MSO430 MCUs. CCS Cloud is also integrated with the LaunchPad development kit, and supports Energia IDE (open-source), TI-RTOS and C/C++ development. The IAR Embedded Workbench Kickstart IDE helps developers build and debug MSP430-based applications, and includes a C-Compiler and an integrated debugger. Finally, the open-source GCC is the product of a collaboration between TI and Redhat, and offers compiling and debugging tools.

TI's software offerings for the MSP MCUs support code generation, GUI peripheral configuration, driver libraries and APIs. Through the RF software offerings, wireless functionality is added as well. While the SmartRF v7 software provides radio evaluation and configuration tools, the SmartRF packet sniffer facilitates displaying and storing radio packets. Finally, the SimpliciTI software protocol stack features energy efficient battery-powered modules with nodes communicating directly with each other.

To support a variety of DSPs and ARM MCUs, SYS/BIOS, a real-time OS is available as well.

2.1.3.2.4. Availability of parts

TI has its own store (https://store.ti.com) to order parts, which also can be ordered from Mouser Electronics, Digi-Key. 2,082 parts of MSP430FR6972 are available immediately and there is a lead time of 9 weeks for orders of more than 2,082 units.

2.1.3.2.5. Texas Instruments company profile

Texas Instruments (TI) was incorporated on 23 December 1938. The company designs, manufactures and sells semiconductors to electronics designers and manufacturers worldwide (Figure 8). The company operates through three segments: analog, embedded processing and other products. The company has design, manufacturing or sales operations in more than 30 countries and has around 29,700 employees worldwide as of 31 December 2016 [23].

- **Analog**—The Analog business, which accounts for about two-thirds of sales, includes high-volume analog and logic products, power management semiconductors, and amplifiers and data converters. The company's analog products are used in the personal electronics, automotive and industrial markets, as well as others.

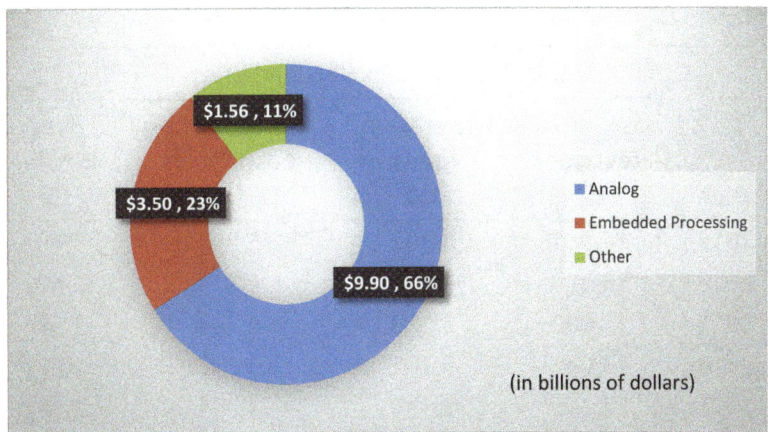

Figure 8. Texas Instruments' 2017 revenues [23].

- **Embedded processing**—The Embedded Processing segment, which generates about a quarter of sales, makes Application Specific Integrated Circuits (ASICs), Digital Signal Processors (DSPs) and microcontrollers. TI's embedded processors range from low-cost microcontrollers used in products such as electric toothbrushes to complex devices used in automotive applications, such as infotainment and advanced driver assistance systems.
- **Other**—The remaining revenue comes from the Other segment, which includes digital light processors (DLPs) used in projectors to create high-definition images, calculators and custom semiconductors [23].
- **Financial performance**—TI followed a 3% revenue increase in 2016 with a 12% jump in 2017 to $14.9 billion. Profit in 2017 rose slightly to $3.7 billion as compared to $3.6 billion in 2016 (Table 2 and Figure 9). TI's cash flow from operations was $5.36 billion in 2017, about $1 billion higher than 2016. Its free cash flow was $4.67 billion, about 31% of its revenue, up from 30.5% in 2016 [24].

Table 2. Texas Instruments past financial performances [24].

	Data in USD millions			
Year	Revenues ($)	Operating Expenses ($)	Net Operating Cash Flows ($)	Net Income ($)
2017	14,961	3,202	5,363	3,682
2016	13,370	3,137	4,614	3,595
2015	13,000	3,028	4,268	2,986
2014	13,045	3,201	3,892	2,821
2013	12,205	3,380	3,384	2,162
2012	12,825	3,681	3,414	1,759

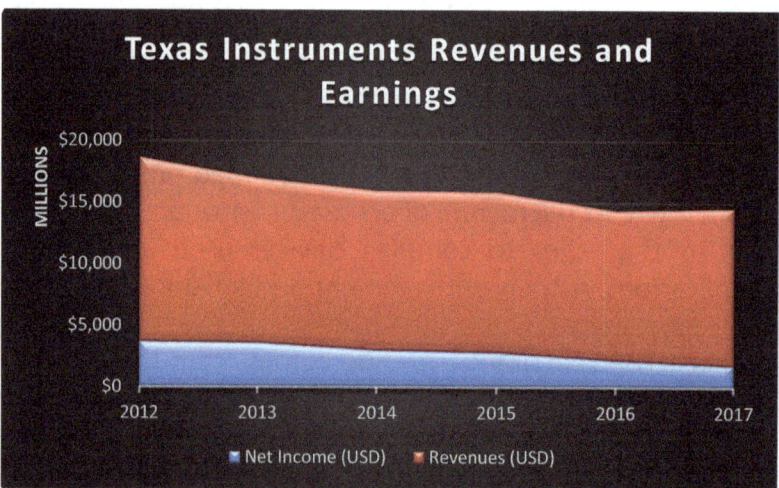

Figure 9. Texas Instruments' earnings and revenues [24].

- **Historical debt**—Texas Instruments' level of debt (38.3%) compared to net worth is satisfactory (less than 40%) (Figure 10). This level has gone down from 51.8% to 38.3% over the last 5 years. Operating cash flow covers debt, while interest payments cover earnings [24].

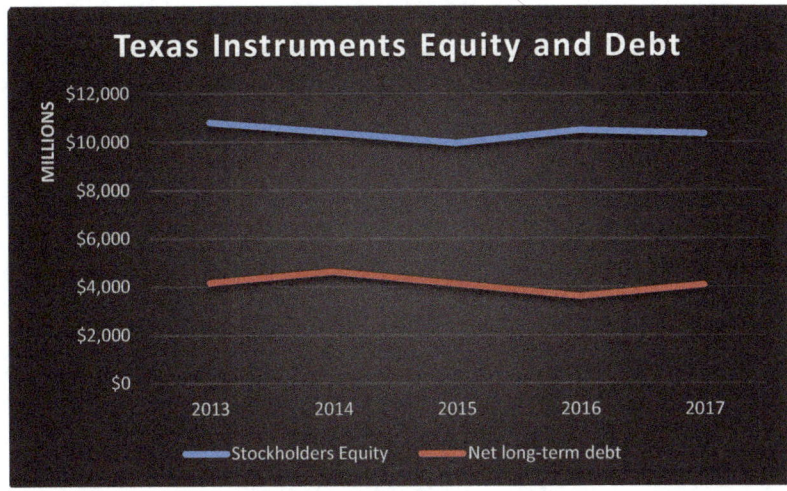

Figure 10. Texas Instruments' historical debt, as of 13 Jan 2018 [24].

2.1.3.3. ATSAML22G16A from Atmel (now Microchip Technologies)

Atmel's SMART SAM L22 ARM Cortex-M0+ based MCU series features a segment LCD controller and large memories (Figure 11). With power consumption rates of 39 µA/MHz in active mode and 490 nA in ultra-low-power backup mode, the microcontrollers are market leaders in terms of energy efficiency. They use the Atmel-developed picoPower technology in conjunction with energy efficient peripherals that can operate independently of the main processor to achieve very low power consumption rates. Hence, this allows the CPU to remain inactive. The series' LCD controller can deliver up to 320 segments and also supports a smart card interface, a full speed USB device, Event System and Sleepwalking, 12-bit Analog, AES, Capacitive Touch Sensing, etc. Hence, these MCUs are applicable in a variety of devices such as thermostats, electric/gas/water meters, home control, and medical and access systems. The SAM L22 series features the 32-bit ARM® Cortex® -M0+ processor, ranging from 48- to 100-pins with up to 256 KB Flash and 32 KB of

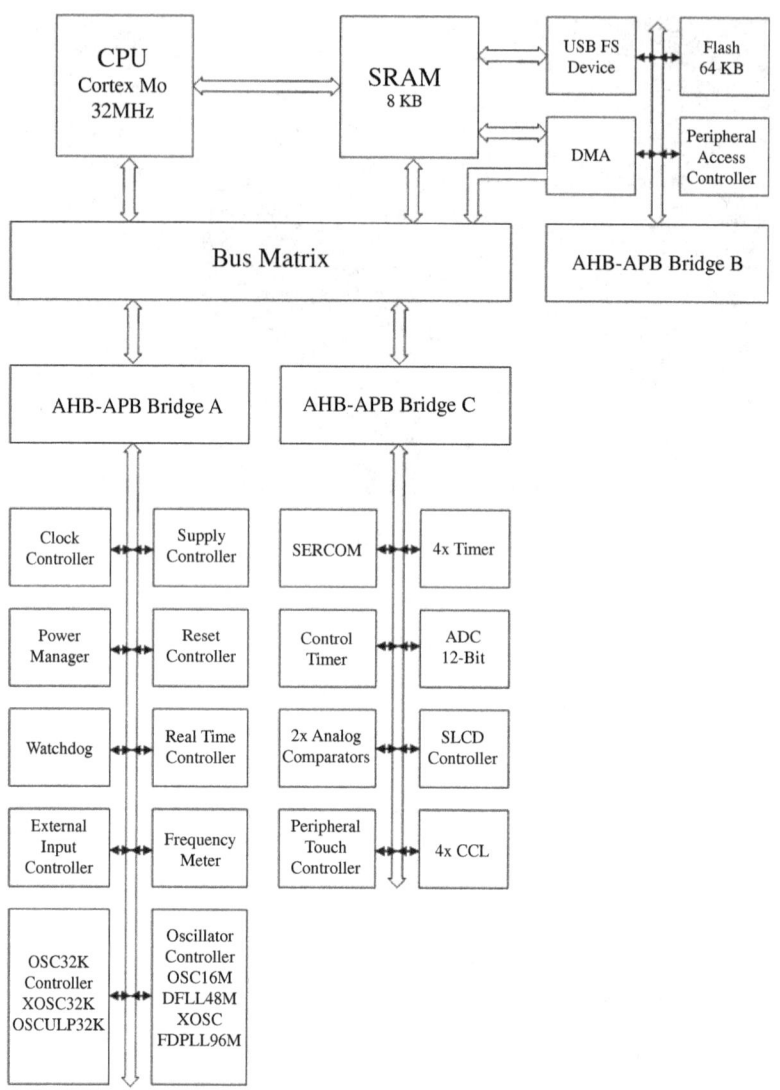

Figure 11. ATSAML22 block diagram [25].

SSRAM [25, 26]. It delivers advanced security features with a 256-bit AES, Cyclic Redundancy Check (CRC), True Random Number Generator (TRNG), Flash Protection and Tamper Detection to securely store, deliver and provide access to information.

An evaluation board containing an on-board debugger and standardized extension connectors is offered with the SAM L22 Atmel Xplained Pro; this will help streamline the design process. The device is also supported by Atmel Studio, a free IDE that features power-profiling tools that make it easy to optimize applications and calculate that average power consumption. Atmel Xplained development kits are inexpensive and easy-to-use, and facilitate rapid prototyping and evaluation for Atmel AVR and Atmel | SMART ARM-based microcontrollers. The kit is ideal for demonstrating Atmel MCU/MPU features and capabilities and is customizable via a wide range of expansion boards [26].

2.1.3.3.1. Availability of parts

Atmel (now Microchip Technologies) has its own website (https://www.microchipdirect.com) where orders for parts can be placed. ATSAML22G16A has 25,000 units available immediately. For orders of more than 25,000, there is a lead time of ~26 weeks.

2.1.3.4. R5F10NLEDFB#30 from Renesas

Renesas Electronics RL78 16-bit microcontrollers deliver ultra-low power consumption, enhanced performance, high integration and an extensive range of powerful peripheral functions. These features make the RL78 MCUs ideal for several applications, including battery-operated devices and household applications [27].

The RL78/I1C Group of MCUs is suitable for smart meters as they possess DLMS support, enhanced security functionality, improved arithmetic operation and numerous on-chip peripheral functions for a reduced system cost. Like their predecessor, the RL78/I1B microcontroller series (of which 30 million were sold over 3 years)—the RL78/I1C MCUs—is highly accurate as well as energy efficient [28].

106 Digital Transformation

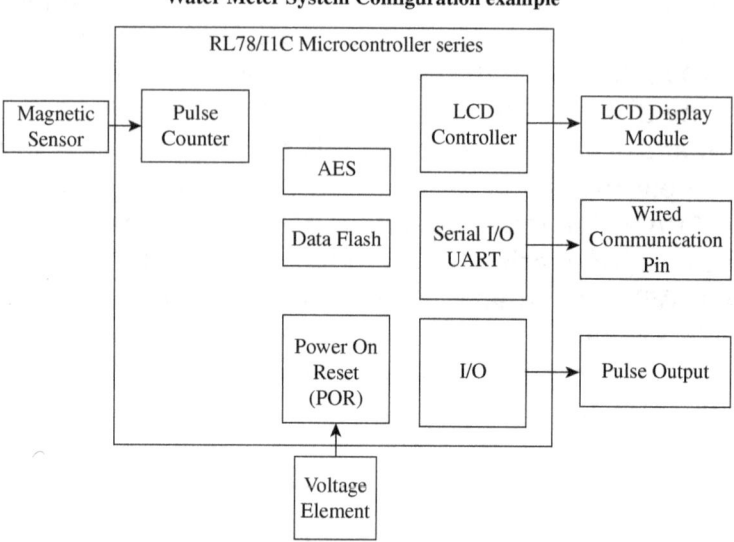

Figure 12. A water meter configuration example from Renesas [29].

RL78/I1C microcontrollers feature integrated functions such as a 24-bit Delta-Sigma A/D operation, AES H/W integration, up to 256 KB flash line-up, a 32-bit MAC (Multiply and Accumulation) function, independent power supply for RTC and enhanced battery pin (Vbat) function for AC-off operation, and Hardware zero cross detection. Performance is enhanced with a high temperature coefficient internal reference voltage. A segment LCD driver supports 8-COM [27].

Renesas offers a multitude of microcontrollers that combine low power consumption and excellent reliability as solutions for water meters [29] (Figure 12).

2.1.3.4.1. Key features

- **Low power consumption**—Three low-power modes maximize battery life, either by putting on-chip functions such as the CPU, clock and peripherals on standby, or by turning them off when they are not being used (Figure 13).

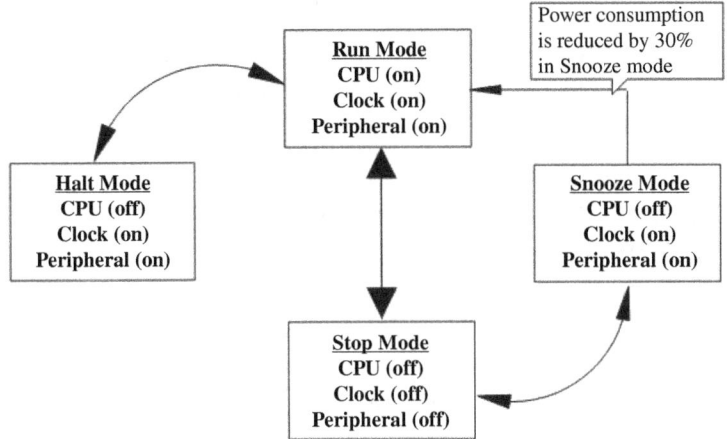

Figure 13. RL 78 family low-power modes [30].

When in standard modes such as RUN and STOP, the RL78 microcontrollers operate at 65.5 µA/MHz (at 32 MHz operation) and at 0.57 µA when in standby/HALT mode with the RTC and LVD modules active. In addition to these modes, a new mode called SNOOZE has been implemented, during which the CPU is on "standby" while the ADC and serial communication are still active independently. The CPU becomes active only when required, hence reducing power consumption and improving efficiency.

Broad scalability
- *Extensive memory size and package options*—With memories ranging from 1 to 512 KB and package pin counts from 10 pins to 128 pins, more than 300 versions of the product are available. This diverse array supports a wide range of application fields, including consumer, automotive, industrial and communications [31].
- *Excellent pin compatibility*—The location of peripheral and input/output pins remains the same even when the pin count changes. Therefore, customers can continue to use the RL78 MCUs in the future [31].

- *High performance*—RL78/I1C microcontrollers with a CPU core employing a three-stage pipeline and operating frequency of 24 MHz, and a processing performance of 1.39 DMIPS/MHz [27] (Figure 14).
    - On-chip security AES H/W with Cipher modes of operation: GCM/ECB/CBC and an encryption key length of 128/192/256 bits.
    - On-chip analog features that supports a 24-bit 4ch Delta-Sigma ADC, multiple ADC channels to support input from a variety of sensors, internal reference voltage (1.45 V) and a temperature sensor.
    - Supported peripherals:
        - Data transfer controller to transfer data without CPU intervention.
        - Event link controller (√) routes interrupt event signals from one peripheral to another while the CPU is processing other tasks. Reductions in interrupts improve real-time performance and reduce program size and average power consumption.
        - Timers/counters/CRC/PWM modules/hardware Real-Time Clock (RTC)/watchdog timer.
        - Serial peripheral interfaces SPI/I2C/UART/IrDA.
- *Reduced system cost*—Helping customers reduce system size and cost, on-chip peripheral functions include a high precision (±1%) high-speed on-chip oscillator, background operation data flash supporting 1 million erase/program cycles, a temperature sensor and multiple power supply interface ports. The RL78 Family is fabricated using a newly developed 130 nm process that enables customers to achieve reduced system cost and smaller overall system size [31].
- *Reliable safety functions*—Safety functions are built into the microcontroller to enhance system reliability. The RL78 series of MCUs have several safety functions like error detection, memory guard, fault detection, etc., that check for problems. Customers can use these functions to run diagnostics

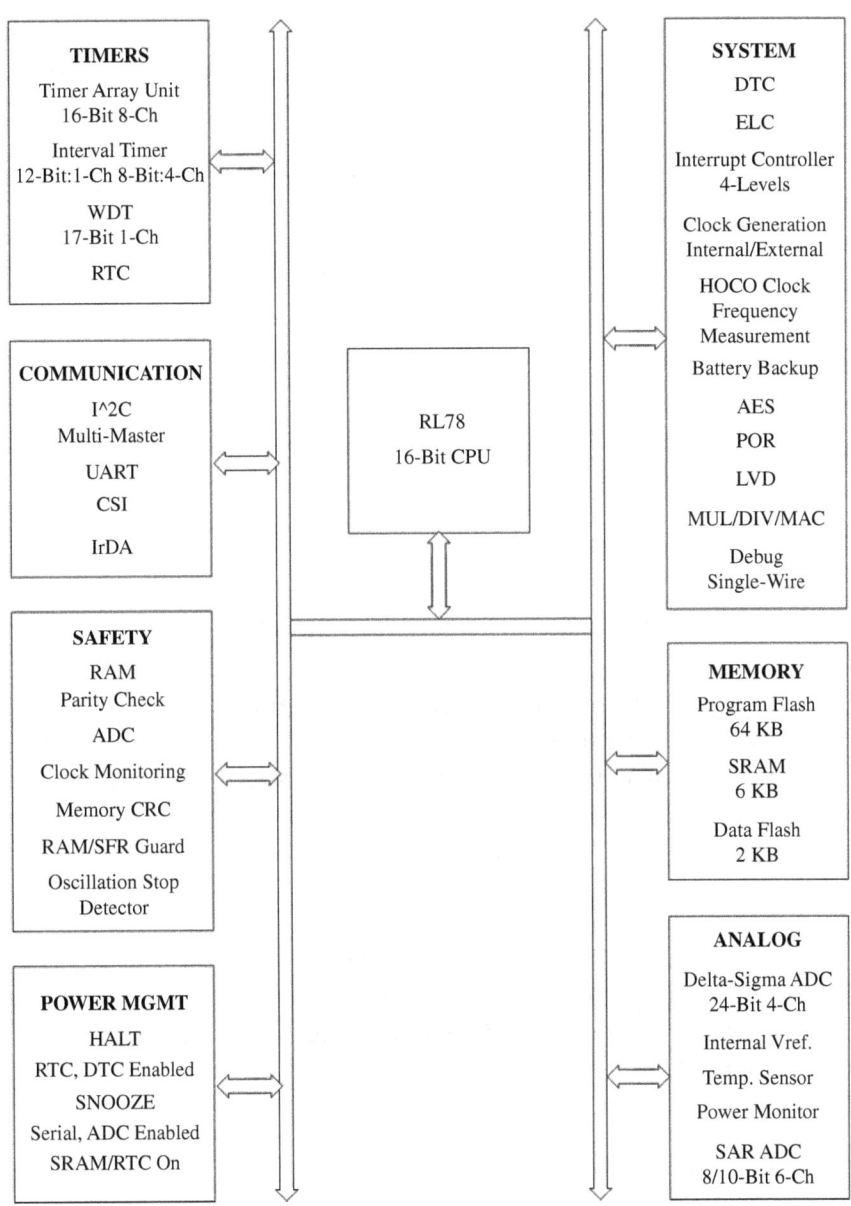

Figure 14. RL78/I1C block diagram [27].

and identify problems with the MCU. This self-diagnostic capability of the RL78 series of MCUs enhances the reliability of the system [31].
   - ECC memory.
   - Compliancy with Safety Standard for Household Appliances (IEC 60730).
   - Supports high operating temperatures (up to 150°C).
   - Detects/avoids system errors.
- *Comprehensive development tools*—Renesas provides a full lineup of tools that facilitate efficient and rapid development, and supports all stages of the development process. These tools include CS+ (a fully integrated development IDE) and e2 Studio (an Eclipse-based IDE). Additional tools include real-time OSes, evaluation and debugging software, and custom flash programming libraries. Renesas' emulator offerings include the E1 and E2 Lite, and support three distinct levels of debugging. To change circuit configurations and features via software settings, Renesas also offers smart analog development tools. Furthermore, third party providers have a diverse range of offerings that have been developed in partnership with Renesas [31].

2.1.3.4.2. Availability of parts

Renesas Electronics has tie-ups with distributors Mouser Electronics, Digi-Key, Arrow, Avnet and Future Electronics to order parts; R5F10NLEDFB#30 parts are non-stocked and backordered.

2.1.3.4.3. Renesas Electronics company profile

Renesas Electronics Corporation was founded in 2002 and is headquartered in Tokyo, Japan. It engages in research, development, design, manufacturing, sale and servicing of semiconductor products in Japan for both domestic and international audiences. The company offers microcontrollers and microprocessors, smart analog ICs, secure microcomputer units, software and tools, power MOSFETs, insulated-gate bipolar transistors, intelligent power devices, triac and thyristor devices, transistors, diodes, power management ICs, analog ICs for automotive, analog and mixed signal ICs and graphic controllers, general-purpose linear ICs, and general-purpose logic devices. It also offers USB power delivery, USB ASSP, RF devices, optoelectronics, memory products, package technology solutions and ROM ordering solutions, as well as LSIs for automotive, factory automation and communications and mobile devices.

- **Financial performance**—Renesas followed up a 11% revenue decrease in 2016 with a 24% jump in 2017 to JPY 788,527 million, i.e., equivalent to US$7.1 billion (Table 3 and Figure 15). Its profits in 2017 rose to $0.7 billion as compared to $0.53 billion in 2016. The company's cash flow from operations was $1.4 billion in 2017, about $0.3 billion higher than 2016 [32].
- **Historical debt**—Renesas' level of debt (44.4%) compared to net worth is high (greater than 40%). However, its level of

Table 3. Renesas' past financial performances [32].

Fiscal Year	Data in JPY billions		
	Revenues	Net Income	Net Operating Cash Flows
2012/03	883.1	−62.6	−9.7
2013/03	785.8	−167.6	−54.1
2014/03	833.0	−5.3	93.7
2015/03	791.1	82.4	116.7
2016/03	693.3	86.3	126.3
2016/12	471.0	44.1	95.0
2017/12	780.3	77.2	164.2

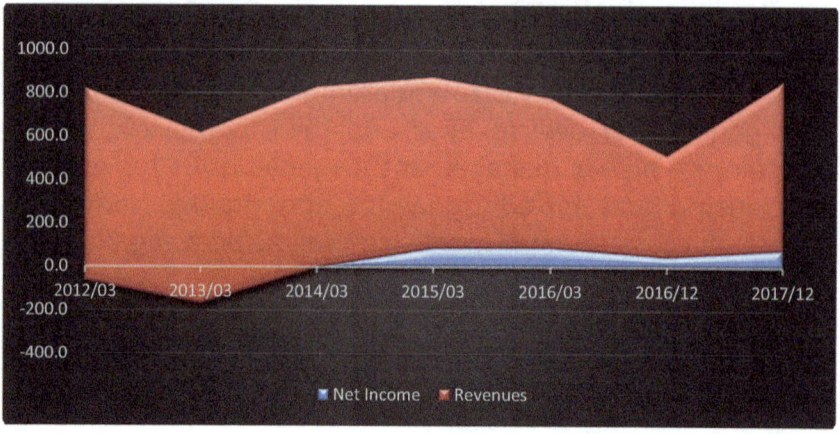

Figure 15. Renesas' revenues and earnings.

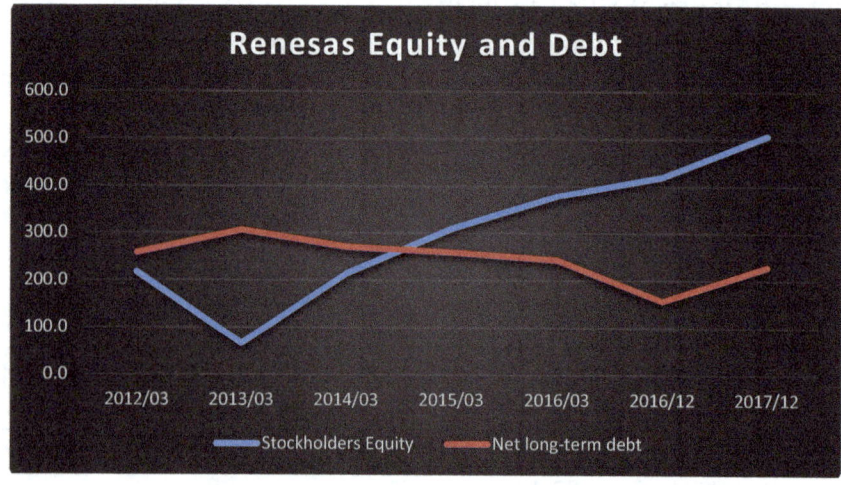

Figure 16. Renesas' historical debt levels [32].

debt has been reduced over the past 5 years (393% vs. 44.4% today) (Figure 16). Its debt is well covered by its operating cash flow. Interest payments on debts are well covered by earnings [32].

3. Data and Data Source(s)

Data was gathered from a panel of experts as well as from an extensive literature review of the IoT, smart water meters, microcontrollers, vendor data sheets, application notes, embedded forums, etc.

3.1. *Panel of experts*

To build and evaluate the model, a panel of experts was established. They had extensive experience in the semiconductor industry, specifically in the areas of embedded product development, Smart Water Meter IoT and microcontroller/microprocessor design. Table 4 contains more information about these experts.

Table 4. The experts panel.

Expert Details		Experience		
Expert	Current Position	Embedded Product Development (S/W, H/W, Application)	Smart Water Meter IoT	Microcontroller/ Microprocessor Development
Expert 1	Marketing Director Embedded Products			
Expert 2	Founder and CEO of an IoT			
Expert 3	Engineering Director			
Expert 4	Hardware Architect			
Expert 5	Corporate Applications Engg			
Expert 6	Product Manager			
Expert 7	Staff Design/Application Engg			
Expert 8	Staff Applications Engg			

These experts were asked to evaluate the model. They were contacted via general communication platforms such as text messages and e-mail, and given an explanation of the objective and what was required of them. Then, each of them received an e-mail with specific details of the model plus consolidated data from Table 8, along with a guide to evaluate the data. Finally, the experts were instructed to use the ETM HDM tool to evaluate the model. Their final evaluation and analysis of the model, derived from studying the criteria and subcriteria, is detailed in the results section.

4. Data

Data was gathered from the experts, as well as from a literature review in the following areas:

- **IoT:** The IoT was created to address the challenges of bridging the gap between the physical world and the information world [33]. It allows electronic devices to wirelessly communicate with each other and store information on the internet. This information is analyzed and used to optimize various systems, making them more efficient and hence saving energy, time, money and even lives. From a more technological standpoint, IoT is a network of smart devices interacting and communicating with each other and with objects, environments and infrastructures. They aggregate the data generated by each device individually, and process this massive data bank to create useful commands and functions for both home and business environments [3].
- **IoT use cases/applications:** Listed below are some applications of IoT technology [3]:
  - Machine-to-machine communication,
  - Machine-to-infrastructure communication,
  - Real-time monitoring of patients coupled with diagnosis and drug delivery,
  - Continuous monitoring and upgrading of vehicles,

- Tracking mobile goods/assets,
- Automated traffic management,
- Remote security and control,
- Environmental monitoring and control,
- Home and industrial automation, applicable in a variety of settings including cities, water, agriculture, buildings, grid, meters, broadband, cars, appliances, tags, animal farming and the environment.

- **Smart Water Meter:** Water is a vital resource for human survival; hence, solutions that improve the conservation and management of water resources are important. Smart Water Meters are electronic measuring devices that perform three basic functions: they automatically and electronically capture, collect and communicate real-time water usage readings to facilitate monitoring and billing [34].
- **Microcontrollers:** A microcontroller is a microprocessor that has been optimized for embedded control applications, which typically involve monitoring numerous single-bit control signals but do not require large data computations. Several peripheral components required for these applications, such as serial communication peripherals, timers, counters, pulse-width modulators and analogue-digital converters are integrated into the microcontroller. This integration of peripherals enables single-chip implementations and results in smaller and lower-cost products [4].
- **Microcontroller classification based on architecture:** Based on the number of bits, MCUs are of four different types: 4-bit, 8-bit, 16-bit and 32-bit. 4-bit microcontrollers are typically used in electronic toys, while 8-bit microcontrollers are used in various control applications. 16-bit microcontroller are designed and developed specifically for high-speed control applications, and are programmed either by a high-level programming language or an assembly language [16].
- **Smart Water Meter IoT system requirements** [3]: Figure 17 provides the functional view of IoT technologies [3].

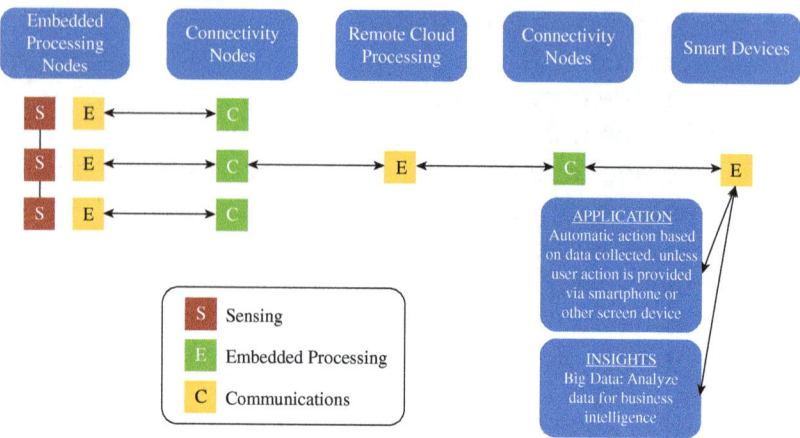

Figure 17. Functional view of IoT [3].

The following requirements are common too:

- **Sensing and data collection capability (sensing nodes):** Flow meter sensor for a Smart Water IoT that converts the water flow into an electrical signal, which can be accepted and processed by the MCU.
- **Layers of local embedded processing capability (embedded processing nodes):** Embedded processing is at the heart of the IoT and integrated MCUs that provide "real-time" embedded processing is a key requirement.
- **Wired and/or wireless communication capability (connectivity nodes):** Low power RF radios are typically used to facilitate communication between a battery-powered water meter and another meter, either in a mesh network or a data collector. Requirements for communication functions, such as cost-effectiveness, low power, quality, reliability and security are almost the same as those for a microcontroller.
- **Software for automating tasks and enabling new services:** Software plays a key role in connecting all segments of the IoT and making them work together to successfully rollout new technology, products and services.

○ **Remote network/cloud-based embedded processing capability (remote embedded processing nodes):** Accessing remote supercomputing nodes (cloud computing) for heavy-duty data processing and analysis.
  ○ **Full security across the signal path:** Ensuring the security of the information that gets passed around by various parts of the IoT system, as well the information from each of these parts, at the device-level.

- **MCU requirements for Smart Water Meter IoT [3]:**

   The block diagram in Figure 18 from TI [34] highlights the different connectivity options for an MCU in a generic flow meter topology.
   The following are the most important:

   ○ **Energy efficiency:** Microcontrollers need to have low power consumption rates as their sensors are battery-operated satellite nodes. Hence, good energy efficiency would minimize battery replacement.

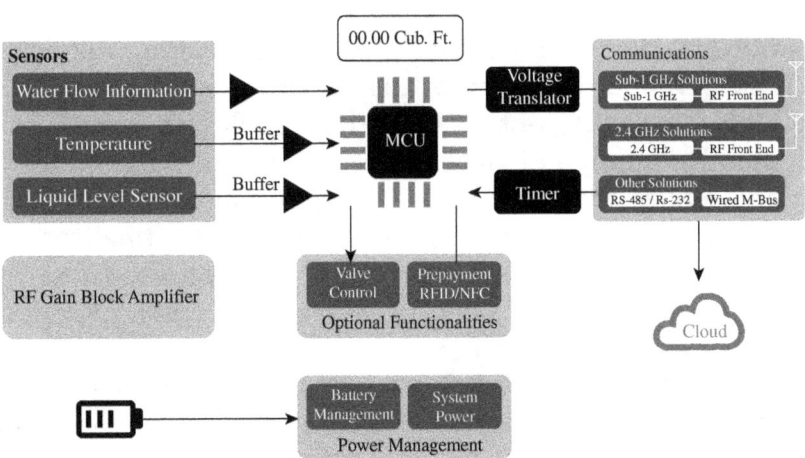

Figure 18. A TI-enabled Smart Flow Meter solution featuring various connectivity options [34].

- ○ **Embedded architecture with a rich software ecosystem:** The MCU's supporting IoT projects need to have a broad software development environment available. The software ties together the applications, commands, controls, routing processes and the security of the system.
- ○ **Portfolio breadth that cost-effectively enables different levels of performance and a robust mix of I/O interfaces:** There are many different MCUs available on the market, each with different tiers of devices and diverse I/Os. The MCU that is cost-performance-optimized for the specific application needs to be chosen.
- ○ **Cost-effectiveness:** The overall product cost involves the costs of both the system's parts as well the services required for maintenance, and needs to be minimized.
- ○ **Quality and reliability:** Product life cycles in the industrial market are typically 10–15 years long. Hence, the long-term quality, reliability and longevity of a microcontroller are critical to the success of the application.
- ○ **Security:** A variety of cryptographic engines and security accelerators to support data encryption and authentication, such as DES and AES for the former and SHA for the latter, are available.

- **Criteria and subcriteria:** After conducting extensive research and reviewing of the IoT ecosystem, Smart Water Meter IoTs, different MCU vendors' application notes, various literature and datasheets, different semiconductors, and embedded forums and blogs, and also by taking into account the panel of experts' opinions, we finalized the most important selection criteria and factors. These are shown in Table 5, specifically when selecting an MCU for a Smart Water Meter IoT [3, 4, 6].
- **Microcontroller Alternatives:** Once again, after conducting extensive research of different MCU vendors and

Table 5. Criteria and subcriteria for selecting an MCU for a Smart Water Meter IoT.

Objective	Criteria	Subcriteria
	Suitability	Voltage
		I/O pins
		Peripherals
		CPU throughput
		Affordability
	Availability	Quantity
		Production status
		Future
	Development support	Assemblers/compilers
		Debug tools
		On-line support
		Applications support
	Manufacturers track record	Competency in design
		Reliability in silicon
		Years in business
		Financials

their product offerings, the available options are shown in Table 6 [35].

Table 7 lists the most important factors in choosing an MCU, which are based on research and the opinions of the panel of experts.

Finally, after analyzing their features as described in their datasheets, the following four microcontroller options were finalized:

- PIC24FJ64GA204 from Microchip Technologies.
- MSP430FR6972 from Texas Instruments.

Table 6. List of manufacturers and their products.

MCU Manufacturers/ Vendors	Products
Microchip	PIC10, PIC12, PIC16 series, PIC18 series (8-bit), PIC24, dsPIC (16-bit), PIC32MX series
Atmel (now Microchip)	AT89 series (Intel 8051 architecture), AT90, ATtiny, ATmega, ATxmega series (AVR architecture), AT91SAM (ARMarchitecture), AVR32 (32-bit AVR architecture), MARC4
Freescale Semiconductor (now NXP)	68HC05, 68HC08, 68HC11 (8-bit), 68HC12, 68HCl6 (16-bit), 683XX, MCF5xxx, M-core, MPC500, MPC860 (32-bit)
NXP Semiconductor	80C51 (8-bit), XA (16-bit), ARM7/LPC2000, ARM9/LPC3000, ARM Cortex-M0/LP C800, LPC 1100, LPC 1200, ARM Cortex-M3/ LPC1300, LPC1700, LPC1800, ARM Cortex-M4/ LPC4300
Texas Instruments	TMS370 (8-bit), MSP430 (16-bit), TMS 320, ARM Cortex-R4/TMS570 (32-bit)
Renesas	RL 78 16-bit MCU; RX 32-bit MCU; SuperH; V850 32-bit MCU; H8; R8C 16-bit MCU
ST Microelectronics	ST6, ST7, STM8, uPSD (8-bit), ST10 (16-bit), ST20, ARM7/STR7, ARM9/STR9, ARM Cortex-M0/STM32 F0, ARM Cortex-M3/STM32 F1, F2, ARM Cortex-M4/STM32 F4

- ATSAML22G16A from Atmel.
- R5F10NLEDFB#30 from Renesas.

For the final step of selecting an MCU, Table 8 was constructed with each microcontroller listed on one axis, with the important criteria and subcriteria on the other. For a fair

Table 7. Most important MCU requirements for the Smart Water IoT.

MCU Requirements for Smart Water Meter IoT
Low Power – 2.5 v – 3.6 v
16-bit Architecture
CPU Throughput – 16 Mhz or less
Flash – 64 KB
RAM – 4 KB
Digital Sensor Interface (Flow sensor & Temp Sensor) – GPIO with interrupt
RFE Interface – UART/I2C/ SPI
Battery Interface – PWM and ADC
TIMERS
RTC
CRYPTO
UART for Debug
LCD Interface
ADC – 8 ch
Comparator
I/Os – 32 pins or 44 pins or 64 pins max with package size of 64 (max)
Operating Temp Range — 20C to 85C

comparison, blanks were filled in with data from the manufacturer's data sheets.

5. Analysis and Key Findings

After analyzing and evaluating the results of the HDM, which included taking into consideration the opinions of the eight experts who have knowledge of and long experience in the fields of Embedded Product Development, Smart Water IoTs and microcontrollers, a conclusion can be reached through the final calculation results shown in Appendix A.

Table 8. Criteria and subcriteria data of Alternatives.

Criteria	Sub-Criteria		MicroChip PIC24FJ64GA204 (16-bit)	Texas Instruments MSP430FR6972 (16-bit)	Atmel ATSAML22G16A (32-bit)	Renesas R5F10NLEDFB#30 (16-bit)
Suitability	Voltage		2 to 3.6 V	1.8 to 3.6 V	1.62 to 3.63 V	1.7 to 5.5 V
	I/O Pins (Package Size)		35 pins (44 size)	51 pins (44 size)	36 pins (48 size)	35 pins (64 size)
	Peripherals	DATA RAM	8 KB	2 KB	8 KB	6 KB
		PROGRAM FLASH	64 KB	64 KB	64 KB	64 KB
		DATA FLASH			2 KB	2 KB
		Watchdog Timer	Yes	Yes	Yes	17-bit 1 ch
		PWM	Yes	Yes	Yes	Yes
		Comparators	3	8 ch Analog	2 Analog	Yes
		OSC (On Chip Oscillator Clock)	Yes	Yes	Yes	Yes
		DMA	6 ch	3 ch	16 ch	Yes
		RTC	Yes	Yes	Yes	Yes
		UART	4 UART w/IrDA	2	Yes	3
		ADC – Channels	13 ch, 10/12-bit ADC	8 ch, 12-bit ADC	10 ch 12-bit ADC	4 ch 10-bit ADC

Technical Transformation: Internet of Things

			5, 16-bit Timers	5, 16-bit Timers	4 16-bit, 1 24-bit	4 8-bit, 8 16-bit
		TIMER				
		SPI	3	4	3	
		I2C	2	2	2	3
		CRYPTO	Yes	Yes	Yes	Yes
		LCD	No	116-segment LCD	Yes	Yes
		CPU Throughput	32 MHz	16 MHz	32 MHz	24 MHz
		Affordability	2.96	2.62 (>4000 units)	2.46	2.86 (>2560 units)
	Availability	Quantity	1600 units available now, 16 weeks lead time >1600 units	2082 units available now, 9 weeks lead time >2082 units	25,000 units available now, 26 weeks lead time >25,000 units	Backordered
		Production Status	Yes	Yes	Yes	Yes
		Future availability	Yes	Yes	Yes	Yes
Development Support		Assemblers/Compilers	Yes	Yes	Yes	Yes
		Debug Tools	Yes	Yes	Yes	Yes
		On-line Support	Yes	Yes	Yes	Yes
		Applications Support	Yes	Yes	Yes	Yes
Manufacturers Track Record		Competency in Design	Yes	Yes	Yes	Yes
		Reliability in Silicon	Yes	Yes	Yes	Yes
		Years in Business	29 Yrs	80 Yrs	29 Yrs	15
		Financials (12/31/2017) – (Net Income)	246 M	3.6 B	246 M	0.7 B

5.1. Relative weights of Alternatives towards the Objective

Appendix B illustrates the relative weights of Alternatives towards the Objective (Figure 19). The higher the weight, the more important that particular Alternative was for satisfying the decision level. Thus it can be seen from the data in Appendix B that the Atmel ATSAML22G16A is the best microcontroller choice with a weight of 0.28, followed by both the Microchip PIC24FJ64GA406 and TI MSP430FR6972 microcontrollers. The Alternatives were chosen to have competing features, so the results were very close.

The inconsistency in an individual expert's judgmental value is within limit. For each expert the inconsistency is <0.1, as shown in Figure 20. This is an acceptable value, as a value near to zero indicates close consensus. Disagreement values help identify consensus among experts in pairwise comparisons, and the disagreement value for the data obtained from this panel of experts was only 0.015. Low inconsistency and disagreement values indicate that the experts are in close agreement. The disagreement and inconsistency results for this data highly support the model and illustrate its reliability.

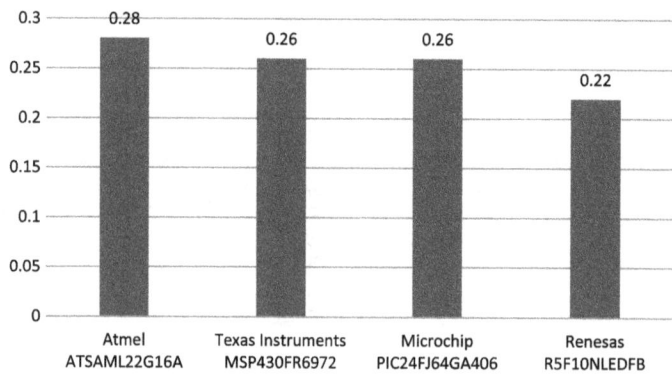

Figure 19. Relative weights of Alternatives towards the Objective.

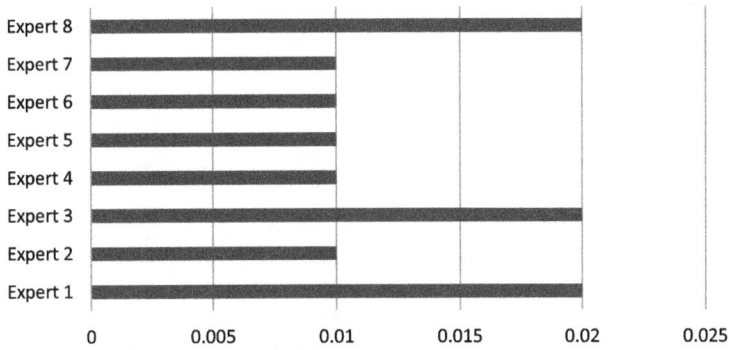

Figure 20. Inconsistency measurement of Experts' Judgment of Alternatives towards the Objective.

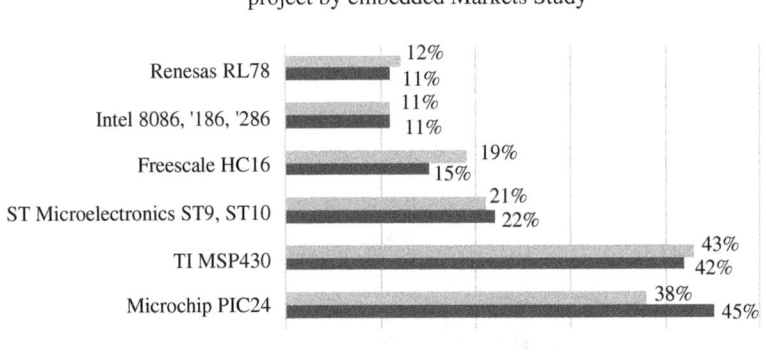

Figure 21. The 2017 Embedded Markets Study—16-bit chip families to consider [36].

According to a 2017 Embedded Markets Study as shown in Figure 21, the top two 16-bit chips under consideration are the Microchip PIC 24 (dsPIC) and the TI MSP430 [36]. In our model, the 16-bit chips selected are the Microchip PIC24FJ64GA406, the TI MSP430FR6972 and the Renesas R5F10NLEDFB. Our model results are comparable to this study.

According to another 2017 Embedded Markets Study (see Figure 22), Microchip and TI have the best ecosystems [36].

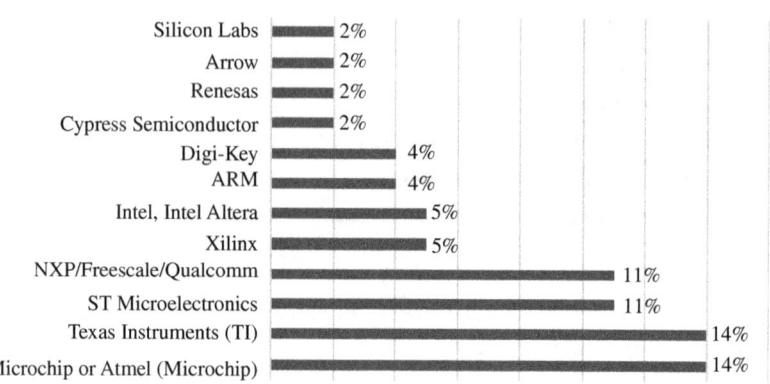

Figure 22. The 2017 Embedded Markets Study—best ecosystems [36].

Hence, our model's results are comparable to this study, as Microchip (Atmel) and TI have the highest weightage based on Expert Judgments.

5.1.1. *Level 1: Relative weights of each criteria towards the Objective*

Table A.1 illustrates the relative weights of the Criteria towards the Objective. From the table we can conclude that Suitability, with a weight of 0.33, is the most important Criteria towards the Objective, followed by Availability, Manufacturer's Track Record and Development Support, with weights of 0.24, 0.22 and 0.21, respectively, as shown in Figure 23.

5.1.2. *Level 2: Relative weights of each Subcriteria towards the Objective via the Suitability criteria*

Table A.2 illustrates the relative weights of each Subcriteria towards the Objective via the Suitability Criteria (Figure 24).

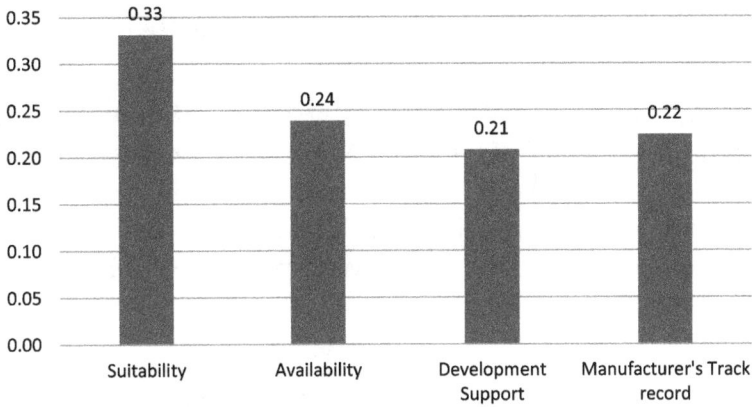

Figure 23. Level 1: Relative weights of criteria towards the Objective.

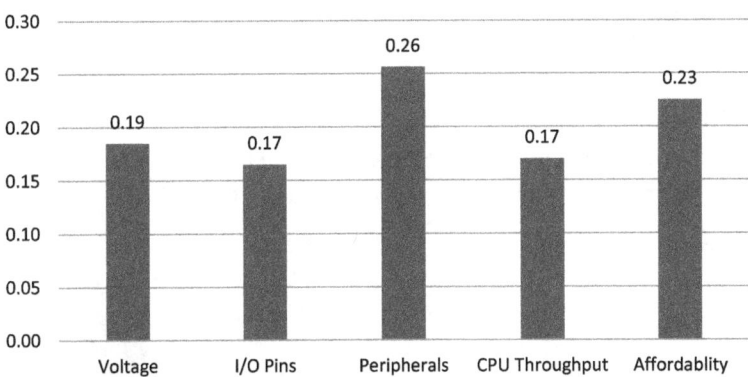

Figure 24. Relative weights of the Subcriteria towards the Suitability perspective.

From the table we can conclude that Peripherals, with a weight of 0.26, make up the most important Subcriteria towards the objective, followed by Affordability, Voltage, I/O Pins and CPU Throughput, with weights of 0.23, 0.19, 0.17 and 0.17, respectively.

5.1.3. Level 2: Relative weights of each Subcriteria towards the Objective via the Availability Criteria

Table A.3 illustrates the relative weights of the Subcriteria towards the Objective via the Availability Criteria. From the table we can conclude that Future Availability is the most important Subcriteria, with a weight of 0.35. It is followed by Quantity and Production Status with weights of 0.33 and 0.32, respectively (see Figure 25).

5.1.4. Level 2: Relative weights of each Subcriteria towards the Objective via the Development Support criteria

Table A.4 illustrates the relative weights of each Subcriteria towards the Objective via the Development Support Criteria (Figure 26). From the table we can conclude that Future Availability is the most important Subcriteria with a weight of 0.35, followed by Quantity and Production Status with weights of 0.33 and 0.32, respectively.

According to the 2017 Embedded Markets Study, the most important software/hardware tools (see Figure 27) are

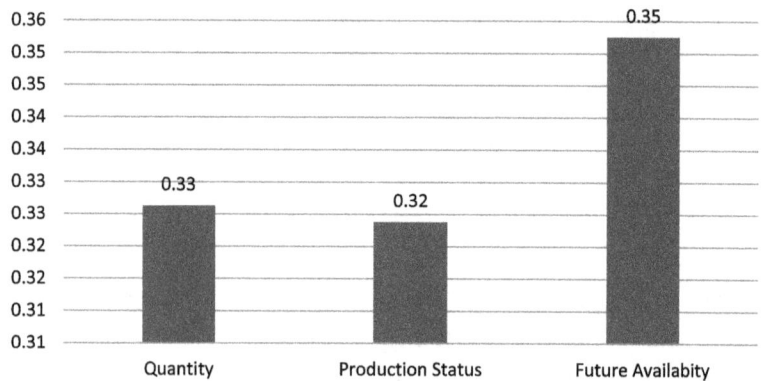

Figure 25. Relative weights of the Subcriteria towards the Availability perspective.

Technical Transformation: Internet of Things 129

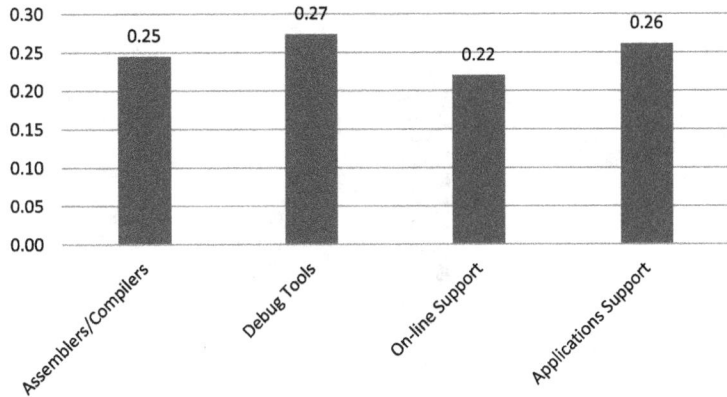

Figure 26. Relative weights of the Subcriteria towards the Development Support perspective.

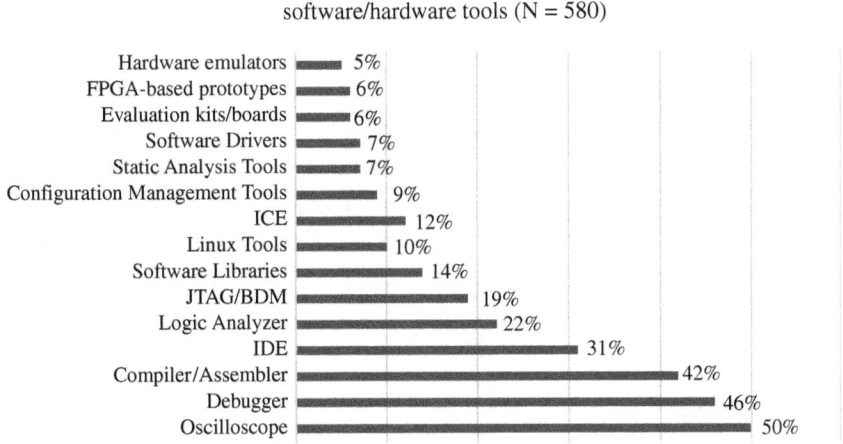

Figure 27. The 2017 Embedded Markets Study—important software/hardware tools [36].

oscilloscopes, debuggers, compilers/assemblers, etc. [36]. Our model results indicate the availability of Debug Tools as being the most important subcriteria.

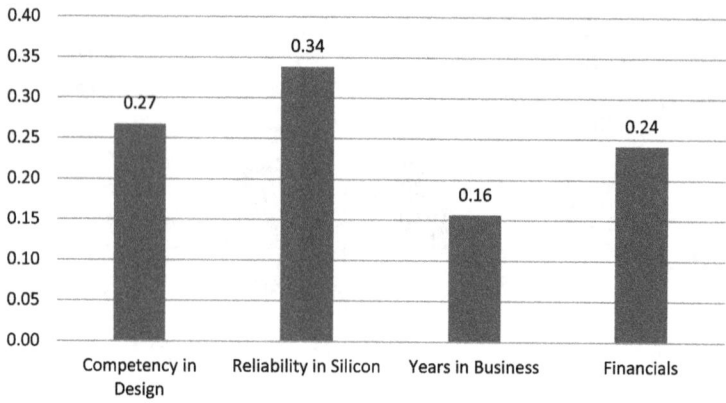

Figure 28. Relative weights of the Subcriteria towards the Manufacturers Track Record perspective.

5.1.5. *Level 2: Relative weights of each Subcriteria towards the Objective via the Manufacturer's Track Record Criteria*

Table A.5 illustrates the relative weights of each Subcriteria towards the Objective via the Manufacturer's Track Record Criteria (Figure 28). From the table we can conclude that Reliability in Silicon is the most important Subcriteria, with a weight of 0.34. This is followed by Competency in Design, Financials, and Years in Business with weights of 0.27, 0.24 and 0.16, respectively.

5.2. *Discussion*

The results from the experts' evaluation of the model indicate that the most important criteria for selecting an MCU for the Smart Water Meter IoT are Suitability and Availability. The most important subcriteria are Peripherals, Affordability, Voltage, Future Availability, Debug Tools, Applications Support and Reliability in Silicon. The following consist of a discussion of those criteria and subcriteria based on existing literature review and the feedback received from experts.

The MCU's suitability for the application is one of the main criteria in selecting a microcontroller, since this covers important aspects such as Operating Voltage, required number of I/O (input/output) pins/ports, and various peripherals including serial I/O, RAM, ROM, A/D converters, CPU throughput and affordability. Thus, Suitability has the highest weightage according to the experts' judgment.

Availability is an important consideration for commercial designs. There will be different requirements for different stages of development, for instance, smaller quantities for engineering samples and larger ones during mass production. Also, make sure that the MCU is available in the future by ensuring that it is not nearing its EoL [12].

Battery powered applications such as the Smart Water Meter IoT require low power consumption for the embedded MCU. Power consumption is the product of operating voltage (Vcc) multiplied by the current consumption (Icc). Decreasing Vcc directly reduces the current consumption and the overall power drain. So, low operating voltage is one of the key areas of concern when selecting an ultra-low power MCU [37].

Many vendors are offering chips with similar features and at almost similar prices across various segments. So, the support and development tools they offer play a very important role when selecting an MCU. "The availability and richness of the development tools is a key aspect. As these tools are expensive and mostly tied to the microcontroller, they need to be evaluated clearly, ensuring flexibility of the tool in supporting multiple families with ease. The debugging capabilities and an easy graphic user interface (GUI) for configuring the various peripherals of the MCU helps a lot [12]".

According to the 2017 Embedded Markets Study, the important factor in choosing a processor is the availability of software and hardware development tools, along with the chip's performance, cost, the on-chip I/O or peripherals,

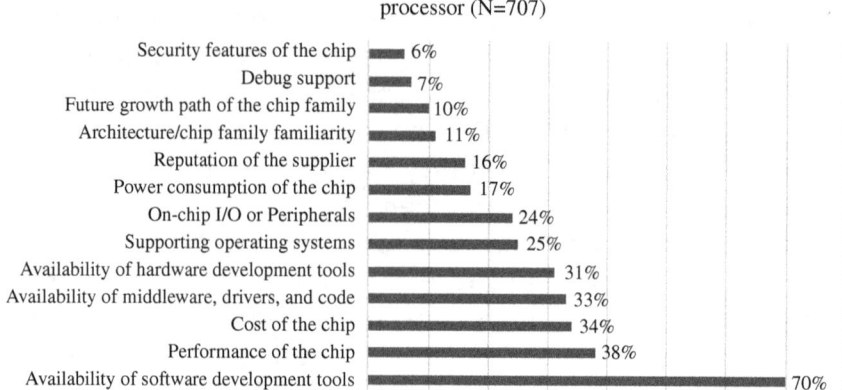

Figure 29. 2017 Embedded Market Study—most important factors in choosing a processor [36].

power consumption, future growth path, etc., which are comparable to our model's most important Subcriteria (Figure 29).

6. Future Research

Listed below are some of the limitations of this paper that can be addressed in future works:

- In the literature review there are a number of scholarly papers explaining the criteria for choosing a microcontroller for a suitable application. However, there is no literature that explains how to choose the right microcontroller for an IoT System.
- Microcontrollers have many different competing requirements to consider. Even though this model tried to cover most of the important criteria and subcriteria, there is room to expand. Additional criteria such as Maintainability, Flexibility, Scalability, Security, Low Power Modes, etc., can be addressed in future works.

- This model considered only four MCU alternatives, but there are more alternatives on the market, including a few wireless MCUs. These too can be taken into consideration in future works.
- This model focused on the Smart Water Meter IoT application. In the future, this model can either be generalized or developed specifically for other IoT applications like automotives, wearables or any other embedded applications.
- Multiple expert panels at different levels can be used.
- Real case studies can be considered to better evaluate the reliability and robustness of this model.

References

1. "Gartner says 8.4 billion connected 'things' will be in use in 2017, up 31 percent from 2016", *Gartner*, February 2017. https://www.gartner.com/newsroom/id/3598917.
2. J. Chase, "The evolution of the internet of things—from connected things to living in the data, preparing for challenges and IoT readiness", *Texas Instruments,* September 2013. http://www.ti.com/lit/ml/swrb028/swrb028.pdf.
3. "What the Internet of Things (IoT) needs to become a reality", NXP Semiconductor. http://www.nxp.com/assets/documents/data/en/white-papers/INTOTHNGSWP.pdf.
4. F. Takawira and D. S. Dawoud, "Selecting the right microcontroller unit", *Zimbabwe Journal of Science and Technology,* 2010, 4(1) 39–46.
5. T. U. Daim and D. F. Kocaoglu, *Hierarchical Decision Modeling: Essays In Honor of Dundar F. Kocaoglu* (Cham: Springer, 2016).
6. "Selecting the microcontroller unit", Freescale Semiconductor. http://www.freescale.com/files/microcontrollers/doc/app_note/AN1057.pdf.
7. Y. W. Lee, S. Eun and S. H. Oh, "Wireless digital water meter with low power consumption for automatic meter reading", in *IEEE International Conference for Convergence and Hybrid Information Technology,* 2008, pp. 639–645.
8. F. Feckl, "Sensing the IoT: IoT long-life wireless sensors require ultra-low power architectures", *ECN-Electronic Component News,* December 2015, p. 12.

9. I. Khajenasiri, P. Zhu, M. Verhelst and G. Gielen, "A low-energy ultra-wideband internet-of-things radio system for multi-standard smart-home energy management", *IEIE Transactions on Smart Processing and Computing* 4(5) (2015) 354–365.
10. J. J. Vaglia and P. S. Gilmour, "How to select a microcontroller", *IEEE Spectrum* 27 (1990) 106–109.
11. "Selecting the right MCU can squeeze nanoamps out of your next Internet of Things application", SiliconLabs, 2014. http://pages.silabs.com/rs/silabs/images/select-the-right-mcu-for-iot-app.pdf.
12. "How to select a microcontroller", *Electronics For You*, December 2012, pp. 54–60.
13. B. Gundry and T. Hampton, "Don't neglect design and development tool support when choosing an MPU", *Computer Design*, 12 February 1990, p. 16.
14. "Popular development tools from IAR systems support complete Atmel/SMART MCU and MPU portfolio", *PR Newswire*, 13 November 2014.
15. "Complete development tools platform further simplifies MCU design process", *ENP Newswire*, 24 April 2013.
16. M. K. Parai, B. Das and G. Das, "An overview of microcontroller unit: from proper selection to specific application", *International Journal of Soft Computing and Engineering* 2(6) (2013) 2231–2307.
17. "PIC24FJ128GA204 FAMILY datasheet", Microchip Technologies. http://ww1.microchip.com/downloads/en/DeviceDoc/30010038c.pdf.
18. "PIC24F curiosity development board", Microchip Technologies. http://www.microchip.com/DevelopmentTools/ProductDetails/PartNO/DM240004.
19. "What investors should know about microchip technology incorporated's (NASDAQ:MCHP) financial strength", *Simply Wall St.* https://simplywall.st/stocks/us/semiconductors/nasdaq-mchp/microchip-technology.
20. "MSP430FR6972 datasheet", Texas Instruments. http://www.ti.com/product/MSP430FR6972/datasheet.
21. "MSP430FR6989 LaunchPad™ development kit (MSP-EXP430FR6989)", Texas Instruments. http://www.ti.com/lit/ug/slau627a/slau627a.pdf.
22. "TI designs ultrasonic water flow measurement", Texas Instruments. http://www.ti.com/lit/ug/tidua10a/tidua10a.pdf.

23. "Investor overview", Texas Instruments. http://www.ti.com/corp/docs/investor_relations/downloads/investor_overview.pdf.
24. "What investors should know about Texas Instruments Incorporated's (NASDAQ:TXN) financial strength", *Simply Wall St.* https://simplywall.st/stocks/us/semiconductors/nasdaq-txn/texas-instruments.
25. "ATSAML22G16A datasheet", Atmel, now Microchip Technologies. http://www.microchip.com/wwwproducts/en/ATSAML22G16A.
26. "Atmel expands line of new innovative ultra-low-power ARM Cortex M0+ MCUs with segment LCD controller", *ENP Newswire*, 19 August 2015.
27. "R5F10NLEDFB#30 datasheet", *Renesas*. https://www.renesas.com/en-us/products/microcontrollers-microprocessors/rl78/rl78i1x/rl78i1c.html.
28. "Renesas Electronics launches RL78/I1C group of microcontrollers supporting meter international standards for smart meters", *Wireless News*, 30 August 2016.
29. "Smart meter solutions catalog", *Renesas*, September 2017. https://www.renesas.com/en-us/doc/application/r30ca0164ej0200-meter.pdf.
30. "RL78 microcontrollers, featuring snooze mode for energy-efficient applications", *Renesas*. https://www.renesas.com/en-us/docs/products/microcontrollers-microprocessors/rl78/RL78_20page_080116a.pdf.
31. "RL78 family, Renesas microcontrollers", *Renesas*. https://www.renesas.com/en-us/doc/products/mpumcu/doc/rl78/r01cp0003ej0800-rl78.pdf.
32. "What investors should know about Renesas Electronics financial strength", *Simply Wall St.* https://simplywall.st/stocks/us/semiconductors/otc-rnec.f/renesas-electronics.
33. P. Suresh, J. V. Daniel, V. Parthasarathy and R. H. Aswathy, "A state of the art review on the Internet of Things (IoT) history, technology and fields of deployment", in *International Conference on Science Engineering and Management Research (ICSEMR)*, 2014, pp. 1–8.
34. E. Idris, "Smart metering: a significant component of integrated water conservation system", *Proceedings of the 1st Australian Young Water Professionals Conference*, International Water Association, Sydney, 2006.
35. "List of common microcontrollers", *Wikipedia*. http://en.wikipedia.org/wiki/List_of_common_microcontrollers.

36. AspenCore Network, "2017 embedded markets study, integrating IoT and advanced technology designs, application development and processing environments", *Eetimes/embedded.com,* April 2017.
37. R. Dasgupta, "Low power MCU selection criteria and sleep mode implementation using hardware/software CoDesign technique", *Techonline,* 2009. https://www.techonline.com/electrical-engineers/education-training/tech-papers/4137624/low-power-mcu-selection-criteria-and-sleep-mode-implementation-using-hardware-software-codesign-technique?cid=nl_embedded.

Appendix A: Final, Quantified Model

Computation of the average of the weights on Levels 1 and 2 criteria (Figure A.1 and Tables A.1–A.5).

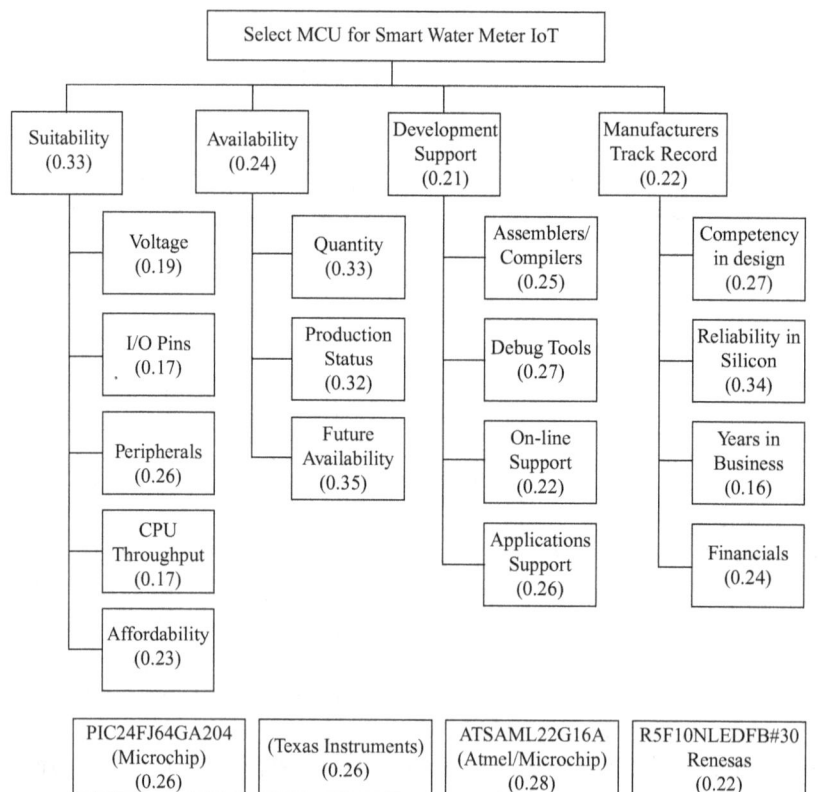

Figure A.1. Final quantified model.

Table A.1. Level 1: Relative weights of each criteria towards the objective.

Level 1	Suitability	Availability	Development Support	Manufacturer's Track Record	Inconsistency
Expert 1	0.52	0.26	0.15	0.07	0.03
Expert 2	0.43	0.12	0.18	0.28	0.03
Expert 3	0.43	0.13	0.28	0.16	0.02
Expert 4	0.26	0.29	0.27	0.19	0.19
Expert 5	0.27	0.25	0.26	0.22	0.04
Expert 6	0.25	0.32	0.13	0.29	0.01
Expert 7	0.24	0.29	0.14	0.33	0.01
Expert 8	0.25	0.25	0.25	0.25	0
Mean	0.33	0.24	0.21	0.22	
Minimum	0.24	0.12	0.13	0.07	
Maximum	0.52	0.32	0.28	0.33	
Std. Deviation	0.11	0.07	0.06	0.08	

Table A.2. Level 2: Relative weights of each subcriteria to the objective through the Suitability criteria.

Level 2— Suitability	Voltage	I/O Pins	Peripherals	CPU Throughput	Affordability	Inconsistency
Expert 1	0.06	0.17	0.35	0.21	0.22	0.07
Expert 2	0.09	0.3	0.3	0.18	0.13	0.03
Expert 3	0.17	0.14	0.29	0.1	0.3	0.03
Expert 4	0.2	0.14	0.31	0.27	0.08	0.01
Expert 5	0.25	0.17	0.25	0.2	0.14	0.01
Expert 6	0.2	0.16	0.14	0.16	0.34	0.02
Expert 7	0.17	0.09	0.15	0.13	0.45	0.09
Expert 8	0.34	0.15	0.26	0.11	0.14	0.13
Mean	0.19	0.17	0.26	0.17	0.23	
Minimum	0.06	0.09	0.14	0.1	0.08	
Maximum	0.34	0.3	0.35	0.27	0.45	
Std. Deviation	0.09	0.06	0.08	0.06	0.13	

Table A.3. Level 2 Criteria—Availability.

Level 2—Availability	Quantity	Production Status	Future Availability	Inconsistency
Expert 1	0.38	0.46	0.16	0.01
Expert 2	0.08	0.1	0.82	0
Expert 3	0.13	0.28	0.59	0
Expert 4	0.33	0.43	0.25	0
Expert 5	0.29	0.43	0.29	0
Expert 6	0.33	0.43	0.25	0
Expert 7	0.74	0.13	0.13	0
Expert 8	0.33	0.33	0.33	0.18
Mean	0.33	0.32	0.35	
Minimum	0.08	0.1	0.13	
Maximum	0.74	0.46	0.82	
Std. Deviation	0.20	0.14	0.24	

Table A.4. Level 2: Development Support.

Level 2—Development Support	Assemblers/Compilers	Debug Tools	On-line Support	Applications Support	Inconsistency
Expert 1	0.43	0.29	0.08	0.19	0.06
Expert 2	0.17	0.36	0.19	0.27	0.03
Expert 3	0.09	0.28	0.32	0.32	0.01
Expert 4	0.22	0.2	0.27	0.3	0
Expert 5	0.25	0.27	0.18	0.3	0
Expert 6	0.29	0.25	0.19	0.27	0.03
Expert 7	0.21	0.21	0.37	0.22	0
Expert 8	0.3	0.33	0.16	0.22	0.01
Mean	0.25	0.27	0.22	0.26	
Minimum	0.09	0.2	0.08	0.19	
Maximum	0.43	0.36	0.37	0.32	
Std. Deviation	0.10	0.05	0.09	0.05	

Table A.5. Level 2: Manufacturers Track Record.

Level 2—Manufacturers Track Record	Competency in Design	Reliability in Silicon	Years in Business	Financials	Inconsistency
Expert 1	0.39	0.31	0.12	0.19	0.07
Expert 2	0.22	0.32	0.17	0.3	0.07
Expert 3	0.22	0.24	0.1	0.44	0.08
Expert 4	0.29	0.36	0.19	0.16	0.01
Expert 5	0.29	0.36	0.16	0.19	0.01
Expert 6	0.16	0.33	0.23	0.28	0.03
Expert 7	0.21	0.47	0.14	0.18	0.16
Expert 8	0.36	0.32	0.14	0.19	0
Mean	0.27	0.34	0.16	0.24	
Minimum	0.16	0.24	0.1	0.16	
Maximum	0.39	0.47	0.23	0.44	
Std. Deviation	0.08	0.06	0.04	0.09	

Appendix B: AHP/HDM PCM Data Tables

Table B.1 shows the final results of the online HDM tool for the decision model.

Table B.1. Relative value of each Alternative towards the Objective.

Objective	Microchip PIC24FJ64GA406	Texas Instruments MSP430FR6972	Atmel ATSAML22G16A	Renesas R5F10NLEDFB	Inconsistency
Expert 1	0.25	0.28	0.27	0.2	0.02
Expert 2	0.26	0.26	0.27	0.21	0.01
Expert 3	0.27	0.25	0.29	0.19	0.02
Expert 4	0.25	0.25	0.28	0.22	0.01
Expert 5	0.25	0.25	0.26	0.23	0.01
Expert 6	0.24	0.24	0.32	0.21	0.01
Expert 7	0.27	0.28	0.23	0.23	0.01
Expert 8	0.25	0.23	0.29	0.23	0.02
Mean	0.26	0.26	0.28	0.22	
Minimum	0.24	0.23	0.23	0.19	
Maximum	0.27	0.28	0.32	0.23	
Std. Deviation	0.01	0.02	0.02	0.01	
Disagreement					0.015

Chapter 5

Technical Transformation: IT in Disaster Management

Namitha Shetty* and Tugrul Daim*,†,‡

*Portland State University, Portland, Oregon, USA
†Higher School of Economics, Moscow, Russia
‡Chaoyang University of Technology, Taiwan

Abstract

To develop an integrated framework to understand the field of disaster management and help nonprofit organizations take effective technology management decisions in terms of availability of technology, effective response to disruptions, mitigate risk and improve their overall disaster management plan by using IT as a tool.

Keywords: Technology assessment, disaster management, information technology.

1. Introduction

According to the World Health Organization, a disaster is defined as "a sudden ecologic phenomenon of sufficient magnitude to require external assistance". Disasters can be classified into either natural or man-made disasters. Natural disasters constitute earthquakes, volcanoes, hurricanes, floods and fires, while man-made disasters consist of war, pollution, nuclear explosions, fires, hazardous materials exposures and transportation accidents [1]. China, the United States (US), the Philippines, India and Indonesia are countries most affected by natural disasters [2]. In 2012, the US faced a total 34 natural disasters, including the Colorado fire ($500 million in damages), Hurricane Sandy ($65 billion) and the Midwest drought ($35 million), which led to thousands of casualties (11,000) and economic losses of at least $140 billion [3–5].

Natural disasters are inevitable, yet collaborative efforts thorough public and private partnerships, in terms of planning, technology and resource management, can mitigate to a large extent the casualties of disaster [6]. Hence it is very crucial for organizations to devise a disaster management plan well ahead of time, as calamities or catastrophic events of any magnitude can occur at any time. A disaster response plan is a predefined contingency plan of action to be implemented after a potentially detrimental situation occurs [7]. A good disaster response plan not only helps organizations recover from emergency situations but also helps them to reduce wastages in terms of infrastructure, revenue, valuable time and human effort. Nonprofit organizations have always taken the initiative in disaster/crisis planning throughout the globe. A few prominent names in the field of disaster response and recovery are the American Red Cross (International Federation of Red Cross and Crescent Society (IFRC)), Salvation Army, United Methodist Committee on Relief (UMCOR), Lutheran Disaster Response, Christian Reformed World Relief Committee and Mennonite Disaster Services [8]. The IFRC are pioneers in the field of disaster crisis management. Up till today, the IFRC has successfully handled more than 70,000 disasters with the help of indigenous crisis-oriented programs [9]. In 2010, Red Cross volunteer services provided what was estimated to be over $6 billion in aid and reached out to over 30 million people affected by disasters [10]. With such high stakes in terms of property, finances and life, any organization (either in the nonprofit or corporate world) related to disaster management and recovery efforts can learn valuable lessons from the IFRC.

The evolution, advancement and increasing prominence of technology in developed nations have led to a large-scale embedment of IT into disaster response and recovery and management efforts by providing access to instant, reliable, sophisticated real-time information through weather monitoring devices, Geographic Information Systems (GIS), crisis management software and social media [11]. For example, technology has played a crucial role in disaster response in these following disasters: Haiti earthquake (2010)—the use of "crowdsourcing" via free texting to get help, the San Bruno Fire (2010)—a "volunteer mapper" to help managers locate volunteers, and the Pakistan flood (2010)—"geo-locate messages" to help locate flood-affected areas in maps offsite [12]. Social media platforms like Twitter was a primary source of information during the South East Queensland Floods (2011), where individuals could share via tweets their flood experience, call for donations and spread awareness of the situation by posting videos and photographs [13].

2. Literature Review

2.1. *Social media*

Social media is a web-based medium to encourage people to create, interact and exchange data in the form of texts, pictures and videos [14]. Some examples of social media include blogs, discussion forums, chat rooms, wikis, YouTube channels, LinkedIn, Facebook and Twitter. From the time the first social media platform has been in existence nearly two decades ago, social media has undergone a series of evolutions to now offer people new and productive ways to engage with events, brands and social causes [15]. According to a recent estimate in 2013, nearly 150 million US web users use social networks via any device on at least a monthly basis, bringing the reach of social media sites to 63.7% of the online population [16].

In last five years popular social media platforms like Facebook and Twitter have played an increasingly significant role in emergency and disaster efforts. Universities, nonprofit organizations and governments use Facebook to broadcast information, communicate with each other, recruit volunteers and coordinate activities [17]. According an American Red Cross study in 2009, social media sites are the fourth-most commonly used source to access news or updates on emergencies and disasters. Individuals and communities also commonly use social media to warn others of unsafe areas, inform friends and family that someone is safe, and raise funds for disaster relief. However, social media can also be used to spread rumors and unconfirmed speculations. In an ideal world, social media should be used in a methodic fashion to send warnings, locate victims and spread awareness. For instance, during Hurricane Sandy (2012), volunteers and help teams used social media to speed up insurance claims payment and encourage insurers to better support their customers [18].

2.2. *Crowdsourcing software*

Crowdsourcing is one of dynamic and innovative trends on the internet, where a task to be performed is outsourced to a crowd. It has been in existence since the 1990s and has been widely applicable in business and social causes [18]. For example, Wikipedia, Threadless, iStockphoto, InnoCentive, Crowdspring, etc., are based on a crowdsourcing model to achieve their business objectives. For social purposes software like Samasource, txtEagle, Ushahidi, Peer Water Exchange, mCollect, the Community Knowledge Worker initiative and Babel are being widely used as disaster management software tools [16]. In disaster management, crowdsourcing software relies on information collected by volunteers to respond during a disaster [19]. One of the most popular model implemented during disaster management in recent times is Ushahidi, an open source customizable volunteer-based mapping tool that collates disaster/crisis information through SMS, e-mail or web entry [20]. All the data received gets timestamped and geo-located to generate a crisis map to facilitate crisis communication and response [21].

Even though Ushahidi has been around only since 2008, it is already widely implemented in projects like Unsung Peace

Heroes and Building Bridges, Uchaguzi in Kenya and Tanzania, as well as in the Haiti Earthquake (2010) and during Hurricane Sandy (2012) [22].

2.3. *GIS and remote sensing devices*

Advanced technology products like the GIS and remote sensing have played a significant role in disaster management planning, communication and response and mitigation efforts [23]. Remote Sensing is a tool that helps to identify disaster-prone areas, monitor Earth for changes on a real time basis, and give early warnings on impending catastrophic events [24].

The GIS is an information system to capture, store, manipulate, analyze, manage, interpret data, and identify relationships and patterns in a geographical area, thus helping decision makers in identifying disasters in preliminary stages, guiding disaster development activities and the implementation of emergency preparedness and response action. The data collected is used to design a model, which helps in determining potential impacts and the appropriate mitigation requirements. The GIS facilitates the viewing of disaster-prone locations along with its infrastructure. This helps to identify high-risk entities whenever calamity strikes and guides governments to better design their response preparedness. Hence, the GIS and remote sensing devices management analyze risks and hazards to determine the values at risk and the operations necessary to reduce exposure, respond more effectively, and recover quickly [24, 25].

GIS has an extensive data management system that enables efficient storage and updating, and faster information access, thus leading to a better decision-making process. The cost of the GIS constitutes three major areas of data, and the software and hardware cost components relative to the cost of a whole project will be illustrated later [26]. When compared to other information systems, the costs for the software and hardware are much lesser than the data cost, which constitutes 80% of the total cost of a GIS [27].

GIS are customized according to an organization's requirements, adding on further to implementation costs. Thus, this makes GIS a very complex information system with significant computing power [25, 28].

2.4. *Disaster Management Information System*

The Disaster Management Information System (DMIS) is a web-based working tool initiated in February 2001 and made available only to Red Cross and Red Crescent staff working in national societies, delegations and the Red Cross' Geneva headquarters. The DMIS facilitates real-time information on disaster trends, and online internal and external resources tools and databases [27]. The DMIS is also a key component system in the IFRC information management system [26]. Information management during disasters relies on the global network of the IFRC and its field-based staff to build on and maximize the use of IT, along with the existing expertise and experience in the Red Cross.

The DMIS project was launched due to the necessity for cognitive decisions, speed and efficient operational readiness. The DMIS is the result of key efforts made by the IFRC taking into account the intricacy of information exchange in the philanthropic community and supporting an international efficient disaster preparedness and response for the IFRC's Red Cross and Crescent network [29]. An information management system for disasters is a crucial element of international disaster response and relief. Hence it is very important to have right, accurate and real-time information before, during and after disasters to handle disaster situation. The DMIS consists of coordination, delivery of relief assistance, beneficiary involvement, marketing and external relations, monitoring and evaluation during a disaster. Hence the advantages of such a system are [29, 30]: (1) it saves lives: people residing in disaster-prone areas can be saved through early warning systems and emergency alert systems, (2) relief efforts reduce suffering and improve relief activities during disasters by providing tracking services, concise information on disaster assistance and relief packages (e.g., shelter assistance), and (3) media coverage facilitates awareness by covering the world's less known disasters and disaster-prone areas, hence promoting international acknowledgment and assistance to disasters [26, 27].

2.5. *Mobile and HAM radio*

Amateur radios like the HAM radio are the most commonly and reliable mode of communications during disasters. Amateur radios have been successfully implemented in government bodies for a century now [31]. It has been proven beneficial when wired lines, cell phones and other traditional means of communications are not responsive/helpful during emergency situations. The reason being that HAM radios are not dependent on aerial facilities to carry out communication networks. Apart from providing disaster communications, they also allow emergency agencies to maximize their resources (e.g., facilitating communications), and provide emergency managers with situational awareness and help to manage large-scale events. In addition, amateur radio operators are often the first to respond after any disaster, by broadcasting details of the emergency situation to emergency managers so that they can start framing a disaster response strategy [32]. Recent use of HAM radios includes the attacks on the World Trade Center in Manhattan (2001), the North America blackout (2003) and during Hurricane Katrina (2005), where thousands of HAM radios were used to provide emergency responses. HAM radios have also been successfully used and implemented in Asia, for example, during the great tsunami disaster (2004) in the Indian Ocean when HAM radios were used to provide relief efforts. HAM radios were also used for other applications, such as providing weather forecasters with "live" information on

Hurricane Frances in the Bahamas (2004). Recently, amateur radio operators in China provided emergency communications after the Sichuan earthquake (2008) [33].

3. Methodology

3.1. *Framework*

In this paper, we are mainly concerned with the various stages of managing disaster management in a nonprofit organization and how technology management techniques like the decision-making principle and strategic planning can be leveraged for this purpose. Thus, effective use of reliable and real time technology can help nonprofit organizations draft an effective decision-making process in an emergency situation. Hence, the general model for an investigating approach for disaster management technology selection is shown in Figure 1.

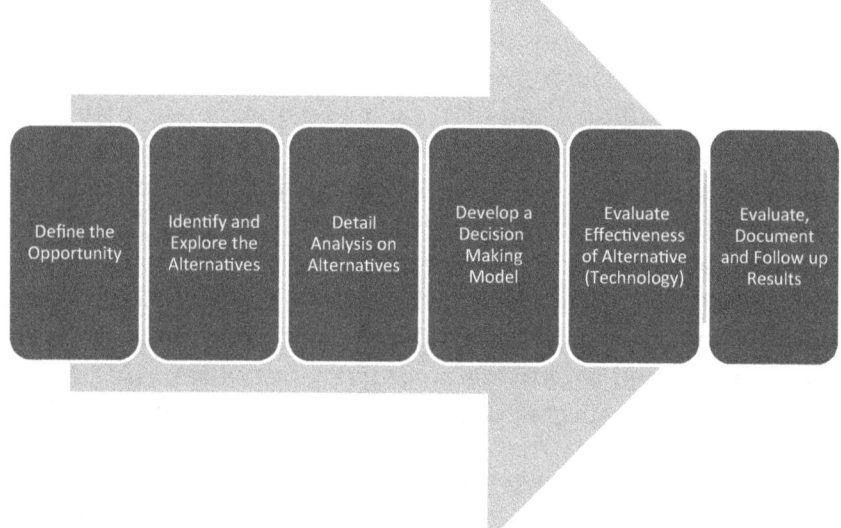

Figure 1. The disaster management technology selection approach.

Listed below are few significant details to realize the scope of the project:

1. Identify and understand various aspects of disaster management.
2. How and what technologies are being used in the disaster management process.
3. Perform a detailed analysis on all the technologies currently implemented.
4. Attempt to develop a model to help a technology manager in a nonprofit organization make an informed decision on the technologies to use.
5. Use the model to evaluate the effectiveness of each technology to handle disaster response and recovery efforts.
6. Documentation of observation and feedback of the framework (system) for future references.

Listed below are the key questions/issues to be addressed in the paper:

1. What are the factors to be considered in disaster response plans in terms of technology management, project management and communication, as well as coordination efforts?
2. Which technology would directly/indirectly influence the impact of the factors identified?
3. Develop a decision-making model (Hierarchical Decision-Making (HDM)/Technology Development Envelope (TDE) model) that can be used to design a framework.

We have approached this paper in two parts. First, we research questions 1 and 2 through a literature review and gathered experts' opinions on the topic. Secondly, we demonstrate how the issues identified in the paper can be effectively mitigated in disaster management efforts through the effective use of technology management principles, as seamlessly as possible.

3.2. Evaluation methodology

Our evaluation included a literature review to evaluate whether the candidate technologies were relevant to the overall disaster management process. We then created and combined two highly symbiotic models to select the best technology. Table 1 explains the application of a candidate to the phases of disaster management like planning, mitigation, preparedness, response and recovery.

Table 1. Application of technology in every stage of disaster management.

Technology	Stages of Disaster Management				
	Planning	Mitigation	Preparedness	Response	Recovery
1. Social media (Facebook, Twitter, blog, etc.)	X		X	X	X
2. Geographic Information System	X	X	X	X	X
3. Web-based crowdsourcing software (information collaboration and publishing tools like Ushahidi, IBM's Sahana and Microsoft Disaster Portal)	X		X	X	X
4. Weather monitoring and remote sensing devices (GPS)	X	X	X	X	X
5. Radio communication (HAM radio, satellite radio and community radio)	X		X	X	X
6. Mobile technology (geo-located text messages, phone calls)				X	
7. Disaster Management Information System (governments along with the Red Cross)	X	X	X	X	X

The two decision-making models implemented in the study include the HDM model and the Technology Valuation (TV) model. The identification and the usage of these models have been carefully evaluated to suit the purpose of the project while considering prospects to future modifications as needed.

3.2.1. HDM

In this report, the HDM model is used to evaluate five candidate technologies (social media, crowdsourcing software, GIS and remote sensing devices, the DMIS, and mobile and HAM radios) for disaster management by nonprofit organizations in combination with the TV model, which will quantify the contribution of each technology option to the overall objective of finding the most effective technology for disaster management for nonprofit organizations [30] (Figure 2).

The first level in this model contains five criteria: Political Ecology, Communication and Coordination, Data Management,

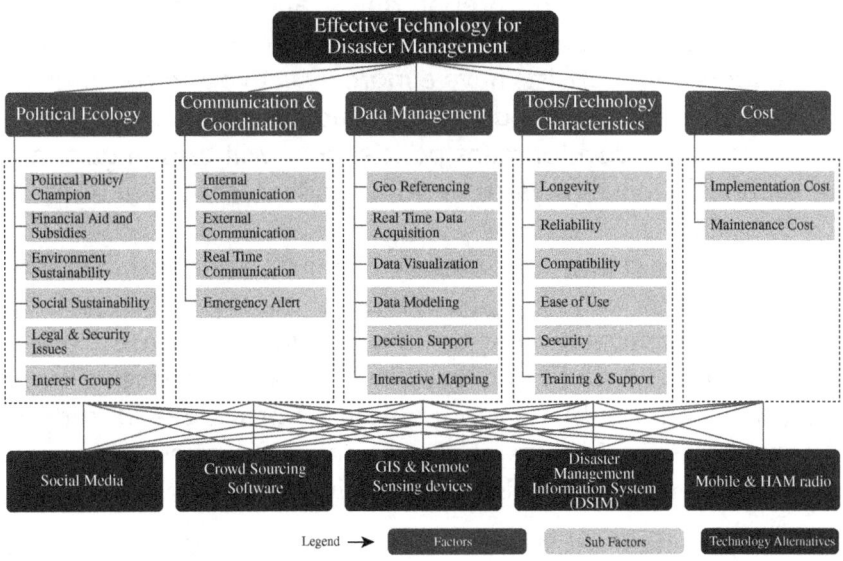

Figure 2. The HDM model helps to select an effective technology for disaster management.

Technology/Tool Characteristics and Cost. Any subsequent level consists of subcriteria associated with the criteria. Consequently, the weights for the criteria and subcriteria levels were calculated by using pairwise comparisons. Since this required deeper understanding and knowledge of the subject matter, two non-profits experts with emergency response backgrounds were invited to carry out the pairwise comparisons for the subcriteria/factors. After determining the weights for the criteria and subcriteria levels using the PCM software, it was possible to find the contribution of each candidate technology to the overall objective of this project by using the TV equation.

3.2.1.1. Criteria and subcriteria

Criteria 1: Political ecology

1. **Political policy/champion:** Political support in the form of a policy/champion is required to address holistic management of disasters by incorporating pre-disaster issues like prevention, mitigation and preparedness, and post-disaster issues relating to response, recovery and reconstruction. Political leaders and lobbyists have a more effective and comprehensive role to play in addressing, implementing and building sustainable development policies that will bridge gaps and build resilience towards national and international disasters [34, 35].
2. **Financial aid and subsidies:** Availability of financial aid and subsidies to fulfill long term disaster management planning and implementation in the United States. It consists of the allocation of contingency reserve funds, disaster relief funds (DRFs) and public saving policy to provide and support resources to aid disaster management activities and restore livelihoods. This consists of, namely, [36, 37].
    i. Individual assistance—Disaster housing, disaster grants, low-interest disaster loans and other disaster-aid programs.

ii. Public assistance—Aid to state or local governments to pay part of the costs of rebuilding a community's damaged infrastructure.

iii. Hazard mitigation—The DRF is an appropriation against which the Federal Emergency Management Agency (FEMA) can direct, coordinate, manage and fund eligible response and recovery efforts associated with domestic major disasters and emergencies.

3. **Environment sustainability:** Defining strategic collaborations between disaster management, resource management and maintenance of the factors and practices that contribute to the quality of the environment on a long-term basis. There are a few constraints imposed on disaster management activities due to their direct/negative impacts on the sustainability of ecosystems and resource systems [38].
4. **Social sustainability:** Designing effective collaboration of processes and abilities to respond better to (local) communities for present and future generations [39].
5. **Legal and security issues:** Legal implications in international and state collaborations, partnerships, contracts, ethics and human resources, including working with volunteers, planning responsibilities and declaring an emergency (response and recovery issues) [35, 38].
6. **Interest groups:** Public opinion and interest groups may have a substantial influence on elected representatives, sequentially leading to drafting and following up with disaster policies and emergency management in general [32, 37].

Criteria 2: Communication and coordination

1. **Internal communication:** Communication and coordination activities within nonprofit/government/private organizations to respond productively to a disaster situation [40].
2. **External communication:** Information gathering (from sources like telecommunication satellites, radar, telemetry,

meteorology and remote sensing, early warning systems, etc.) and communications between individuals/communities in disaster and rescue/relief/government organizations [40].
3. **Real-time communication:** The ability to gather and share information instantaneously to communicate assistance, resources and coordinate emergency efforts [41].
4. **Emergency alert:** An Emergency Alert System is used by state and local authorities to announce vital information such as weather information for any specific area [42]. The communication process during disaster management will be illustrated later.

Criteria 3: Data management

1. **Geo-referencing:** Effective use of geo-referenced/geospatial information tools for the implementation of disaster risk preparedness and recovery activities [35].
2. **Real-time data acquisition:** The well-coordinated satellite remote sensing programs allow the acquisition of real-time low-resolution data, and detailed identification of objects and processes. This gathers information instantaneously about emergency areas, assistance and resources, and also helps to coordinate emergency efforts [43].
3. **Decision support:** Knowledge extracted from data processing and analysis to understand disaster scenes and aid critical decision-making processes, which helps emergency response teams (nonprofit/private/government organizations) to efficiently plan disaster relief efforts to mitigate property loss, reduce injuries, save lives and restore amenities like water and electricity to affected disaster regions [44].
4. **Data visualization:** The ability to facilitate viewing the location of current and forecasted emergency locations on a map, and allows users to easily retrieve detailed reports [45].
5. **Data modeling:** Provides the information needed for disaster management. This may consist of existing information (e.g., information for buildings and road and utility networks)

and information collected during disasters (e.g., location of a disaster incident, extent and possible escalation, and the number of victims) [46].
6. **Interactive mapping:** Interactive maps are well timed and precise information used by survivors and relief responders to check the disaster-affected areas from a remote location, without interrupting response efforts or putting themselves in extreme conditions [47].

Criteria 4: Tools/technology characteristics

1. **Longevity:** The ability of a software application to perform consistently and remain reliable over a long period of time [48].
2. **Reliability:** Capability of a computer program to perform its functions and operations in a system's environment, without any failures in hardware or software in the form of system crashes (i.e., hangs, functionally incorrect response, untimely responses—too fast or too slow, etc.), planned events (i.e., updates relating to configuration changes that require a reboot) and configuration failures (i.e., application/system incompatibility errors, installation/setup failures, etc.) [25].
3. **Compatibility:** Software that runs on one of the systems in the family can also be run on all the other member systems with no alterations required [48].
4. **Ease of use:** The simplicity of use of the software/system, which does not require the user to have any particular skillset or undergo training to perform tasks [25].
5. **Security:** Includes a full range of integrated security and emergency management services—physical, cyber and telecommunications security, facility management, process control and emergency response, and responses to threats and vulnerabilities. It develops plans to ensure business continuity, emergency response and safeguard national security [49].
6. **Training and support:** Availability/requirement of assistance and training to users of technology products [50].

Criteria 5: Cost

1. **Implementation cost:** Implementation consists of all the processes involved in getting new software or hardware to operate properly in its new environment. This includes installation, configuration, running, testing, and making the necessary changes [51].
2. **Maintenance cost:** Software maintenance costs include corrective maintenance, adaptive maintenance, perfective maintenance, and enhancements [23].

3.2.2. TV Model

According to Gerdsri [52], one of the investors of the TV model, "The TV model enables the evaluation of emerging technologies though semi-absolute values instead of the relative values". The TV is calculated using this equation:

$$TV_n \equiv \sum_{k=1}^{k} \sum_{jk=1}^{jk} wk.f_{jk,k}.V(t_{n,jk,k}),$$

where,

TV_n: Technology value of technology (n) determined according to a company's objective.
Wk: Relative priority of criterion (k) with respect to company objective.
$f_{jk,k}$: Relative importance of factor (jk) with respect to criterion (k).
$t_{n,jk,k}$: Performance and physical characteristics of technology (n) along with factor (jk) for criterion (k).
$V(t_{n,jk,k})$: Desirability value of performance and physical characteristics of technology (n) along with factor (jk) for criterion (k).

In order to find the relative priority values of the criteria and subcriteria, pairwise comparison input sheets were provided to experts in nonprofit organizations engaged in international disaster management efforts.

4. Data Analysis and Results

For the purpose of this project a combination of both the HDM and TV models was used. The TV model was used to evaluate five candidate technologies using the HDM. Then, inputs for the pairwise comparisons at the criteria level were obtained from experts from nonprofit organizations engaged in disaster management. These experts compared all five criteria with each other and provided values on the relative importance level of each criterion.

Pairwise inputs for the subcriteria (Factors) level were obtained from two industry experts over a month. These industry experts have an extensive experience in national and international emergency services. The desirability values were also sent to industry experts for the purpose of this project. These values were plotted on a graph to find the desirability value (in percentage form) for each alternative technology. Based on these curves and using the TV equation the final technology values would be computed for each candidate technology. The final conclusions will be made based on the TV values to pick the most effective technology among the candidate technologies for disaster management.

4.1. *Expert input*

The inputs for the analysis process were performed in three different stages. The first stage included a pairwise comparison for the criteria level (see Figures 3 and 4). Two experts from nonprofit organizations were contacted to evaluate the five criteria in the first level by performing a pairwise comparison. Appendices A, B, C and D provide tools and results of this study.

In the second level included pairwise comparison for subcriteria level (see Figures 5 and 6), the complete list is available under Appendix C. Within each criterion experts provide pairwise analysis for all 24 subcriteria. The third and final input stage was for the desirability values for each of the 24

158 Digital Transformation

Criteria 1	C1%	Vs	Criteria2	Total
Political Ecology	40	Vs	Communication & Coordination	60
Political Ecology	60	Vs	Data Management	40
Political Ecology	60	Vs	Tools/ Technology Characteristics	40
Political Ecology	70	Vs	Cost	30
Communication & Coordination	70	Vs	Data Management	30
Communication & Coordination	70	Vs	Tools/ Technology Characteristics	30
Communication & Coordination	70	Vs	Cost	30
Data Management	60	Vs	Tools/ Technology Characteristics	40
Data Management	60	Vs	Cost	40
Tools/ Technology Characteristics	50	Vs	Cost	50

Figure 3. Pairwise comparisons from Industry Expert 1 (Criteria).

Criteria 1	C1%	Vs	Criteria2	Total
Political Ecology	20	Vs	Communication & Coordination	80
Political Ecology	30	Vs	Data Management	70
Political Ecology	40	Vs	Tools/ Technology Characteristics	60
Political Ecology	10	Vs	Cost	90
Communication & Coordination	80	Vs	Data Management	20
Communication & Coordination	70	Vs	Tools/ Technology Characteristics	30
Communication & Coordination	40	Vs	Cost	60
Data Management	50	Vs	Tools/ Technology Characteristics	50
Data Management	40	Vs	Cost	60
Tools/ Technology Characteristics	40	Vs	Cost	60

Figure 4. Pairwise comparisons from Industry Expert 2 (Criteria).

Political Ecology				
Sub Criteria 1	C1%	Vs	Sub Criteria2	Total
Political Policy/Champion	40	Vs	Financial Aid and Subsidies	60
Political Policy/Champion	60	Vs	Environment Sustainability	40
Political Policy/Champion	30	Vs	Social Sustainability	70
Political Policy/Champion	40	Vs	Legal & Security Issues	60
Political Policy/Champion	70	Vs	Interest Groups	30
Financial Aid and Subsidies	60	Vs	Environment Sustainability	40
Financial Aid and Subsidies	40	Vs	Social Sustainability	60
Financial Aid and Subsidies	60	Vs	Legal & Security Issues	40
Financial Aid and Subsidies	60	Vs	Interest Groups	40
Environment Sustainability	40	Vs	Social Sustainability	60
Environment Sustainability	40	Vs	Legal & Security Issues	60
Environment Sustainability	40	Vs	Interest Groups	60
Social Sustainability	70	Vs	Legal & Security Issues	30
Social Sustainability	70	Vs	Interest Groups	30
Legal & Security Issues	70	Vs	Interest Groups	30

Figure 5. Level 2: Pairwise comparisons from Industry Expert 1 (Subcriteria).

Political Ecology				
Sub Criteria 1	C1%	Vs	Sub Criteria2	Total
Political Policy/Champion	60	Vs	Financial Aid and Subsidies	40
Political Policy/Champion	40	Vs	Environment Sustainability	60
Political Policy/Champion	40	Vs	Social Sustainability	60
Political Policy/Champion	30	Vs	Legal & Security Issues	70
Political Policy/Champion	80	Vs	Interest Groups	20
Financial Aid and Subsidies	60	Vs	Environment Sustainability	40
Financial Aid and Subsidies	50	Vs	Social Sustainability	50
Financial Aid and Subsidies	30	Vs	Legal & Security Issues	70
Financial Aid and Subsidies	80	Vs	Interest Groups	20
Environment Sustainability	45	Vs	Social Sustainability	55
Environment Sustainability	40	Vs	Legal & Security Issues	60
Environment Sustainability	80	Vs	Interest Groups	20
Social Sustainability	40	Vs	Legal & Security Issues	60
Social Sustainability	80	Vs	Interest Groups	20
Legal & Security Issues	90	Vs	Interest Groups	10

Figure 6. Level 2: Pairwise comparisons from Industry Expert 2 (Subcriteria).

subcriteria. The desirability values were sent but we are yet to receive any responses from the experts. The responses for desirability values will be between a scale of 10 to 100% for the factor values, ranging between the worst and the best scale provided to them for each of the 24 subcriteria. Some of these metrics were quantitative while others were qualitatively defined on a five-point scale ranging from Excellent to Poor. These values will be later projected onto a line graph called the "desirability graph" to identify the metric values for the final TV calculation for each candidate technology.

Besides providing the values necessary for analysis purposes, the industry experts also provided valuable inputs that were based on their study and analysis of the criteria, and their valuable experience in the field of disaster management. According to one expert in a strategic response global emergency team.

"The cost of responding to the tsunami in Japan and the cost of responding to the tsunami in Chile were a lot different.

The supplies needed and supports required were completely different. Even if we needed some of the same things, the costs were different. In an emergency nothing is consistent. Here are our priorities, 1. Life saving interventions: Does this help people; help them survive without doing long term harm? 2. Security: We cannot endanger the folks we are helping, and the risks we take must be in proportion to the life saving interventions we are providing. 3. Resources: Do we have the funds and the expertise to help people?"

According to other experts with more than decades of experience in disaster management, "During emergency situation, developed countries (US) have organized planning and technology, we can depend on. But in underdeveloped nations situations may get so [bad] that people will have to be contingent on traditional approaches like face-to-face meetings—to understand, plan and cope with situations".

4.2. *Results (Figures 7–12)*

Criteria pairwise results:

- Political Ecology and Communication and Coordination were the most important followed by Data Management.
- Political Support and Will is a crucial element for an NGO to successfully respond to disasters.
- Political Ecology and Communication and Coordination play a pivotal role in selecting a technology for disaster management.

Subcriteria for political ecology:

- Within Political Ecology, government policies and financial aids were the most important followed by Environment Sustainability.

- A Political Policy/Champion is vital to the holistic management of disasters by incorporating pre-disaster issues like prevention, mitigation and preparedness, and post-disaster issues of response, recovery and reconstruction.

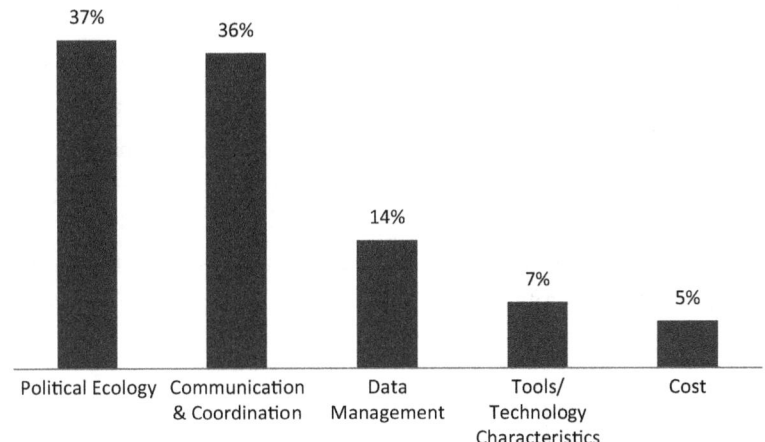

Figure 7. Overall Criteria Level-based pairwise comparisons from both experts.

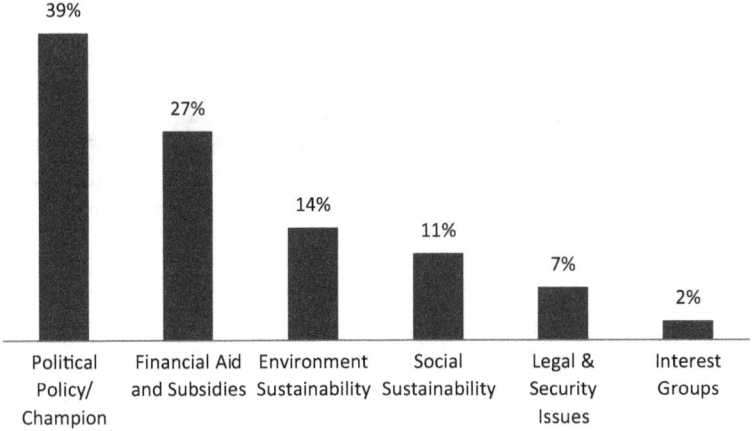

Figure 8. Relative importance of the subcriteria of the Political Ecology-based pairwise comparisons from both experts.

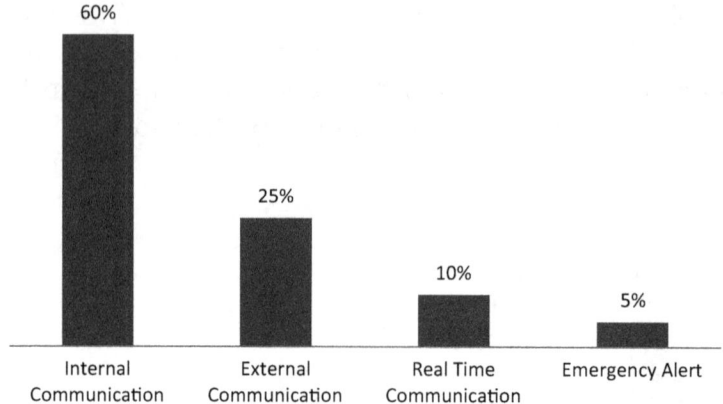

Figure 9. Relative importance of the subcriteria of Communication and Coordination, based on pairwise comparisons from both experts.

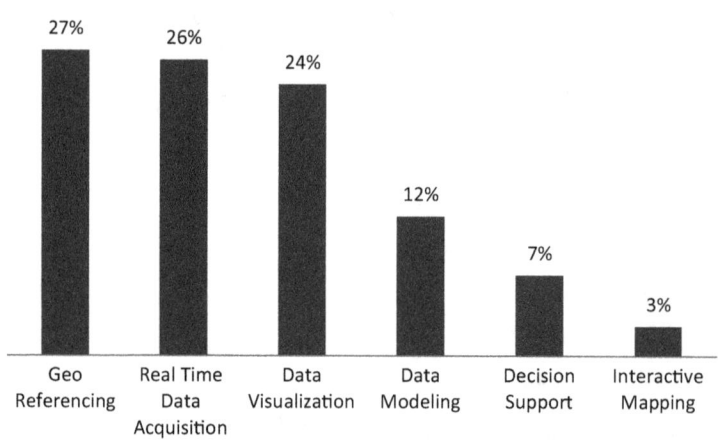

Figure 10. Relative importance of the subcriteria of the Data Management-based pairwise comparisons from both experts.

This stems from the fact that many organizations rely on aid and donations to fund their operations. For example, tax-exempted charitable donations are a function of tax policy.

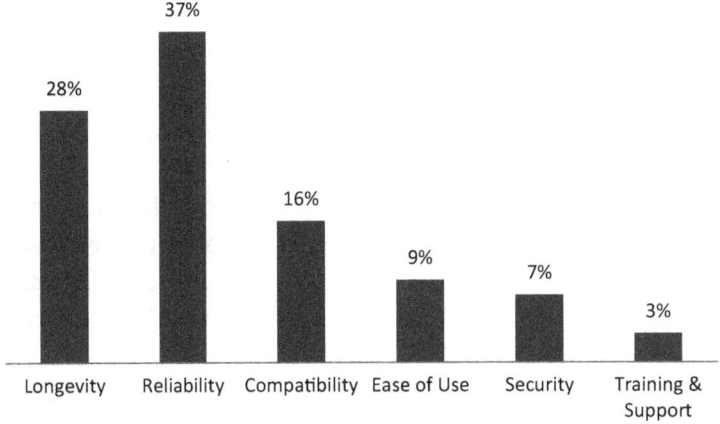

Figure 11. Relative importance of the subcriteria of Tools/Technology Characteristics-based pairwise comparisons from both experts.

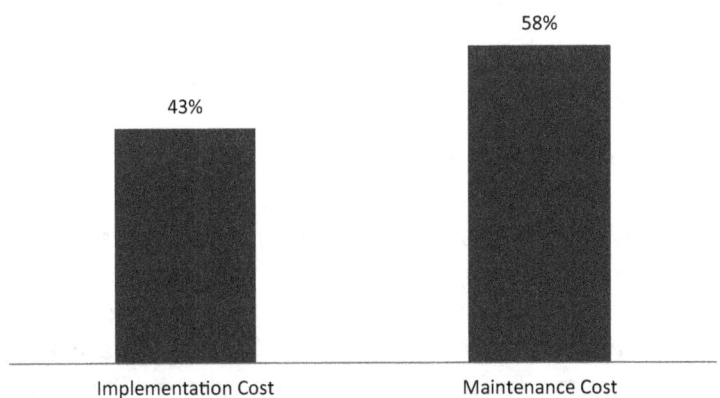

Figure 12. Relative importance of subcriteria of the Technology of Cost-based pairwise comparisons from both experts.

Subcriteria for communication and co-ordination:

- Within Communication and Coordination, Internal Communication was the most important factor followed by External Communication.
- Communication within organizations or volunteer groups for data gathering and coordinating activities is very essential in communication and coordination efforts.

Subcriteria for data management:

- Within Data Management, Geo Referencing, Real Time Data Acquisition and Data Visualization were more or less equally important, followed by Data Modeling.
- In this age of "Big Data", these activities have become an integral part of any business, be it for profit or NGOs. These results highlight that the ability to acquire data in real time and then being able to visualize it makes disaster management more efficient.
- The use of geo-referenced/geospatial information tools for the implementation of disaster risk preparedness and timely early recovery efforts is an effective data management segment.

Subcriteria for tools/technology characteristics:

- Longevity and Reliability were the most important factors followed by Compatibility.
- Given that these systems require considerable investment and are not used that frequently, it makes sense that Longevity and Reliability were the most important factors.
- The ability of a computer program to perform its functions and operations in a system's environment, without any hardware/software failures in the form of crashes, planned events and configuration failures. It is a key subcriteria in the Tools/Technology Characteristics.

Subcriteria for costs:

- Although both implementation and maintenance cost are important, our experts rank maintenance cost slightly higher than implementation cost.
- Software maintenance costs include corrective maintenance, adaptive maintenance and perfective maintenance. Enhancements is the most significant subcriteria in cost.

5. Conclusion

Nature is a highly unpredictable entity. There can be situations when data analysis and modeling cannot capture all the intrinsic elements and magnitude of a catastrophic event. Whenever an emergency occurs there is usually a limited window of time when one needs to act in order to contain the extent and impact of a disaster. A technology that is not intuitive, flexible and transferable is not very useful in these situations. There are many considerations in selecting the right technology, so that it can be effectively deployed across multiple scenarios and geographies. A disaster has no boundaries and can happen anywhere, anytime, and this in itself poses a challenge for an NGO—multiple geographies have their own unique political hurdles. Based on the survey's results and the model that we developed, Political Ecology and Communication & Coordination play a pivotal role in selecting the right technology for disaster management. Although technology captures enough of patterns and trends and helps to alert people in the wake of a disaster, it tends to over-rely on the infrastructure that supports its platform. Hence technology must not be the only alternative for disaster management efforts; there should be back up measures in case of events like power outages and loss of internet connection. As technologists, we need to be optimistic that in this age of technology, it is easier to manage, recover and prevent both natural and man-made disasters.

6. Future Recommendations

Although sufficient research was done in identifying the factors used in this study, there is an opportunity to consider other factors that will make the study exhaustive, for example, global coordination, literacy rate, availability of power supply, back-up technology and opportunity cost. Along with existing technologies, emerging technologies like the Intelligent Deployable Augmented Wireless Gateway (iDAWG), Simulation Deck and Intellistreets are also recommended as alternate technology options to be implemented in future disaster management scenarios.

The model used for this study does not take into account the expected impact of risk to each factor over time. For instance, if the software updates are expected to last for a longer time horizon, then this needs to be captured appropriately. The model has only used limited expert input for the pairwise comparison and we are yet to receive responses for the desirability values. Once received, they will be updated in the analysis accordingly. We recommend including additional expert inputs (for the pairwise comparison) and desirability values as part of a more robust evaluation process.

References

1. J. Zibulewsky, "Defining disaster: the emergency department perspective". http://www.ncbi.nlm.nih.gov/pmc/articles/PMC1291330/.
2. "Annual disaster statistical review 2011: the numbers and trends", *Reliefweb*. http://reliefweb.int/report/world/annual-disaster-statistical-review-2011-numbers-and-trends.
3. "2012 natural disasters: a year in review", *Relief.org*. http://www.relief.org/2012-natural-disasters-a-year-in-review/.
4. D. Rice, "Hurricane Sandy, drought cost U.S. $100 billion", *USA Today*. http://www.usatoday.com/story/weather/2013/01/24/global-disaster-report-sandy-drought/1862201/.
5. Swiss Re Group, "Sigma preliminary estimates for 2012: insurers to pay for close to half of the USD 140 billion in economic losses caused by natural catastrophes and man-made disasters", *Reliefweb*. https://reliefweb.int/report/world/sigma-preliminary-estimates-2012-insurers-pay-close-half-usd-140-billion-economic.
6. Frost and Sullivan, "Planning is the key to successful disaster recovery", p. 2.
7. L. Do, "Disaster recovery: helping non-profits to plan, prepare and recover by", p. 2.
8. J. J. Stys, "Non-profit involvement in disaster response and recovery", pp. 4–6.

9. L. Poston, "Shining examples of excellent social media crisis management". http://www.salesforcemarketingcloud.com/blog/2012/09/shining-examples-of-excellent-social-media-crisis-management/.
10. International Federation of Red Cross and Red Crescent Societies (IFRC), "Red Cross Red Crescent approach to disaster and crisis management". http://www.ifrc.org/en/who-we-are/vision-and-mission/disaster-and-crisis-management/.
11. V. M.-Y. Ng, "Evolving technologies for disaster planning in US cities", Master in City Planning, Massachusetts Insitute of Technology, 2011.
12. R. W. M. Narvaez, "Carnegie Mellon University". http://www.cmu.edu/silicon-valley/news-events/news/2011/novel-social-networking.html.
13. A. Bruns, J. E. Burgess, K. Crawford and F. Shaw, "# qldfloods and @ QPS Media: crisis communication on Twitter in the 2011 South East Queensland floods", ARC Centre of Excellence for Creative Industries and Innovation Queensland University of Technology, January 2012, p. 40.
14. "Digital communications", Tufts University. http://webcomm.tufts.edu/web-resources-tufts/social-media-overview/.
15. "State of the media: the social media report 2012", *Nielsen*. http://www.nielsen.com/us/en/reports/2012/state-of-the-media-the-social-media-report-2012.html.
16. A. Sharma, "Crowdsourcing critical success factor model strategies to harness the collective intelligence of the crowd", *Semantic Scholar*, 2010. https://www.semanticscholar.org/paper/Crowdsourcing-Critical-Success-Factor-Model-to-the-Sharma-Sharma/7e47d791397f1b78ca4ced2dcfddd182e5abddc0.
17. N. Garun, "Hurricane Sandy by the social media numbers", *Digital Trends*, 30 October 2012. https://www.digitaltrends.com/photography/hurricane-sandy-by-social-media-numbers/.
18. H. Gowdy, A. Hilderbrand, D. L. Plana and M. M. Campos, *How five trends will reshape the social sector* (La Piana Consulting: Los Angeles, 2009).
19. R. W. Narvaez, "Crowdsourcing for disaster preparedness: realities and opportunities", PhD dissertation, Graduate Institute of International and Development Studies, Geneva, 2012. https://repository.graduateinstitute.ch/record/14852/?ln=en.

20. "Ushahidi". http://www.ushahidi.com/services.
21. P. Greenough, J. Chan, P. Meier, L. Bateman and S. Datta, "Applied technologies in humanitarian assistance: report of the 2009 applied, prehospital and disaster medicine", *2009 Humanitarian Action Summit* **24**, S2 (2009) s206–s209.
22. "Ushahidi—Evaluation blog series", *Harvard Humanitarian Initiative*, Vol. 1. http://www.knightfoundation.org/media/uploads/media_pdfs/blogseries-121024124807-phpapp02.pdf.
23. T. Vyas and A. Desai, "Information technology for disaster management", *Proceedings of National Conference; INDIACom-2007*, 23 to 24 February 2007, pp. 1–6. http://www.bvicam.ac.in/news/NRSC%202007/pdfs/papers/st_230_03_02_07.pdf.
24. "Analyze and model your world". Esri. http://www.esri.com/what-is-gis.
25. "The nature of GIS". http://gisweb.massey.ac.nz/topic/NatureofGIS/lectures/benefitscosts.html#_costs.
26. "Regional and international disaster response tools and systems". http://www.ifrc.org/en/what-we-do/disaster-management/responding/disaster-response-system/dr-tools-and-systems/.
27. "Parallel architectures laboratory for Geographical Information Systems".http://gisweb.massey.ac.nz/topic/webreferencesites/whatisgis/edinburghwhatisgis/www.geo.ed.ac.uk/home/research/gispal/gispal.htm.
28. L. Daniel, "SDSS for location planning, or the seat of the pants is out". http://gisweb.massey.ac.nz/topic/webreferencesites/whatisgis/sdsswhatisgis/sdss/sdss.htm.
29. "DMIS Disaster Management Information System". https://www-secure.ifrc.org/DMISII/Pages/00_Home/login.aspx.
30. "International Red Cross and Red Crescent Movement". https://www-secure.ifrc.org/dmis/login.asp.
31. Federal Emergency Management Agency, "Federal Emergency Management Agency (FEMA)", *Amateur Radio in Disaster Preparedness*. http://fema.ideascale.com/a/dtd/Amateur-Radio-in-Disaster-Preparedness/322143-14692.
32. C. McKenna, "Amateur radio operators provide a critical communications link during emergencies". *Emergency Management*, 23 June 2010. http://www.emergencymgmt.com/safety/Amateur-Radio-Operators-Communications.html?page=2&.

33. "Role of HAM radio in disaster management?" *WikiAnswers*. http://wiki.answers.com/Q/Role_of_ham_radio_in_disaster_management.
34. P. G. D. Chakrabarti, "Challenges of disaster management in India: implications for the economic, political, and security environments", NBR Special Report No. 34, The National Bureau of Asian Research, 6 November 2011. http://www.nbr.org/publications/element.aspx?id=549#.Ug2Qlhb_RUQ.
35. R. S. Tiwaree, "Geospatial information for disaster risk management in Asia-Pacific region". *19th United Nations Regional Cartographic Conference for Asia and the Pacific*, 29 October to 1 November 2012. http://unstats.un.org/unsd/geoinfo/RCC/docs/rccap19/ip/E_Conf.102_IP15_GEO-REF-cartographic%20conf12_revised.pdf.
36. Federal Emergency Management Agency, "The disaster process and disaster aid programs". http://www.fema.gov/disaster-process-disaster-aid-programs.
37. M. Phaup and C. Kirschner, "Budgeting for disasters: Focusing on the good times", OECDi-library, 16 June 2010. https://doi.org/10.1787/budget-10-5kmh5h6tzrns.
38. S. Dovers, "Sustainability and disaster management", *Australian Journal of Emergency Management*, **19**, 1 (2004): 21–25.
39. Center for History and New Media, "Zotero quick start guide". http://zotero.org/support/quick_start_guide.
40. L. J. Perez-Calderon, "Emergency communications for disaster management", pp. 11–13.
41. "Communications: Staying Connected to the World", HISG. http://hisg.org/training-and-models/articles/disaster-communications/.
42. Federal Communications Commission, "Emergency alert system". http://transition.fcc.gov/pshs/services/eas/.
43. V. Gershenzon, "Remote sensing data acquisition for disaster management", *Geospatial World*, 9 January 2009. http://www.geospatialworld.net/Paper/Application/ArticleView.aspx?aid=948#sthash.I1WEdoMz.dpuf.
44. S. Parks, "Disaster Management using LiDAR". http://www.imagingnotes.com/go/article_freeJ.php?mp_id=294.
45. "Visualizing data with OnTheMap for emergency management". http://blogs.census.gov/2011/06/06/visualizing-data-with-onthemap-for-emergency-management/.

46. A. Dilo and S. Zlatanova, "Spatiotemporal data modeling for disaster management in the Netherlands", January 2008. http://www.gdmc.nl/publications/2008/Spatiotemporal_Modeling_Disaster_Management.pdf.
47. Federal Emergency Management Agency, "The big picture: The role of mapping in assessing disaster damages". *FEMA*. http://www.fema.gov/blog/2013-06-07/big-picture-role-mapping-assessing-disaster-damages.
48. "Lifespan of software: how often do you expect to do start from scratch?" *stackoverflow*. http://stackoverflow.com/questions/360297/lifespan-of-software-how-often-do-you-expect-to-do-start-from-scratch.
49. "Security and Emergency Management", CH2MHILL. http://www.ch2m.com/corporate/services/security-emergency-management/default.asp.
50. P. Rubens, "Technical support for the neighbours", *BBC News*. http://news.bbc.co.uk/2/hi/uk_news/magazine/4387525.stm.
51. C. Thomas, "How much does social media cost?" http://www.bluecloudsolutions.com/blog/social-media-cost/.
52. N. Gerdsri, "An analytical approach to building a Technology Development Envelope (TDE) for roadmapping of emerging technologies", Portland State University, 2005.

Appendix A: Survey HDM

Criteria 1	C1%	Vs	Criteria2	Total
Political Ecology		Vs	Communication & Coordination	100
Political Ecology		Vs	Data Management	100
Political Ecology		Vs	Tools/ Technology Characteristics	100
Political Ecology		Vs	Cost	100
Communication & Coordination		Vs	Data Management	100
Communication & Coordination		Vs	Tools/ Technology Characteristics	100
Communication & Coordination		Vs	Cost	100
Data Management		Vs	Tools/ Technology Characteristics	100
Data Management		Vs	Cost	100
Tools/ Technology Characteristics		Vs	Cost	100

Please input data here

Political Ecology

Sub Criteria 1	C1%	Vs	Sub Criteria2	Total
Political Policy/Champion		Vs	Financial Aid and Subsidies	100
Political Policy/Champion		Vs	Environment Sustainability	100
Political Policy/Champion		Vs	Social Sustainability	100
Political Policy/Champion		Vs	Legal & Security Issues	100
Political Policy/Champion		Vs	Interest Groups	100
Financial Aid and Subsidies		Vs	Environment Sustainability	100
Financial Aid and Subsidies		Vs	Social Sustainability	100
Financial Aid and Subsidies		Vs	Legal & Security Issues	100
Financial Aid and Subsidies		Vs	Interest Groups	100
Environment Sustainability		Vs	Social Sustainability	100
Environment Sustainability		Vs	Legal & Security Issues	100
Environment Sustainability		Vs	Interest Groups	100
Social Sustainability		Vs	Legal & Security Issues	100
Social Sustainability		Vs	Interest Groups	100
Legal & Security Issues		Vs	Interest Groups	100

Please input data here

Communication & Coordination

Sub Criteria 1	C1%	Vs	Sub Criteria2	Total
Internal Communication		Vs	External Communication	100
Internal Communication		Vs	Real Time Communication	100
Internal Communication		Vs	Emergency Alert	100
External Communication		Vs	Real Time Communication	100
External Communication		Vs	Emergency Alert	100
Real Time Communication		Vs	Emergency Alert	100

Please input data here

Data Management

Sub Criteria 1	C1%	Vs	Sub Criteria2	Total
Geo Referencing		Vs	Real Time Data Acquisition	100
Geo Referencing		Vs	Data Visualization	100
Geo Referencing		Vs	Data Modeling	100
Geo Referencing		Vs	Decision Support	100
Geo Referencing		Vs	Interactive Mapping	100
Real Time Data Acquisition		Vs	Data Visualization	100
Real Time Data Acquisition		Vs	Data Modeling	100
Real Time Data Acquisition		Vs	Decision Support	100
Real Time Data Acquisition		Vs	Interactive Mapping	100
Data Visualization		Vs	Data Modeling	100
Data Visualization		Vs	Decision Support	100
Data Visualization		Vs	Interactive Mapping	100
Data Modeling		Vs	Decision Support	100
Data Modeling		Vs	Interactive Mapping	100
Decision Support		Vs	Interactive Mapping	100

172 Digital Transformation

Please input data here ↓

Cost						
Sub Criteria 1		C1%	Vs	Sub Criteria2		Total
Implementation Cost			Vs	Maintenance Cost		100

Tools/ Technology Characteristics						
Sub Criteria 1		C1%	Vs	Sub Criteria2		Total
Longevity			Vs	Reliability		100
Longevity			Vs	Compatibility		100
Longevity			Vs	Ease of Use		100
Longevity			Vs	Security		100
Longevity			Vs	Training & Support		100
Reliability			Vs	Compatibility		100
Reliability			Vs	Ease of Use		100
Reliability			Vs	Security		100
Reliability			Vs	Training & Support		100
Compatibility			Vs	Ease of Use		100
Compatibility			Vs	Security		100
Compatibility			Vs	Training & Support		100
Ease of Use			Vs	Security		100
Ease of Use			Vs	Training & Support		100
Security			Vs	Training & Support		100

Appendix B: Survey Results

Political Ecology						
Sub Criteria 1		C1%	Vs	Sub Criteria2		Total
Political Policy/Champion		40	Vs	Financial Aid and Subsidies		60
Political Policy/Champion		60	Vs	Environment Sustainability		40
Political Policy/Champion		30	Vs	Social Sustainability		70
Political Policy/Champion		40	Vs	Legal & Security Issues		60
Political Policy/Champion		70	Vs	Interest Groups		30
Financial Aid and Subsidies		60	Vs	Environment Sustainability		40
Financial Aid and Subsidies		40	Vs	Social Sustainability		60
Financial Aid and Subsidies		60	Vs	Legal & Security Issues		40
Financial Aid and Subsidies		60	Vs	Interest Groups		40
Environment Sustainability		40	Vs	Social Sustainability		60
Environment Sustainability		40	Vs	Legal & Security Issues		60
Environment Sustainability		40	Vs	Interest Groups		60
Social Sustainability		70	Vs	Legal & Security Issues		30
Social Sustainability		70	Vs	Interest Groups		30
Legal & Security Issues		70	Vs	Interest Groups		30

Pairwise Analysis Expert 1: Subcriteria Expert 1

Political Ecology				
Sub Criteria 1	C1%	Vs	Sub Criteria2	Total
Political Policy/Champion	40	Vs	Financial Aid and Subsidies	60
Political Policy/Champion	60	Vs	Environment Sustainability	40
Political Policy/Champion	30	Vs	Social Sustainability	70
Political Policy/Champion	40	Vs	Legal & Security Issues	60
Political Policy/Champion	70	Vs	Interest Groups	30
Financial Aid and Subsidies	60	Vs	Environment Sustainability	40
Financial Aid and Subsidies	40	Vs	Social Sustainability	60
Financial Aid and Subsidies	60	Vs	Legal & Security Issues	40
Financial Aid and Subsidies	60	Vs	Interest Groups	40
Environment Sustainability	40	Vs	Social Sustainability	60
Environment Sustainability	40	Vs	Legal & Security Issues	60
Environment Sustainability	40	Vs	Interest Groups	60
Social Sustainability	70	Vs	Legal & Security Issues	30
Social Sustainability	70	Vs	Interest Groups	30
Legal & Security Issues	70	Vs	Interest Groups	30

Pairwise Analysis Expert 2: Subcriteria Expert 2

Communication & Coordination	Please input data here			
Sub Criteria 1	C1%	Vs	Sub Criteria2	Total
Internal Communication	70	Vs	External Communication	30
Internal Communication	80	Vs	Real Time Communication	20
Internal Communication	70	Vs	Emergency Alert	30
External Communication	70	Vs	Real Time Communication	30
External Communication	60	Vs	Emergency Alert	40
Real Time Communication	40	Vs	Emergency Alert	60

Pairwise analysis Expert 1: Subcriteria Expert 1

Communication & Coordination				
Sub Criteria 1	C1%	Vs	Sub Criteria2	Total
Internal Communication	70	Vs	External Communication	30
Internal Communication	80	Vs	Real Time Communication	20
Internal Communication	70	Vs	Emergency Alert	30
External Communication	70	Vs	Real Time Communication	30
External Communication	60	Vs	Emergency Alert	40
Real Time Communication	40	Vs	Emergency Alert	60

Pairwise Analysis Expert 2: Subcriteria Expert 2

Data Management

Sub Criteria 1	C1%	Vs	Sub Criteria2	Total
Geo Referencing	10	Vs	Real Time Data Acquisition	90
Geo Referencing	10	Vs	Data Visualization	90
Geo Referencing	10	Vs	Data Modeling	90
Geo Referencing	15	Vs	Decision Support	85
Geo Referencing	50	Vs	Interactive Mapping	50
Real Time Data Acquisition	40	Vs	Data Visualization	60
Real Time Data Acquisition	30	Vs	Data Modeling	70
Real Time Data Acquisition	40	Vs	Decision Support	60
Real Time Data Acquisition	50	Vs	Interactive Mapping	50
Data Visualization	80	Vs	Data Modeling	20
Data Visualization	75	Vs	Decision Support	25
Data Visualization	90	Vs	Interactive Mapping	10
Data Modeling	85	Vs	Decision Support	15
Data Modeling	80	Vs	Interactive Mapping	20
Decision Support	80	Vs	Interactive Mapping	20

Pairwise analysis Expert 1: Subcriteria Expert 1

Data Management

Sub Criteria 1	C1%	Vs	Sub Criteria2	Total
Geo Referencing	70	Vs	Real Time Data Acquisition	30
Geo Referencing	45	Vs	Data Visualization	55
Geo Referencing	60	Vs	Data Modeling	40
Geo Referencing	40	Vs	Decision Support	60
Geo Referencing	75	Vs	Interactive Mapping	25
Real Time Data Acquisition	40	Vs	Data Visualization	60
Real Time Data Acquisition	30	Vs	Data Modeling	70
Real Time Data Acquisition	35	Vs	Decision Support	65
Real Time Data Acquisition	60	Vs	Interactive Mapping	40
Data Visualization	60	Vs	Data Modeling	40
Data Visualization	45	Vs	Decision Support	55
Data Visualization	70	Vs	Interactive Mapping	30
Data Modeling	40	Vs	Decision Support	60
Data Modeling	60	Vs	Interactive Mapping	40
Decision Support	80	Vs	Interactive Mapping	20

Pairwise Analysis Expert 2: Subcriteria Expert 2

Data Management				
Sub Criteria 1	C1%	Vs	Sub Criteria2	Total
Geo Referencing	70	Vs	Real Time Data Acquisition	30
Geo Referencing	45	Vs	Data Visualization	55
Geo Referencing	60	Vs	Data Modeling	40
Geo Referencing	40	Vs	Decision Support	60
Geo Referencing	75	Vs	Interactive Mapping	25
Real Time Data Acquisition	40	Vs	Data Visualization	60
Real Time Data Acquisition	30	Vs	Data Modeling	70
Real Time Data Acquisition	35	Vs	Decision Support	65
Real Time Data Acquisition	60	Vs	Interactive Mapping	40
Data Visualization	60	Vs	Data Modeling	40
Data Visualization	45	Vs	Decision Support	55
Data Visualization	70	Vs	Interactive Mapping	30
Data Modeling	40	Vs	Decision Support	60
Data Modeling	60	Vs	Interactive Mapping	40
Decision Support	80	Vs	Interactive Mapping	20

Pairwise Analysis Expert 2: Subcriteria Expert 2

Cost				
Sub Criteria 1	C1%	Vs	Sub Criteria2	Total
Implementation Cost	25	Vs	Maintenance Cost	75

Pairwise analysis Expert 1: Subcriteria Expert 1

Cost				
Sub Criteria 1	C1%	Vs	Sub Criteria2	Total
Implementation Cost	60	Vs	Maintenance Cost	40

Pairwise Analysis Expert 2: Subcriteria Expert 2

Tools/Technology Characteristics

Sub Criteria 1	C1%	Vs	Sub Criteria2	Total
Longevity	10	Vs	Reliability	90
Longevity	50	Vs	Compatibility	50
Longevity	60	Vs	Ease of Use	40
Longevity	10	Vs	Security	90
Longevity	30	Vs	Training & Support	70
Reliability	90	Vs	Compatibility	10
Reliability	95	Vs	Ease of Use	5
Reliability	50	Vs	Security	50
Reliability	75	Vs	Training & Support	25
Compatibility	90	Vs	Ease of Use	10
Compatibility	30	Vs	Security	70
Compatibility	50	Vs	Training & Support	50
Ease of Use	20	Vs	Security	80
Ease of Use	30	Vs	Training & Support	70
Security	80	Vs	Training & Support	20

Pairwise Analysis Expert 2: Subcriteria Expert 2

Tools/Technology Characteristics

Sub Criteria 1	C1%	Vs	Sub Criteria2	Total
Longevity	30	Vs	Reliability	70
Longevity	40	Vs	Compatibility	60
Longevity	20	Vs	Ease of Use	80
Longevity	45	Vs	Security	55
Longevity	45	Vs	Training & Support	55
Reliability	60	Vs	Compatibility	40
Reliability	40	Vs	Ease of Use	60
Reliability	70	Vs	Security	30
Reliability	75	Vs	Training & Support	25
Compatibility	30	Vs	Ease of Use	70
Compatibility	60	Vs	Security	40
Compatibility	60	Vs	Training & Support	40
Ease of Use	70	Vs	Security	30
Ease of Use	80	Vs	Training & Support	20
Security	60	Vs	Training & Support	40

Pairwise analysis Expert 1: Subcriteria Expert 1

Appendix C: Desirability Table Description

Factors	5-point scale	Description
1. Political Policy/ Champion	Excellent	Immense political support to fulfill majority of the disaster management (resource) requirements in the United States.
	Very Good	Adequate political support to fulfill majority of the disaster management (resource) requirements in the United States.
	Good	Decent political support to fulfill partial disaster management requirements (resource) in the United States.
	Acceptable	Adequate availability of political support to fulfill peak scenarios/ immediate emergency situation of disaster management in the United States.
	Poor	Inadequate availability of political support to fulfill peak scenarios/ emergency situations of disaster management in the United States.
2. Financial Aid and Subsidies	Excellent	Surplus availability of financial aid and subsidies to fulfill long term disaster management planning and implementation in the United States.
	Very Good	Adequate availability of fuel to fulfill majority of the long term availability of financial aid and subsidies to fulfill long term disaster management planning and implementation in the United States.

(Continued)

178 *Digital Transformation*

(Continued)

Factors	5-point scale	Description
	Good	Adequate availability of fuel to fulfill partial demand of the short term availability of financial aid and subsidies to fulfill disaster management planning and implementation in the United States.
	Acceptable	Fair availability of financial aid and resources to fulfill peak scenarios/emergency situation of disaster management planning and implementation in the United States.
	Poor	Inadequate availability of financial aid and resources to fulfill peak scenarios/emergency situation of disaster management planning and implementation in the United States.
3. Social Sustainability	Excellent	Effective collaboration of processes and abilities to respond better to (local) communities for present and future generation.
	Very Good	Adequate collaboration of processes and abilities to respond better to (local) communities for present and future generation.
	Good	Decent collaboration of processes and abilities to partially respond to (local) communities for present and future generation.

(Continued)

Factors	5-point scale	Description
	Acceptable	Fair collaboration of processes and abilities to respond to (local) community's emergency situation for present generation.
	Poor	Inadequate collaboration of processes and abilities to respond to (local) community's emergency situation for present generation.
4. Environment Sustainability	Excellent	Effective long term strategic collaborations between disaster management, resource management and maintenance of the factors and practices that contribute to the quality of environment.
	Very Good	Adequate long term strategic collaborations between disaster management, resource management and maintenance of the factors and practices that contribute to the quality of environment.
	Good	Adequate short term strategic collaborations between disaster management, resource management and maintenance of the factors and practices that contribute to the quality of environment.

(Continued)

(Continued)

Factors	5-point scale	Description
	Acceptable	Adequate short term strategic collaborations between disaster management, resource management and maintenance of the factors and practices that contribute to the quality of environment.
	Poor	Inadequate short term strategic collaborations between disaster management, resource management and maintenance of the factors and practices that contribute to the quality of environment.
5. Legal and Security Issues	Excellent	No legal issues in international and state collaborations, partnerships, contracts, ethics and human resources, including working with volunteers, planning responsibilities and declaring an emergency (response and recovery issues).
	Very Good	Minimum legal issues in contracts, ethics and human resources, including working with volunteers, planning responsibilities and declaring an emergency (response and recovery issues).
	Good	Manageable national and international issues in partnerships, contracts and human resource planning responsibilities, and declaring an emergency (response and recovery issues).

(*Continued*)

Factors	5-point scale	Description
	Acceptable	Manageable national and state issues in partnerships, contracts and human resource planning responsibilities, and declaring an emergency (response and recovery issues).
	Poor	Maximum legal issues in contracts, ethics and human resources, including working with volunteers, planning responsibilities and declaring an emergency (response and recovery issues) on a national/international level.
6. Interest Groups	Excellent	Actively involved large number of interest groups with significant influence on policymakers.
	Very Good	Actively involved large number of interest groups with good influence on policymakers.
	Good	Satisfactorily involved large number of interest groups with fair influence on policymakers.
	Acceptable	Passive interest groups with a little bit of influence on policymakers.
	Poor	Absence of interest groups and no influence on policymakers.
7. Internal Communication	Excellent	Well-coordinated communication and activities within the organization to respond productively in a disaster situation.

(*Continued*)

182 *Digital Transformation*

(Continued)

Factors	5-point scale	Description
	Very Good	Adequately coordinated communication and activities within the organization to respond productively in a disaster situation.
	Good	Adequately coordinated communication and activities within the organization to respond to a disaster situation.
	Acceptable	Satisfactorily coordinated communication and activities within the organization to respond to a disaster situation only during peak scenarios.
	Poor	Inadequately coordinated communication and activities to respond to a disaster situation during peak scenarios.
8. External communication	Excellent	Admirable information gathering (sources like telecommunication satellites, radar, telemetry, meteorology and remote sensing, early warning systems) and communication between individuals/communities in disaster and rescue/relief/government organizations.
	Very Good	Adequately coordinated information gathering (sources like telecommunication satellites, radar, telemetry, meteorology and remote sensing, early warning systems) and communication

(Continued)

Factors	5-point scale	Description
		between individuals/communities in disaster and rescue/relief/ government organizations.
	Good	Manageable coordination and information gathering (sources like telecommunication satellites, radar, telemetry, meteorology and remote sensing, early warning systems) and communication between individuals/communities in disaster and rescue/relief/ government organizations.
	Acceptable	Suitably coordinated information gathering (sources like telecommunication satellites, radar, telemetry, meteorology and remote sensing, early warning systems) and communication between individuals/communities in disaster and rescue/relief/ government organizations during peak scenarios.
	Poor	Inadequately coordinated information gathering (sources like telecommunication satellites, radar, telemetry, meteorology and remote sensing, early warning systems) and communication between individuals/communities in disaster and rescue/relief/ government organizations during peak scenarios.

(Continued)

(Continued)

Factors	5-point scale	Description
9. Real Time Communication	Excellent	Well-coordinated ability to gather and share information instantaneously to communicate assistance, resources and coordinate emergency efforts.
	Very Good	Adequate ability to gather and share information instantaneously to communicate assistance, resources and coordinate emergency efforts.
	Good	Manageable ability to gather and share information instantaneously to communicate assistance, resources and coordinate emergency efforts.
	Acceptable	Little bit of delay in ability to gather and share information instantaneously to communicate assistance, resources and coordinate emergency efforts.
	Poor	Absence of ability to gather and share information instantaneously to communicate assistance, resources and coordinate emergency efforts.
10. Emergency Alert	Excellent	Well-coordinated emergency alert system used by state and local authorities to announce vital information such as emergency weather information for any specific area.

(*Continued*)

Factors	5-point scale	Description
	Very Good	Adequate emergency alert system used by state and local authorities to announce vital information such as emergency weather information for any specific area.
	Good	Manageable emergency alert system used by state and local authorities to announce vital information such as emergency weather information for any specific area.
	Acceptable	Suitable emergency alert system used by state and local authorities to announce vital information such as emergency weather information for any specific area.
	Poor	Lack of emergency alert system used by state and local authorities to announce vital information such as emergency weather information for any specific area.
11. Geo Referencing	Excellent	Effective use of geo-referenced/geospatial information tools for the implementation of disaster risk preparedness and recovery activities.
	Very Good	Adequate use of geo-referenced/geospatial information tools for the implementation of disaster risk preparedness and recovery activities.

(*Continued*)

(Continued)

Factors	5-point scale	Description
	Good	Manageable use of geo-referenced/geospatial information tools for the implementation of disaster risk preparedness and recovery activities.
	Acceptable	Sparse use of geo-referenced/geospatial information tools for the implementation of disaster risk preparedness and recovery activities.
	Poor	Absence use of geo-referenced/geospatial information tools for the implementation of disaster risk preparedness and recovery activities.
12. **Real Time Data Acquisition**	Excellent	Creditable coordination of satellite remote sensing programs that allow acquisition of real-time low-resolution data and detailed identification of objects and processes to gather information instantaneously about emergency area, assistance, resources and to coordinate emergency efforts.
	Very Good	The well-coordinated satellite remote sensing programs allow acquisition of real-time low-resolution data and detailed identification of objects and processes to gather information instantaneously about emergency area, assistance, resources and to coordinate emergency efforts.

(Continued)

Factors	5-point scale	Description
	Good	Sufficient satellite remote sensing programs allow acquisition of real-time low-resolution data and detailed identification of objects and processes to gather information instantaneously about emergency area, assistance, resources and to coordinate emergency efforts.
	Acceptable	Average satellite remote sensing programs allow acquisition of real-time low-resolution data and detailed identification of objects and processes to gather information instantaneously about emergency area, assistance, resources and to coordinate emergency efforts.
	Poor	Lack of satellite remote sensing programs allow acquisition of real-time low-resolution data and detailed identification of objects and processes to gather information instantaneously about emergency area, assistance, resources and to coordinate emergency efforts.
13. Decision Support	Excellent	Creditable knowledge extracted from data processing and analysis to understand disaster scenes and aid critical decisions-making processes. In the process helping

(Continued)

(Continued)

Factors	5-point scale	Description
		emergency response teams (nonprofit/private/government organizations) to efficiently plan disaster relief efforts to mitigate property loss, reduce injuries, save lives and restore amenities like water and electricity to disaster-affected regions.
	Very Good	Well-coordinated knowledge extracted from data processing and analysis to understand disaster scenes and aid critical decisions-making processes. In the process helping emergency response teams (nonprofit/private/government organizations) to efficiently plan disaster relief efforts to mitigate property loss, reduce injuries, save lives and restore amenities like water and electricity to disaster-affected regions.
	Good	Adequate knowledge extracted from data processing and analysis to understand disaster scenes and aid critical decisions-making processes. In the process helping emergency response teams (nonprofit/private/government organizations) to efficiently plan disaster relief efforts to mitigate property loss, reduce injuries, save lives and restore amenities like water and electricity to disaster-affected regions.

(Continued)

Factors	5-point scale	Description
	Acceptable	Satisfactory knowledge extracted from data processing and analysis to understand disaster scenes and aid critical decisions-making processes. In the process helping emergency response teams (nonprofit/private/government organizations) to efficiently plan disaster relief efforts to mitigate property loss, reduce injuries, save lives and restore amenities like water and electricity to disaster-affected regions.
	Poor	Lack of knowledge extracted from data processing and analysis to understand disaster scenes and aid critical decisions-making processes. In the process helping emergency response teams (nonprofit/private/government organizations) to efficiently plan disaster relief efforts to mitigate property loss, reduce injuries, save lives and restore amenities like water and electricity to disaster-affected regions.
14. Data Visualization	Excellent	Efficient ability to facilitate viewing the location of current and forecasted emergency locations on a map, and allows users to easily retrieve detailed reports.

(Continued)

(Continued)

Factors	5-point scale	Description
	Very Good	Well-coordinated ability to facilitate viewing the location of current and forecasted emergency locations on a map, and allows users to easily retrieve detailed reports.
	Good	Adequate ability to facilitate viewing the location of current and forecasted emergency locations on a map, and allows users to easily retrieve detailed reports.
	Acceptable	Satisfactory ability to facilitate viewing the location of current and forecasted emergency locations on a map, and allows users to easily retrieve detailed reports.
	Poor	Lack of ability to facilitate viewing the location of current and forecasted emergency locations on a map, and allows users to easily retrieve detailed reports.
15. Data Modeling	Excellent	Efficient availability to provide information needed for disaster management consists of existing information, e.g., information for buildings, road and utility networks, and information collected during disasters, e.g., location of a disaster incident, extent and possible escalation, and number of victims.

(*Continued*)

Factors	5-point scale	Description
	Very Good	Well-coordinated ability to provide information needed for disaster management consists of existing information, e.g., information for buildings, road and utility networks, and information collected during disasters, e.g., location of a disaster incident, extent and possible escalation, and number of victims.
	Good	Adequate ability to provide information needed for disaster management consists of existing information, e.g., information for buildings, road and utility networks, and information collected during disasters, e.g., location of a disaster incident, extent and possible escalation, and number of victims.
	Acceptable	Average ability to provide information needed for disaster management consists of existing information, e.g., information for buildings, road and utility networks, and information collected during disasters, e.g., location of a disaster incident, extent and possible escalation, and number of victims.

(*Continued*)

(Continued)

Factors	5-point scale	Description
	Poor	Lack of ability to provide information needed for disaster management consists of existing information, e.g., information for buildings, road and utility networks, and information collected during disasters, e.g., location of a disaster incident, extent and possible escalation, and number of victims.
16. Interactive Mapping	Excellent	Efficiently well timed and precise information that could be used by survivors and relief responders to check the disaster-affected areas from a remote location, without interrupting response efforts or putting themselves in extreme conditions.
	Very Good	Actively timed and precise map information that could be used by survivors and relief responders to check the disaster-affected areas from a remote location, without interrupting response efforts or putting themselves in extreme conditions.
	Good	Well-coordinated interactive maps that provide information that could be used by survivors and relief responders to check the disaster-affected areas from a remote location, without interrupting response efforts or putting themselves in extreme conditions.

(Continued)

Factors	5-point scale	Description
	Acceptable	Fairly coordinated interactive maps that provide information that could be used by survivors and relief responders to check the disaster-affected areas from a remote location, without interrupting response efforts or putting themselves in extreme conditions.
	Poor	Absence of interactive maps that provide any information that could be used by survivors and relief responders to check the disaster-affected areas from a remote location, without interrupting response efforts or putting themselves in extreme conditions.
Reliability	Excellent	Incomparable ability of a computer program to perform its functions and operations in a system's environment, without any hardware or software failures in form of crashes, planned events and configuration failures.
	Very Good	Suitable ability of a computer program to perform its functions and operations in a system's environment, without any hardware or software failures in form of crashes, planned events and configuration failures.

(Continued)

(Continued)

Factors	5-point scale	Description
	Good	Estimable ability of a computer program to perform its functions and operations in a system's environment, without any hardware or software failures in form of crashes, planned events and configuration failures.
	Acceptable	Adequate ability of a computer program to perform its functions and operations in a system's environment, without any hardware or software failures in form of crashes, planned events and configuration failures.
	Poor	Lack of ability of a computer program to perform its functions and operations in a system's environment, without any hardware or software failures in form of crashes, planned events and configuration failures.
17. Compatibility	Excellent	Efficient software that runs on one of the system can also be run on all other systems of the family with no alterations required.
	Very Good	Suitable software that runs on one of the system can also be run on all other systems of the family with a little bit of alterations required.
	Good	Estimable software that runs on one of the system can also be run on all other systems of the family with a little bit of alterations required.

(*Continued*)

Factors	5-point scale	Description
	Acceptable	Adequate software that runs on one of the system can also be run on all other systems of the family with a little bit of alterations required.
	Poor	Lack of software that runs on one of the system can also be run on all other systems of the family with no alterations required.
18. Ease of Use	Excellent	Simple and ease of use of software/system.
	Very Good	Not so easy or complicated software though usage can be learned easily.
	Good	Uncomplicated use of software and requires a little bit of training.
	Acceptable	Complicated use of software but could be learned with a little bit of formal training.
	Poor	Complicated and requires a special skillset to use the software.
19. Security	Excellent	It includes no security issues in any range of integrated security and emergency management services—physical, cyber and telecommunications security, facility management, process control and emergency response, and responses to threats and vulnerabilities. It also develops plans to ensure business continuity and emergency response and safeguards national security.

(*Continued*)

(*Continued*)

Factors	5-point scale	Description
	Very Good	It includes no major but security issues in any range of integrated security and emergency management services—physical, cyber and telecommunications security, facility management, process control and emergency response, and responses to threats and vulnerabilities. It also develops plans to ensure business continuity and emergency response and safeguards national security.
	Good	It includes no major but few minor security issues in a limited range of integrated security and emergency management services—physical, cyber and telecommunications security, facility management, process control and emergency response, and responses to threats and vulnerabilities. It also develops limited plans to ensure business continuity and emergency response and safeguards national security.
	Acceptable	It includes no major but few minor security issues in a very limited range of integrated security and emergency management services—physical, cyber and telecommunications security, facility management, process control and emergency response, and responses to threats and vulnerabilities. It also develops very limited plans to ensure

(Continued)

Factors	5-point scale	Description
		business continuity and emergency response and safeguards national security on a limited scale.
	Poor	It includes numerous and high security alert issues in the entire range of integrated security and emergency management services—physical, cyber and telecommunications security, facility management, process control and emergency response, and responses to threats and vulnerabilities. It does not develop plans to ensure business continuity and emergency response and does not safeguard national security.
20. Training and Support	Excellent	Presence of highly experienced and skilled support and training facility to handle system failure or system concerns at any time.
	Very Good	Presence of experienced support and training facility to handle system failure at any time.
	Good	Presence of support and training facility to handle system failure at any time.
	Acceptable	Presence of highly experienced support to handle system failure during peak emergency hours.
	Poor	Absence of support and training facility to handle system failure any time, including peak/ emergency hours.

Appendix D: Desirability Table

Cost: Implementation Cost											
Conditions	Worst										Best
Limiting Metrics (Constant)	120000.00										0
Relative Desirability (%)	10	20	30	40	50	60	70	80	90	100	
120000											
6000											
3000											
2000											
765											
34											
0											

Cost: Maintence Cost											
Conditions	Worst										Best
Limiting Metrics ($/CFt)	6.52										0
Relative Desirability (%)	10	20	30	40	50	60	70	80	90	100	
620											
619											
618											
617											
34											
25											
0											

Political Ecology: Social Sustainability											
Conditions	Worst									Best	
Limiting Metrics (% Value)	24									50	
Relative Desirability (%)	10	20	30	40	50	60	70	80	90		100
Excellent											
Very Good											
Good											
Acceptable											
Poor											

Political Ecology: Political Policy/ Champion											
Conditions	Worst									Best	
Limiting Metrics (5 Point Scale)	Poor									Excellent	
Relative Desirability (%)	10	20	30	40	50	60	70	80	90		100
Excellent											
Very Good											
Good											
Acceptable											
Poor											

Political Ecology: Environment Sustainability	Worst									Best
Conditions Limiting Metrics (5 Point Scale)	Poor									Excellent
Relative Desirability (%)	10	20	30	40	50	60	70	80	90	100
Excellent										
Very Good										
Good										
Acceptable										
Poor										

Political Ecology: Financial Aid and Subsidies	Worst									Best
Conditions Limiting Metrics (5 Point Scale)	Poor									Excellent
Relative Desirability (%)	10	20	30	40	50	60	70	80	90	100
Excellent										
Very Good										
Good										
Acceptable										
Poor										

Technical Transformation: IT in Disaster Management 201

Political Ecology: Legal & Security Issues											
Conditions	Worst									Best	
Limiting Metrics (5 Point Scale)	Poor									Excellent	
Relative Desirability (%)	10	20	30	40	50	60	70	80	90	100	
Excellent											
Very Good											
Good											
Acceptable											
Poor											

Political Ecology: Interest Groups											
Conditions	Worst									Best	
Limiting Metrics (5 Point Scale)	Poor									Excellent	
Relative Desirability (%)	10	20	30	40	50	60	70	80	90	100	
Excellent											
Very Good											
Good											
Acceptable											
Poor											

Communication & Coordination: Internal Communication											Best
Limiting Metrics (5 Point Scale)	Worst										
Relative Desirability (%)	Poor										Excellent
	10	20	30	40	50	60	70	80	90	100	
Excellent											
Very Good											
Good											
Acceptable											
Poor											

Communication & Coordination: External Communication											Best
Conditions	Worst										
Limiting Metrics (5 Point Scale)	Poor										Excellent
Relative Desirability (%)	10	20	30	40	50	60	70	80	90	100	
Excellent											
Very Good											
Good											
Acceptable											
Poor											

Technical Transformation: IT in Disaster Management 203

Communication & Coordination: Emergency Alert												
Limiting Metrics (5 Point Scale)	Worst											Best
Relative Desirability (%)	0.5											15.0
	10	20	30	40	50	60	70	80	90	100		
Excellent												
Very Good												
Good												
Acceptable												
Poor												

Communication & Coordination: Real Time											
Limiting Metrics (5 Point Scale)	Worst										Best
Relative Desirability (%)	Poor										Excellent
	10	20	30	40	50	60	70	80	90	100	
Excellent											
Very Good											
Good											
Acceptable											
Poor											

Data Management: Geo Referencing											
Conditions	Worst										Best
Limiting Metrics (5 Point Scale)	Poor										Excellent
Relative Desirability (%)	10	20	30	40	50	60	70	80	90	100	
Excellent											
Very Good											
Good											
Acceptable											
Poor											

Data Management: Real Time Data Acquisition											
Conditions	Worst										Best
Limiting Metrics (5 Point Scale)	Poor										Excellent
Relative Desirability (%)	10	20	30	40	50	60	70	80	90	100	
Excellent											
Very Good											
Good											
Acceptable											
Poor											

Technical Transformation: IT in Disaster Management 205

Data Management: Data Visualization												
Conditions	Worst									Best		
Limiting Metrics (5 Point Scale)	Poor									Excellent		
Relative Desirability (%)	10	20	30	40	50	60	70	80	90	100		
Excellent												
Very Good												
Good												
Acceptable												
Poor												

Data Management: Data Modeling											
Conditions	Worst									Best	
Limiting Metrics (5 Point Scale)	Poor									Excellent	
Relative Desirability (%)	10	20	30	40	50	60	70	80	90	100	
Excellent											
Very Good											
Good											
Acceptable											
Poor											

Data Management: Decision Support											
Conditions	Worst									Best	
Limiting Metrics (5 Point Scale)	Poor									Excellent	
Relative Desirability (%)	10	20	30	40	50	60	70	80	90	100	
Excellent											
Very Good											
Good											
Acceptable											
Poor											

Data Management: Interactive Mapping											
Conditions	Worst									Best	
Limiting Metrics (5 Point Scale)	Poor									Excellent	
Relative Desirability (%)	10	20	30	40	50	60	70	80	90	100	
Excellent											
Very Good											
Good											
Acceptable											
Poor											

Tools/Technology Characteristics: Longevity

Conditions	Worst									Best
Limiting Metrics (5 Point Scale)	3									20
Relative Desirability (%)	10	20	30	40	50	60	70	80	90	100
3										
8										
10										
15										
20										

Tools/Technology Characteristics: Reliability

Conditions	Worst									Best
Limiting Metrics (5 Point Scale)	Poor									Excellent
Relative Desirability (%)	10	20	30	40	50	60	70	80	90	100
Excellent										
Very Good										
Good										
Acceptable										
Poor										

208 Digital Transformation

Tools/Technology Characteristics: Compatibility											
Conditions	Worst										Best
Limiting Metrics (5 Point Scale)	Poor										Excellent
Relative Desirability (%)	10	20	30	40	50	60	70	80	90	100
Excellent											
Very Good											
Good											
Acceptable											
Poor											

Tools/Technology Characteristics: Ease of Use											
Conditions	Worst										Best
Limiting Metrics (5 Point Scale)	Poor										Excellent
Relative Desirability (%)	10	20	30	40	50	60	70	80	90	100
Excellent											
Very Good											
Good											
Acceptable											
Poor											

Technical Transformation: IT in Disaster Management 209

Tools/Technology Characteristics: Security		Worst									Best
Conditions		Poor									Excellent
Limiting Metrics (5 Point Scale)											
Relative Desirability (%)		10	20	30	40	50	60	70	80	90	100
Excellent											
Very Good											
Good											
Acceptable											
Poor											

Tools/Technology Characteristics: Training & Support		Worst									Best
Conditions		Poor									Excellent
Limiting Metrics (5 Point Scale)											
Relative Desirability (%)		10	20	30	40	50	60	70	80	90	100
Excellent											
Very Good											
Good											
Acceptable											
Poor											

Part 2
Personal Transformation

Chapter 6

Personal Transformation: Evaluation of Smart Home Hubs

Ahmed Alzahrani*, Majed Alshamlani*, Wei-Chen Hsu*,
Shreyas Harish* and Tugrul Daim*,†,‡

*Portland State University, Portland, Oregon, USA
†Higher School of Economics, Moscow, Russia
‡Chaoyang University of Technology, Taiwan

Abstract

A smart home is a hardware system which enables connection and communication between other devices on the home automation network. These devices can include thermostats, light bulbs, wall outlets and switches, door locks, energy monitors, window coverings, appliances, motion sensors, etc. With a smart hub, all of these smart devices can be controlled using a single app.

In spite of being very useful devices, smart hubs have their limitations. Certain smart hubs are only compatible with certain smart devices, whereas others support a limited number of languages. Some operate well only over a short distance (about 15 feet). Some hubs support more smart devices than others, while certain hubs can connect to more devices simultaneously than others. These limitations present a question to consumers: which smart hub will best suit their needs?

This project proposes a Hierarchical Decision Model (HDM) to help consumers decide which smart hub alternative is best for them based on their needs. The HDM is coupled with a

Pairwise Comparison Method (PCM) to evaluate all possible alternatives. The model incorporates various criteria for smart hubs such as range of operation, communication protocols supported, number of devices supported, etc., thus providing the consumer with the right information and feedback needed to purchase the smart hub that suits him/her the best from the various smart hubs available in today's market.

Four alternatives for smart hubs were used in this model, namely, Samsung SmartThings, VeraSecure, Wink 2 and Securifi Almond 3. Rankings were assigned to these alternatives based on experts' opinions used in the HDM. Results concluded that the Samsung SmartThings was the best alternative for the model used, followed by Wink 2 and Securifi Almond 3. This indicated that the initial cost, number of protocols supported by the hub and ease of use were the most important criteria considered to achieve the desired objective of selecting the appropriate smart hub.

Keywords: Technology assessment, smart homes, smart devices.

1. Introduction

A smart hub is a hardware system that can be controlled remotely using smartphones, as it is connected to your home via WiFi [1]. Upon successful installation, it can then connect to and remotely control all other smart devices in the house.

The efficiency of the smart hub lies in its simplicity, where it can handle multiple smart devices without needing a separate app for each one. This not only saves space and time on the smartphone, it also makes it simpler and more intuitive to control multiple devices at any one time. However, it can be complicated in some cases because there are many platforms, apps and hardware providers [2].

There are generally two kinds of smart hubs—those connected to the internet and have additional wireless connections, and those with camera units that are more efficient than most security systems. Each smart hub connects with a particular app, which is able to control other devices and apps through a software system on the phone [3]. Connectivity and quality of connection can be a big factor while buying any app. However, in some cases, certain features are sacrificed in order to connect to multiple devices.

In addition, there is a trade-off between WiFi and Bluetooth technologies. WiFi-connected gadgets can be directly networked to the home's internet and can be controlled from anywhere. Hence, it can act as a good security measure if there are many devices, e.g., lights. On the other hand, Bluetooth kits consume lower power and are more affordable. However, they are effective for only a few hundred meters, though Bluetooth devices can be connected to a smart hub and operated via WiFi connection. They also support a variety of protocols like ZigBee and Z-Wave [3].

It is clear that smart hubs are very useful devices. However, they have their limitations. Certain smart hubs are only compatible with certain smart devices, for example, the Panasonic smart hub is only compatible with Panasonic smart devices [4, 5]. Other hubs can only support certain languages like English, Spanish, etc. There are also certain hubs that can only operate well over a short distance (about 15 feet). Some hubs support more smart devices than others, while certain hubs can connect to more devices simultaneously than others. Other smart hubs have regional limitations, for example, the US version of Samsung's SmartThings hub cannot be used in certain countries due to differences in radio frequency.

The aforementioned limitations make it difficult for a consumer to decide which smart hub he/she should purchase. This project proposes a Hierarchical Decision Model (HDM) to help consumers decide which smart hub is best for them, based on their needs. The model incorporates various criteria for smart hubs such as range of operation, communication protocols supported, number of devices supported, etc., thus providing the consumer with the right information and feedback needed to purchase the smart hub that suits him/her the best from the various smart hubs available in today's market.

2. Literature Review

2.1. *Home automation overview*

While the term "smart home" is broad and covers a range of products with a variety of integration scenarios, the term "home automation" is more specific, which refers to devices and appliances in a home that can function automatically [6]. In the past, semi-automated items were as simple as garage doors, stair lifts and thermostats, but advances in smart home technology have allowed for a large number of devices, appliances and furnishings to be programmed for automation in the home [7].

Some common applications range from lights, locks, cameras and other security features, to thermostats, preheating ovens and coffee makers. For example, an automated lamp can turn itself on as people enter a room, an automated thermostat can adjust the temperature when detecting a presence (or absence) of bodies, and coffee makers can be timed or coordinated to brew a pot as soon as a morning alarm goes off [8].

These systems can be as simple as a light or motion sensor on outdoor lighting to more complex integrations of smart systems, such as a selective profile of specific automations for each individual in the home. Currently, mobile phones and bluetooth are used by these systems to identify users, but they can eventually integrate facial recognition or other identifiers to further personalize the experience [7].

These individualized profiles can be programmed for people, rooms, different times of the day, or other factors that will suit the needs of those in the home. And as this technology advances, the automation becomes consolidated [9]. This is where central hubs play a role. A good smart hub can integrate all smart applications in a home into a single system of controls within a single app. Enter the smart hub era [9].

2.2. *Smart hubs overview*

As smart products proliferate, it is very helpful to control them from a single unit. Smart home hubs can connect all smart devices to the cloud, integrate a number of products with third-party compatibility, and allow customized experiences such as the advanced profiles mentioned earlier [9]. Also, smart hubs often have their own integrated WiFi router, radio, voice operation software, and other applications that enhance the user's experience [9].

With the advancement of Internet of Things (IoT), automations have become smart; they are able to communicate with one another and therefore coordinate their responses to sensors [9]. This new device market stems from an increased demand for systems to centrally control home automations, rather than doing a number of different tasks [10].

Benefits: The bottom line: automation controlled in a single device. Benefits include time, convenience, added security, accessibility, consolidation of tasks, interconnectivity of devices, important notifications (such as leak prevention, electrical problem detection, security breaches, etc.), seamless automation (lights, locks, etc.), energy efficiency, enhanced communications and more. Further advances in the industry will continue to provide further benefits to using and upgrading the systems [9, 10].

3. Problem Statement

For most consumers, there are several issues that increase the difficulty level of home automation implementation, including requiring a large number of control apps, power outlets for different hubs, and the number of Ethernet cables. Taking automation home lighting application as an example, Philips Hue, one of the most popular lighting devices, uses ZigBee to control connected bulbs [11]. Users will first need to connect the smartphone app by using WiFi, then they will need to connect the Hue Bridge to the Home Router through the ethernet cable. After that, the Hue Bridge will be able to connect the Hue Bulbs via ZigBee. Different smart devices have different protocols. The more protocols users use, the greater the number of control apps, power outlets for the Hue Brige, and ethernet cables are needed.

4. Project Objective

The project objective is to evaluate the best smart home hub based on technological, economic and personal perspectives by using a multi-leveled HDM. The results derived will help smart home users to select the best technology candidate that fits their requirements.

5. Methodology

5.1. *Perspectives*

The STEEP Analysis is a method used in marketing to analyze the macroeconomy of the firm to determine which factors can influence its success. STEEP is an acronym for [12, 13]:

- Social
- Technological
- Economic
- Ecological
- Political (and legislative)

The STEEP analysis was used to determine which perspectives mattered the most when making the choice of purchasing an appropriate smart hub. Upon performing this analysis, the political and ecological perspectives were eliminated as they were deemed unnecessary. The remaining perspectives are stated and briefly described as follows:

5.1.1. Technical perspective

The technical perspective envelopes the specifications and features of the smart hubs that distinguish them from one another. The number of protocols supported by a hub, its range of operation, the number of devices it can connect to simultaneously, the number of devices it supports, etc., are some of the criteria under this perspective.

5.1.2. Economic perspective

The economic perspective deals with the financial aspects of the smart hubs under consideration. Initial cost of the hub, monthly charges (if any), warranty, etc., are some of the criteria taken into consideration.

5.1.3. Personal (social) perspective

The personal perspective covers some of the less obvious criteria that should be considered before purchasing a smart hub. It includes the ease of use of the smart hub app and regional limitations, such as not being able to use a certain smart hub in other parts of the world due to differences in the radio frequency of operation and battery backup of the smart hub.

5.2. Gap analysis

Gap analysis is an assessment technique to compare an organization's performance to a desired performance. For decision makers, it helps them to identify the gap between the current situation and the future state that they want to be in, along with the tasks that they need to complete to close this gap [14–17].

We developed the gap analysis of the smart hubs from three different perspectives: technological perspective, economic perspective and personal perspective. Based on the gap analysis, we will be able to find the gap between smart home users' current needs or requirements and their desired performance or capabilities.

5.2.1. *Technological perspective*

For the technological perspective, as shown in Table 1, we came out with five points that are associated with smart home users' requirements. Smart home users expect to have one smart hub with the following features:

1. **Support devices:** They can support the smart devices or any number of objects.
2. **Compatible communication protocols:** These can be compatible with different communication protocols such as WiFi, Bluetooth, ZigBee and Z-Wave.
3. **Low energy consumption:** It can control the power consumption to the battery-powered smart devices.
4. **Connectivity range:** It can connect to the smart devices wirelessly and as wide a range as possible.

Table 1. Technological perspective in the gap analysis.

Gap Analysis	Requirement	Capability	Gap
Technological	• Support 2+ communication protocols (Languages) • Support most devices • Ability to control low energy devices • Wide range to transmit data • 1 Mbps data rate	• Support some protocols • Limited number of compatible devices • High power consumption • Different ranges • 250 Kbps data rate	• Support various communication protocols (WiFi, BT, ZigBee, Z-wave) • High number of compatible devices • Low power consumption • Can transmit up to 30 ft • Mesh Networking • Can transmit data up to 1 Mbps

5.2.2. *Economic perspective*

For the economic perspective, as shown in Table 2, the three points that we generated are the affordable price for the smart hub (because the price range varies from $49 to $149), the affordable price for the smart device (because different applications of the smart device varies from a few dozen to thousands of dollars), and the one-time payment (because smart home users prefer not to be charged a monthly or annual fee for cloud services) [18, 19].

5.2.3. *Personal perspective*

For the personal perspective, as shown in Table 3, smart home users would like to have an integrated app on their platform

Table 2. Economic perspective in the gap analysis.

Gap Analysis	Requirement	Capability	Gap
Economic	• Affordable price for smart hub • Affordable price for smart devices • One time cost	• Expensive price for the advanced smart hub • Expensive price for smart devices • Monthly fees for cloud services	• Different prices • Supported smart devices price differ • Charge fixed and variable costs

Table 3. Personal perspective in the gap analysis.

Gap Analysis	Requirement	Capability	Gap
Personal	• One centralized app/platform to control multiple devices • Ease of use • Battery backup power	• Confusing user interface • Complicated setup process • No battery to backup for power outage	• Convenience of management multiple devices • Need to simplify setup process • Need to support battery

to control all smart devices easily, and the ease of use refers to the installation step to implement the wireless connectivity of smart devices without the complicated setup process and ability to backup data for the battery-powered smart devices that can avoid unexpected situations like power outage.

Based on the above gap analysis, we have demonstrated the gap between users' requirements and capabilities from the technological, economic and personal perspectives. Six advanced and latest smart hub products are selected as our alternatives. The technology candidates are Samsung SmartThings, VeraSecure, Wink 2, Securifi Almond 3, Amazon Echo Plus, and smartphones. Each of them are described in the technology candidate section.

5.3. HDM

The HDM is a Multi-Criteria Decision-Making (MCDM) method that represents the problem in a hierarchical disposition to help decision makers visualize the effectiveness of each decision criteria on the main model's objective [20]. It is used to rank and evaluate the available alternatives based on quantitative and qualitative judgments.

The HDM basically converts the results of pairwise comparisons (performed by experts) into relative weights on a ratio scale by using a series of mathematical operations [21]. Using computations on matrices, any judgment by a single or multiple experts can be quantified [21]. The HDM is known to construct consensus among multiple decision makers with different goals [21]. It links the decision elements at different operational levels in support with higher-level goals in order to make sure that all operational decisions/strategies are aligned to the organization's benefits [21]. But it can be difficult without having intermediate levels of decision hierarchies [21].

Hence, the number of hierarchies in the HDM are dependent on the logical sequence of decisions involved. However, measurements can become too large or excessively aggregated with either too many or too few levels, respectively, in the HDM. It is advisable to maintain five hierarchies of the HDM with a multi-criterion mission at the top and actionable strategies or goals fulfilling these objectives at the next levels [21].

5.4. *HDM development*

The perspectives for this model were chosen based on the literature review. They are technical, economic and personal. The gap analysis shows that each perspective has its gaps, and this helps define the criteria of each perspective.

Based on the perspectives and criteria, the HDM was built. The objective of this research is the highest level in the hierarchy of the model that is aimed to evaluate and assess different smart hubs for the home automation application and then find the best for users. The second level is the criteria groups—the perspectives—that will be compared against each other based on their relative importance in serving the objective. The third level is the criteria and this is where criteria in the same group are compared against each other. The last level is the technology candidates—the alternatives—that are evaluated based on the criteria on the upper level (Figure 1).

5.5 *Pairwise comparison*

Pairwise comparison is a fundamental factor in MCDM, where the probability of rank reversal becomes an important measure to estimate the impact of ambiguity on the results [22]. The Pairwise Comparison Method (PCM) will be used to determine the relative importance of each alternative. Relative scores will help us determine the weights of each criterion. These criteria, when compared to each other in pairs, will show us the successful order of the options. A pairwise comparison combined with the HDM will guide us through choosing the absolute relevant option while keeping in mind the selected criteria.

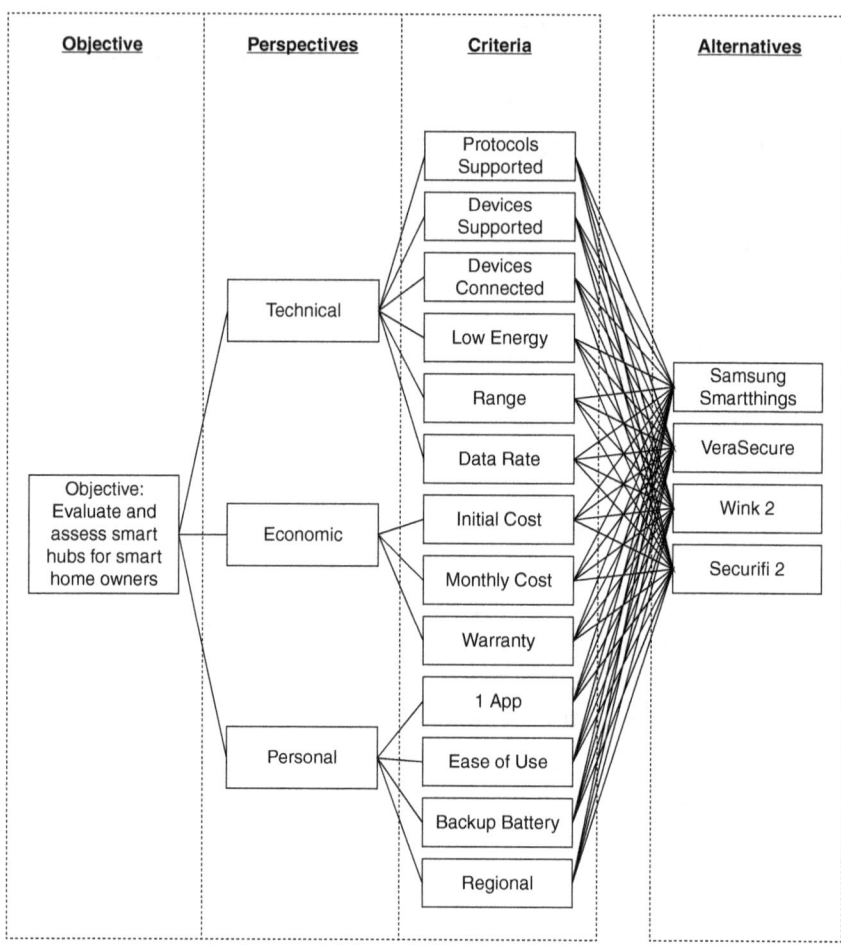

Figure 1. HDM development.

5.6. *Objective level*

The main objective of this project is to help consumers decide which smart hub is best suited to his/her needs based on various criteria, such as cost and affordability, technical specifications, ease of use, etc. In order to accomplish this task, a HDM is used in the project. The model incorporates various criteria for smart hubs, such as range of operation, communication protocols supported, number of devices supported, etc. Thus, the customer can make an appropriate choice of a smart hub based on his/her needs.

5.7. Criteria level

The HDM includes the analysis of a finite set of alternatives described as evaluative criteria. The decision makers can rank the criteria in terms of how important they are whilst considering all the criteria simultaneously. For the HDM used in this project the criteria used for the three previously mentioned perspectives are stated as follows:

Technical perspective

1. **Protocols supported:** This concerns the number of communication protocols supported by the devices (such as ZigBee, Z-Wave, WiFi, Bluetooth, etc.).
2. **Devices supported:** The number of available smart devices in the market that the hub can support.
3. **Devices connected simultaneously:** Number of devices that can be connected to the smart hub at the same time.
4. **Low energy connection:** Ability to control low energy smart objects such smart bulbs, door sensors, etc.
5. **Range:** The maximum distance over which the smart hub can receive/transfer data efficiently.
6. **Data rate:** The amount of data that can be transferred per second.

Economic perspective

1. **Initial cost:** Cost of purchasing the smart hub.
2. **Monthly cost:** Service or repetitive cost. It can be charged on a monthly or annual basis.
3. **Warranty:** Cost covering the warranty of the smart hub.

Personal perspective

1. **One app:** Ability to control all devices using a single app on the user's smartphone.
2. **Ease of use:** This concerns the ease with which the hub can connect to other smart devices.
3. **Battery:** Battery backup (mAh rating) of the smart hub in case of a power outage.
4. **Regional limitations:** This concerns the usage of the smart hub in different regions/countries, i.e., whether it can be operated normally in different countries (due to differences in radio frequencies of operation).

5.8 *Technology alternatives*

An alternative analysis process has been used to evaluate six different smart devices. These different alternatives has been chosen to give a variety of options to critique later. The following sections include information on each alternative.

5.8.1. *Alternative 1: Samsung SmartThings*

Samsung SmartThings allows users to connect, track and control different smart devices such as lights, thermostats and locks. It also improves one home's security, energy consumption, and more [23, 24]. The SmartThings app and Samsung's SmartThings Hub allow users to add as many devices as they want to automate, and they are compatible with ZigBee, Z-Wave and IP-connected devices too. The brands it can compatible with are: Honeywell, Schlage, Yale, First alert, D-Link, OSRAM LIGHTIFY, Leviton, Bose, Cree, and many others [25].

5.8.2. *Alternative 2: VeraSecure*

VeraSecure is an easy-to-install comprehensive home controller, which includes a full-featured alarm security system. It is also more than just a security device—it can be used to adjust temperature, thermostats, control lights and garage doors, etc., through an app on a smartphone, tablet or computer [26, 27].

5.8.3. *Alternative 3*: *Wink 2*

Wink 2 is the first ever home hub created for a conventional customer. It brings together multiple devices and products from popular brands and offers industry-leading smart home protocol support, enhanced connectivity features and a sleek design that packages everything into a simple, intuitive hub [28, 29].

5.8.4. *Alternative 4*: *Securifi Almond 3*

Securifi Almond 3 is the first ever device that incorporates a touchscreen interface in wireless routers. In addition, the Securifi Almond 3 does not require any PC/MAC for setup [30]. Instead, all setup and maintenance options are available from the product's touchscreen interface [30, 31].

5.8.5. *Alternative 5*: *Amazon Echo Plus*

Amazon Echo is a new smart speaker developed by Amazon. This smart speaker has many useful features. It is capable of voice interaction, music playback, playing audiobooks, setting alarms and providing weather updates [32, 33].

5.8.6. *Alternative 6*: *Smartphones*

A smartphone refers to any mobile phone that has the ability to perform similar functions as a computer [27]. Such a phone will have a touchscreen interface, internet access and an operating system that can run downloaded applications [34].

From the above six alternatives, only four were considered for the HDM used in the project, namely: Samsung SmartThings, VeraSecure, Wink 2 and Securifi Almond 3. Smartphones were eliminated due to their inability to support the "one app" feature. Amazon Echo Plus was eliminated because it supports only three protocols. We will use the latter two criteria to define the technology candidates.

5.9. *Expert panel*

After developing the HDM, an expert panel is formed. The expert panel of this project consists of four persons to determine the relative relationship among the decision elements at various levels of the model. There are two steps: data collection and development of the ranking table.

We came out with three important perspectives for smart home users to determine the best smart home hubs based on the developed alternatives. In the data collection stage, we collected information from companies' websites. In the development of the ranking table, we ranked the numbers of each criterion from one to four for experts to measure the relative importance of each alternative, as shown in Table 4.

6. Result Analysis and Discussion

The HDM was used to structure the decision into objective, perspectives and criteria to better understand the important factors to consumers when selecting the best smart hub alternatives. It was then used to identify the best among the alternatives. The following sections show the result of the analysis and discussion. More details about the results are in Appendix B.

6.1. *Perspectives ranking*

The perspectives used in this model are Technical, Economic and Personal. They were ranked using the PCM software provided by Portland State University in order to determine the relative

Table 4. Alternatives ranking table.

Ranking (1–4) 1—Poor 4—Excellent	Criteria	Samsung SmartThings	VeraSecure	Wink 2	Securifi Almond 3
Technical	Protocols supported	3	4	4	4
	Number of devices supported	4	4	2	3
	Devices connected	4	2	3	2
	Ability to control low energy	4	4	4	4
	Range/distance	3	1	2	4
	Data rate	4	4	4	3
Economic	Cost of smart hub	4	1	3	2
	Service cost—monthly/annually	4	4	4	4
	Warranty	4	4	4	4
Personal	One App	4	4	4	4
	Ease of use	4	4	4	4
	Battery backup	3	4	1	1
	Region limitations	1	1	1	1

importance of each perspective to the overall objective of the project (Figure 2). Table 5 shows the experts' weighting for the perspectives regarding the objective. The mean for each perspective's weighting was calculated.

6.2. *Inconsistency and disagreement*

This model shows that the level of inconsistency for all the experts is below 0.10, which is in the acceptable range. The inconsistency level above 0.10 occurs when the choices of

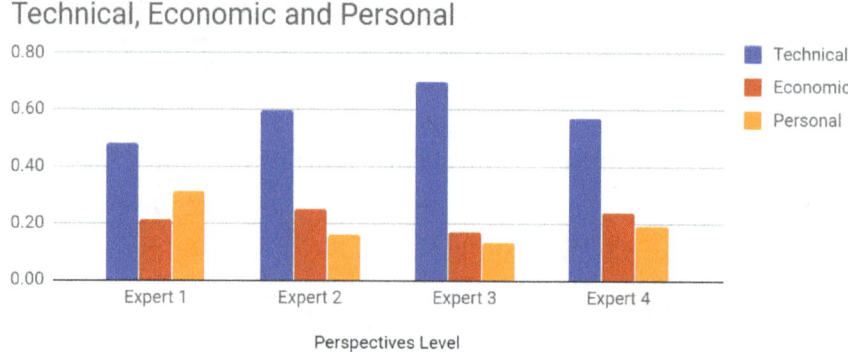

Figure 2. The weighting for each perspective.

Table 5. Perspectives ranking.

Perspectives Level	Technical	Economic	Personal	Inconsistency
Expert 1	0.48	0.21	0.31	0.00
Expert 2	0.60	0.25	0.16	0.00
Expert 3	0.70	0.17	0.13	0.02
Expert 4	0.57	0.24	0.19	0.01
Mean	0.59	0.22	0.20	
Disagreement				0.03

the preferences are not aligned. The experts' answer to each pairwise comparison should be consistent and has an overall level of <= 0.10 for each expert. Inconsistency is considered as a measurement of validation for the results.

6.3. *Criteria ranking*

For this model, 13 criteria were selected to evaluate the alternatives based on three perspectives. The experts ranked each criterion with respect to its corresponding perspective. The higher

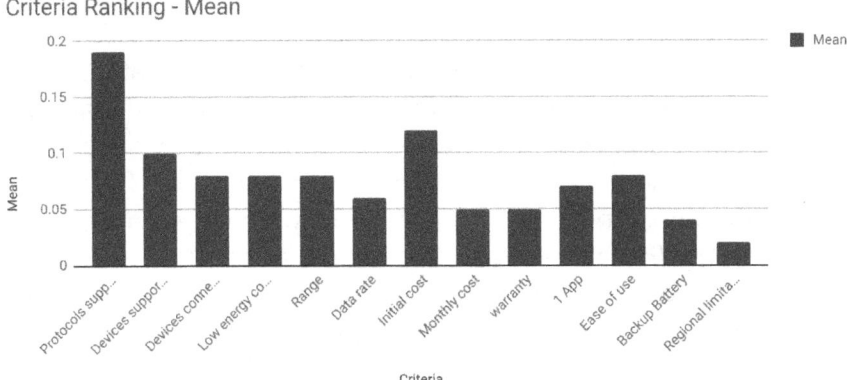

Figure 3. The weighting for each criterion.

value a criterion has, the higher its impact on its perspective. In order to identify the weighting for each criterion, a pairwise comparison was conducted. The weighting for each criterion is shown in Figure 3.

Figure 3 shows each criterion's relative importance to its perspective. From the technical perspective, Protocols Supported has a score of 0.19, which is the most important criterion in the overall model, followed by Devices Supported, Devices Connected, Low Energy Connection, and Range. The least important criterion in this perspective is Data Rate. From the economic perspective, the most important criterion is Initial Cost and the least important criteria are Monthly Cost and Warranty. Initial Cost always plays a big part when considering smart hubs, and depending on the manufacturer of the device, some costs such as cloud services costs and premium accounts are added later onto the user's account. From the Personal perspective, ease of use is the most important criterion; this illustrates how easy it is to use the smart hub to connect new smart objects to the network. The second most important criterion is 1 App, which refers to a single app controlling all connected devices.

6.4. *Alternative ranking*

The experts weighted each technological alternative with respect to the relative importance to each criterion that contributed to its perspective (see Appendix C). Each alternative is assigned a value ranging from one to five, with five being Excellent and one being Poor, with respect to each criterion based on the literature review. These values are then multiplied by the relative importance of each criterion to the consumers. The higher the value, the more important that alternative is to the overall objective. Table 6 and Figure 4 show the ranking values of each perspective for each alternative.

6.5. *Overall HDM results*

The HDM was utilized to structure the decision into the objective, criteria and alternatives. The following figure shows the overall result of the HDM. More details are shown in Appendix A.

Figure 4 shows the top factors from a customer's point of view. These factors should give insight to smart home owners about the criteria they should consider when selecting a smart hub. Table 7 shows the weights of importance of each criterion under each perspective to the overall objective.

The figure also shows that the criteria of ease of use, the ability to connect low energy devices, the number of connected devices simultaneously, and range are highly

Table 6. The ranking values for each perspective.

	Technical	Economic	Personal	Alternative Value
Samsung SmartThings	2.09	0.88	0.74	3.71
VeraSecure	1.96	0.52	0.78	3.26
Wink 2	1.92	0.76	0.66	3.34
Securifi Almond 3	2.04	0.64	0.66	3.34

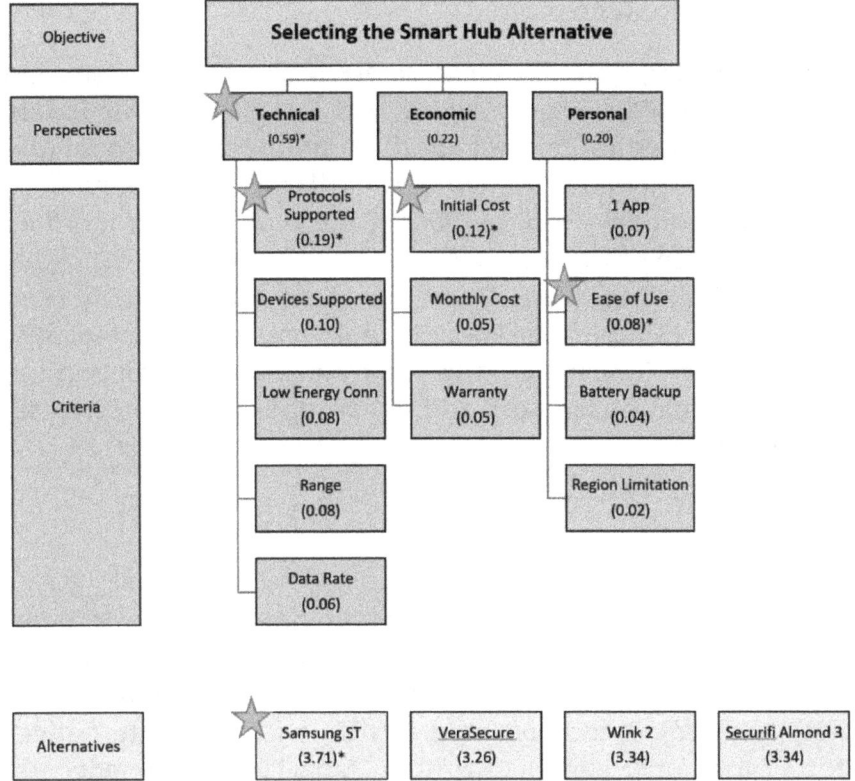

Figure 4. The overall ranking value for each perspective and criterion.
Note: *Highest ranked element at its level.

Table 7. Top three factors.

Criteria	Perspective	Weight
Protocols supported	Technical	0.19
Initial cost of the smart hub	Economic	0.12
Number of devices supported	Technical	0.10

ranked (0.8). These factors should also be taken into consideration by the smart home owner when selecting the best smart hub device.

7. Future Research

In this project, four different smart hubs, namely, Samsung SmartThings, VeraSecure, Wink 2 and Securifi Almond 3 were evaluated based on technological, economic and personal perspectives by using a HDM. Other smart hubs such as Panasonic, Logitech Harmony Hub, VeraEdge, etc., can also be included and evaluated using the HDM.

Including judgments from users who have actually used these smart hub into this model would enhance the credibility of the model. Researchers and professionals specializing in the home automation area can add tremendous inputs via alternatives rankings.

8. Conclusion

The HDM and the PCM will together provide the consumer with the best possible smart hub alternative. This model can be altered to include other smart hubs available in today's market. This project incorporated a HDM with four alternatives for smart hubs, namely, Samsung SmartThings, VeraSecure, Wink 2 and Securifi Almond 3. A ranking was assigned to these alternatives based on the experts' opinions used in the HDM. Based on the weights assigned to the perspectives and criteria, it was concluded that Samsung SmartThings was the best alternative for the model used, followed by Wink 2 and Securifi Almond 3, as shown in Table 6. From Figure 4, Initial Cost, Protocols Supported and Ease of Use were the most important criteria.

References

1. S. Keach, "Best Black Friday Deals 2017: All the latest sales and best deals", *Trusted Reviews*, 26 November 2017. http://www.live-smart.co/review/what-is-a-smart-hub-and-how-do-i-choose-the-right-one-5186. Accessed: 26 November 2017.

2. "7 things to know about a next generation smart home hub", *Homementors—Home Automation Training Courses & Books*, 18 January 2015. http://homementors.com/7-key-features-smart-home-hub/. Accessed: 26 November 2017.
3. "What is smart home hub (home automation hub)? Definition from WhatIs.com", *IoT Agenda*. http://internetofthingsagenda.techtarget.com/definition/smart-home-hub-home-automation-hub. Accessed: 26 November 2017.
4. "AppleHomeKit", *Which? Tech Daily*. https://www.which.co.uk/reviews/smart-home-hubs/article/what-is-a-smart-hub. Accessed: 26 November 2017.
5. H. Walsh, "Smart home hubs—three first look verdicts", *Which? Tech Daily*, 15 February 2016. https://blogs.which.co.uk/technology/app-review/smart-home-hubs-three-first-look-verdicts/. Accessed: 26 November 2017.
6. "Home automation", *Eepartnership.org*. https://eepartnership.org/wp-content/uploads/2014/07/Tom-Kerber.pdf. Accessed: 23 November 2017.
7. "Why the smart home and smart enterprise are more connected than ever before", *Iotjournal.com*, 2017. http://www.iotjournal.com/articles/view?16409. Accessed: 23 November 2017.
8. "Patent US20140324410—Apparatus and method for the virtual demonstration of a smart phone controlled smart home using a website", *Google Books*, 2017. https://www.google.com/patents/US20140324410. Accessed: 23 November 2017.
9. "What is smart home hub (home automation hub)? Definition from WhatIs.com", *IoT Agenda*, 2017. http://internetofthingsagenda.techtarget.com/definition/smart-home-hub-home-automation-hub. Accessed: 23 November 2017.
10. "Smarts Hubs", *Pdfs.semanticscholar.org*, 2017. https://pdfs.semanticscholar.org/6d7f/60b16adead96aafa9e975207980eb32671b5.pdf. Accessed: 23 November 2017.
11. A. Joshi, "Philips Hue: automated home lighting gets colorful", *RSS*, 1 March 2013. https://www.anandtech.com/show/6805/philips-hue-automated-home-lighting-gets-colorful/4. Accessed: 26 November 2017.
12. "The STEEP analysis—what to analyse in the marketing environment", *HubPages*, 25 January 2016. https://hubpages.com/business/The-STEEP-Analysis-What-to-analyse-in-the-marketing-environment. Accessed: 26 November 2017.

13. M. Alipour, R. Hafezi, M. Amer and A. Akhavan, "A new hybrid fuzzy cognitive map-based scenario planning approach for Iran's oil production pathways in the post–sanction period", *Energy* 135 (2017) 851–864.
14. K. Fondren, "Process power tools—conducting a gap analysis while modeling the future state", *Captech Consulting, Inc.*, 12 August 2013. https://www.captechconsulting.com/blogs/process-power-tools-conducting-a-gap-analysis-while-modeling-the-future-state. Accessed: 26 November 2017.
15. "Gap analysis: reaching your ideal future state", *MindTools.com*. https://www.mindtools.com/pages/article/gap-analysis.htm. Accessed: 26 November 2017.
16. "Gap analysis: a simple tool for achieving your business goals—shopify", *Enterprise Business Marketing, News, Tips & More*. https://www.shopify.com/enterprise/102475782-gap-analysis-a-simple-tool-for-achieving-your-business-goals. Accessed: 26 November 2017.
17. K. Bunse, M. Vodicka, P. Schönsleben, M. Brülhart and F. O. Ernst, "Integrating energy efficiency performance in production management—gap analysis between industrial needs and scientific literature", *Journal of Cleaner Production* 19, 6–7 (2011) 667–679.
18. M. Prospero, "Best smart home hubs of 2017", *Tom's Guide*, 20 November 2017. https://www.tomsguide.com/us/best-smart-home-hubs,review-3200.html. Accessed: 26 November 2017.
19. "Learn how much it costs to install a home automation system", *2017 Home Automation Costs | Smart Home Systems Pricing*. https://www.homeadvisor.com/cost/electrical/install-or-repair-a-home-automation-system/. Accessed: 26 November 2017.
20. D. Kocaoglu, "A participative approach to program evaluation", *IEEE Transactions on Engineering Management* 30, 3 (1983) 112–118.
21. D. U. Tugrul, *Hierarchical Decision Modeling, Essays in Honor of Dundar F. Kocaoglu* (Springer International Publishing, 2016), Switzerland.
22. G. Dede, T. Kamalakis and T. Sphicopoulos, "Theoretical estimation of the probability of weight rank reversal in pairwise comparisons", *European Journal of Operational Research*, 252, 2 (2016) 587–600.

23. "Samsung Electronics America", *Smart home—Smartthings | Samsung US*, 2017. https://www.samsung.com/us/smart-home/smartthings/. Accessed: 28 November 2017.
24. "Echo plus", *Amazon.com*, 2017. https://www.amazon.com/Amazon-Echo-Bluetooth-Speaker-with-Alexa-Black/dp/B00X4WHP5E. Accessed: 23 November 2017.
25. Samsung Electronics America", *Samsung SmartThings Hub SmartThings—F-HUB-US-2 | Samsung US*, 2017. https://www.samsung.com/us/smart-home/smartthings/hubs/f-hub-us-2-f-hub-us-2/. Accessed: 28 November 2017.
26. "VeraSecure Advanced Smart Home Security Controller—Vera™", Vera. http://getvera.com/controllers/verasecure/. Accessed: 26 November 2017.
27. "Find the best deal on what is a smartphone", *Exploreshops.net*, 2017. http://exploreshops.net/sch/?phrase=What+Is+A+Smartphone&hash=22fbc7be27bd973c2d8af1d5d1561c17&creative_id=108440997189&keyword_id=kwd-9634772970&keyword=what%20is%20a%20smartphone&match_type=b&device=c&devicemodel=&campaign_id=407866989&adgroup_id=29108693949&target=&placement=&gclid=EAIaIQobChMI16XckO_O1wIVIshkCh27XwOIEAAYAyAAEgKEAfD_BwE. Accessed 23 November 2017.
28. "Wink Hub 2", *Wink.com*. https://www.wink.com/products/wink-hub-2/. Accessed: 28 November 2017.
29. "Wink Hub 2", *Walmart.com*, 2017. https://www.walmart.com/ip/Wink-Hub-2/54803020?wmlspartner=wlpa&selectedSellerId=0&adid=22222222227054358808&wl0=&wl1=s&wl2=c&wl3=154910077106&wl4=pla-267412110675&wl5=9032925&wl6=&wl7=&wl8=&wl9=pla&wl10=8175035&wl11=online&wl12=54803020&wl13=&veh=sem. Accessed: 23 November 2017.
30. "Securifi", *Securifi.com*, 2017. https://www.securifi.com/almond. Accessed: 23 November 2017.
31. "Securifi Almond 3 Smart Home Wi-Fi System", *PCMAG*, 2017. https://www.pcmag.com/review/350459/securifi-almond-3-smart-home-wi-fi-system. Accessed: 23 November 2017.
32. L. Chang, "Despite an email mixup, the Echo Spot should arrive before Christmas", *PC Magazine*, 2017. https://www.digitaltrends.com/home/amazon-echo-plus-second-generation-and-more/. Accessed: 28 November 2017.

33. *Amazon.com*, 2017. https://www.amazon.com/dp/B075RWFCHB?tag=googhydr-20&hvadid=223605111263&hvpos=1t1&hvnetw=g&hvrand=4000390557756188175&hvpone=&hvptwo=&hvqmt=e&hvdev=c&hvdvcmdl=&hvlocint=&hvlocphy=9032925&hvtargid=kwd-365336145051&ref=pd_sl_2hssrnbb2b_e. Accessed: 23 November 2017.
34. "Top 7 smartphones of 2017", *Dignited*, 2017. http://www.dignited.com/23364/top-7-smartphones-2017/ Accessed: 23 November 2017.

Appendix A: References of Alternative Ranking Table

Perspectives	Criteria	Alternatives				
		Samsung SmartThings	VeraSecure	Wink 2	Securifi Almond 3
Technical	Protocols supported	5 WiFi, Bluetooth, ZigBee, Z-Wave, Ethernet port. [1]	7 WiFi, Bluetooth 4.0+BLE, WiFi, ZigBee, Z-Wave, VeraLink, 3G Cellular.	7 Bluetooth LE, Z-Wave®, Security Enabled Z-Wave Plus Device, ZigBee®, WiFi®, Lutron® Clear Connect®, Thread (future), Kidde.	7 WiFi, Bluetooth, Cellular Data Networks (5Ghz), Z-Wave alarms, Zigby radio, Nest Thermo.
	Number of devices supported	2,000+	Up to 2,000	530	1,000+
	Number of devices that can be connected at the same time	Unlimited [2] [3]	100+	350	Up to 100, though slowing occurs at varying thresholds beyond 50–60.
	Ability to control low energy	Yes [1]	Yes	Yes	Controls household sensors by mobile device.

(Continued)

Appendix A: (*Continued*)

Perspectives	Criteria	Alternatives			
		Samsung SmartThings	VeraSecure	Wink 2	Securifi Almond 3
	Range/ distance	50–100 ft. Dependent on home's construction [1]	15 feet optimum, could be affected by obstacles (5–20%). Repeaters can be used to boost signal and improve the range.	Up to 50 ft.	160 ft.
	Data rate	1,000 Mbps (Ethernet). Live video streaming is supported free of charge; video clip recording is available as an optional premium feature within the SmartThings app.	1,000 Mbps (Ethernet)	1,000 Mbps (Ethernet)	867 Mbps [6]
Economic	Cost of smart hub	$79.99 [1]	$299.95	$99	$155

Personal	Service cost monthly/annually	No [3]	No monthly fees, no contracts.	No	Standard WiFi charges.
	Warranty	1 year	12 months [5]	1 year	1 year
	1 App	Yes	Yes	Yes	Yes
	Ease of use	Yes. No wiring or installation needed: anyone with broadband Internet connection can easily set up their hub [1]	Yes	Yes. Auto-discovery and guided setup.	Yes
	Battery backup	4, AA batteries. Up to 2 hrs (backup only) [1]	Internal 2,400 mAh/18Wh.	No. 1 Lithium ion battery required.	No
	Region limitations	Yes	A US device will not work with a Euro Vera controller due to the frequency incompatibility.	Yes	Yes

Appendix B: HDM Result Table

Objective	Protocols supported	Devices supported	Devices connected simultaneously	Low energy connection	Range	Data rate	Initial cost	Monthly cost	Warranty	1 App	Ease of use	Backup Battery	Regional limitations	Inconsistency
Ahmed Alzahran	0.24	0.06	0.1	0.01	0.04	0.02	0.14	0.05	0.02	0.09	0.14	0.06	0.04	0.02
Majed Alshamlar	0.17	0.15	0.1	0.04	0.08	0.05	0.08	0.12	0.05	0.07	0.04	0.03	0.02	0.02
Shreyas Harish	0.1	0.1	0.07	0.08	0.12	0.11	0.14	0	0.1	0.07	0.06	0.04	0.02	0.01
WEI-CHEN HSU	0.26	0.07	0.05	0.2	0.09	0.04	0.13	0.03	0.01	0.03	0.07	0.02	0.01	0.01
Mean	0.19	0.1	0.08	0.08	0.08	0.06	0.12	0.05	0.05	0.07	0.08	0.04	0.02	
Minimum	0.1	0.06	0.05	0.01	0.04	0.02	0.08	0.03	0.01	0.03	0.04	0.02	0.01	
Maximum	0.26	0.15	0.1	0.2	0.12	0.11	0.14	0.12	0.1	0.09	0.14	0.06	0.04	
Std. Deviation	0.06	0.04	0.02	0.07	0.03	0.03	0.02	0.04	0.04	0.02	0.04	0.01	0.01	
Disagreement														0.038

Appendix C: Impact of Alternatives to Overall Objective

Ranking (1-4) 1 – poor 4 – excellent		Alternatives			
	Criteria	Samsung SmartThings	VeraSecure	Wink 2	securifi 2
Technical	Protocols supported	0.57	0.76	0.76	0.76
	Devices supported	0.4	0.4	0.2	0.3
	Devices connected	0.32	0.16	0.24	0.16
	Low energy connection	0.32	0.32	0.32	0.32
	Range/distance	0.24	0.08	0.16	0.32
	Data rate	0.24	0.24	0.24	0.18
	Sum	2.09	1.96	1.92	2.04
Economic	Initial cost	0.48	0.12	0.36	0.24
	Monthly cost	0.2	0.2	0.2	0.2
	Warranty	0.2	0.2	0.2	0.2
	Sum	0.88	0.52	0.76	0.64
Personal	One App	0.28	0.28	0.28	0.28
	Ease of use	0.32	0.32	0.32	0.32
	Battery backup	0.12	0.16	0.04	0.04
	Region limitations	0.02	0.02	0.02	0.02
	Sum	0.74	0.78	0.66	0.66
Alternative Ranking		3.71	3.26	3.34	3.34

Chapter 7

Personal Transformation: Protocols for Home Automation Application

Ahmed Alzahrani* and Tugrul Daim*,†,‡

*Portland State University, Portland, Oregon, USA
†Higher School of Economics, Moscow, Russia
‡Chaoyang University of Technology, Taiwan

Abstract

The idea behind Smart Homes is nothing but all the appliances, lights, locking system, security cameras and other objects fixed in the house are smart and know how to communicate with each other via internet. These objects can be controlled remotely using phone or internet [1]. In order to make all smart devices communicate with each other, they need to be operated by a protocol. Currently, there are variety of protocols available in the market, and they all have their own pros and cons. Each one has its own language. Most of these protocols support wide range of devices/objects and offer interoperability between these devices. Although, there are multiple factors such as data rate, costs associated with building a smart home, and power consumption which matter a lot while choosing any protocol.

This paper uses a methodology Hierarchical Decision Model (HDM) to compare four wireless communication protocols used in building home automation applications. To evaluate these four protocols, the model used different criteria related to these protocols including adoption range, interoperability, encryption rate, data rate and usage, network capabilities.

To identify the best communication protocol, this model uses these four protocols namely- WIFI, Bluetooth, ZigBee, and Z-Wave. By using the information available in literature review and model evaluations, Z-Wave is chosen to be the best communication protocol in general of all. However, the Bluetooth protocol is ranked to be the most flexible protocol, whereas ZigBee is ranked to be the most reliable one. Amongst all the criteria used to choose these protocols, adoption scale, power consumption and encryption seem to be the important ones according to users' opinions.

Keywords: Technology assessment, home automation, smart homes.

1. Introduction

1.1. *Home automation overview*

Home automation, or building a smart home, has been a growing paradigm today. It helps monitor and control the living environment and improves the quality of life by making it more productive, reliable and sustainable [2–4].

Vermesan and Friess [5] define the Internet of Things (IoT) as the networked interconnection of everyday objects by involving self-configuring wireless networks of objects that create a world where everything in it sends information to other objects and to people. The concept of home automation can be defined as making homes smarter by using computers to schedule or automatically control home functions and features remotely.

The process includes varying the state of a device or a system with respect to internal or external occurrences. It can be by a person controlling the device with an action or as a result of movement of time. To achieve this change through automation, some kind of a control message or communication must be sent to the system to be reacted on. This can be a simple on-off switch or can be a complex tweaking in the current state, such as dimming/increasing it to a certain level. In order to make the device act upon the control message, communication signals are sent using home automation protocols. Such protocols are embodied in alliances that which tend to create instant revenue opportunities for members willing to join in demand response with their products or services [6]. Another purpose of such alliances is to maintain industry-wide standards for home automation and avoid interoperability issues [6].

1.2. *Communication protocols*

Communication protocols are networking layer languages. This layer is the middle layer between the physical layer and the application layer. It is responsible for connecting and processing the data gathered from the sensors and other devices in the physical layer, which then transmits and processes the data to the application layer, which will then be delivered to the user.

There is a wide range of technology platforms, or communication protocols. Each protocol speaks its own language to the connected nodes and devices to perform a function. These protocols can be classified into three categories—wire connectivity, wireless hybrid and wireless.

Wired technologies have been around since the 1970s. These technologies transfer signals over the main power lines to control lights, switches, outlets and other objects from IR remote controllers or computer interfaces. On the other side, wireless technologies are more flexible. They send signals wirelessly and can be located wherever needed, as there is no need to change the home wiring.

2. Problem Statement

Most of the home automation devices on the market these days lack interoperability standards. As each protocol has its own functionality and applications, it is hard to maintain commonality in the industry.

On the other side, there is the need to have different communications needs. Security cameras, for example, transfer a high amount of data that needs a high bandwidth. When a high level of data rate is required, devices normally use WiFi. However, WiFi consumes a lot of power. Meanwhile, other protocols like ZigBee and Z-Wave are designed to control smart devices with a long battery life.

This paper focused on home automation applications by analyzing and evaluating wireless communication protocols that can be used for smart home projects. Different criteria and perspectives were considered to find the best communication protocols and to come up some recommendation using the Hierarchy Decision Model (HDM) and pair-wise comparison.

3. Literature Review

IoT applications, including home automation, consists of mainly three layers: the physical layer (the objects that sense and gather the information in the environment); the network layer (the communication protocol that is responsible for connecting the different object with each other); and the application layer, which delivers the gathered data to the end user based on the designed application.

3.1. *Candidate protocols*

An extensive literature review, conducted to find the wireless protocols that can serve the home automation application, found there were many wireless communication protocols. Some of them were considered standards to be followed by smart objects manufacturers, while others were designed for use in a special line of products. This paper will focus on the first group, and therefore, there are four main wireless protocols competing in this field. Namely, they are WiFi, Bluetooth, ZigBee and Z-Wave.

3.1.1. *WiFi*

This communication protocol runs on higher radio bands than other protocols, which allow the transmission of a higher amount of data per second in short distances. However, this protocol consumes a high amount of power for many IoT applications. Most people are familiar with WiFi networks that can be managed locally (LAN) or remotely (internet) [7]. A WiFi network is nothing but a star network in which all the devices communicate with the central hub. Any device moving out of the central hub vicinity cannot be a part of the network [8].

3.1.2. *ZigBee*

Lately, ZigBee is known to be one of the most efficient technologies for wireless home networking, as its applications need significantly less data and consume less battery [9].

ZigBee is a communication protocol developed by the ZigBee alliance. It is a low-cost, low-power and wireless single-mesh network standard developed for long battery life devices. It operates in different radio bands around the world with a variance in the data rate based on the band frequency. ZigBee uses smartphones to control the home automation objects. These objects are made by big brand names including GE, LG, Logitech, Philips and Samsung.

3.1.3. *Z-Wave*

Z-Wave wireless technology, which operates in the 900 MHz spectrum, is primarily used for home automation and has become the de facto home automation wireless standard with widespread support and interoperability by and between vendors [10]. It is a single mesh network that allows communication between devices within the network. It operates at lower radio frequencies than WiFi and ZigBee networks, which makes the data packets at data rates smaller. Z-Wave technology allows the use of smartphones and other mobile devices to control the network. Devices like Amazon Echo and Apple Watch are Z-Wave compatible [11]. It supports a variety of set-ups and sights using different options for lights, switches and door locks [11]. With multiple wireless technologies present in today's market, Z-Wave has been a well-known technology in the resident security channel [12]. Almost all remote home management solutions with virtual security in North America seem to be powered by Z-Wave [12].

3.1.4. *Bluetooth*

Bluetooth Low-Energy, or Bluetooth Smart, is designed to offer reduced power consumption. It communicates a small amount of data over short distances. The latest version is Bluetooth 5, which was announced in 2016, is focused on the emerging IoT. According to the Bluetooth Special Interest Group (SIG), important features in Bluetooth 5 are longer scale, faster speeds, wide broadcast message space, revised interoperability and harmony with other technologies [13]. Bluetooth 5 continues to improve the IoT experience with effective interactions across a wide range of devices [13].

3.2. Mesh networking

Z-Wave and ZigBee are mesh networks that act like a star network with a signal initiated by the central hub [14]. However, devices are not required to communicate with the central hub directly. A mesh network permits all the devices to act as a repeater and pass the signal forward, which makes the network versatile, even with greater distance and obstacles around [8].

4. Methodology

4.1. *Multiple-Criteria Decision Models*

Any environmental project requires a decision-making process to be considerate of socio-political, environmental and economic aspects. In addition, the stakeholders' point of view makes the process even more complicated [15].

The Multiple-Criteria Decision Model (MCDM) is known to be a formal approach that considers both technical data as well as stakeholders' significance [15]. The only concern with using MCDM for obtaining universal preference relation on the set of alternatives is to aggregate preferences with respect to relations with all the criteria [16].

4.1.1. *The Analytic Network Process*

The Analytic Network Process (ANP) theory of multi-criteria measurement is used to obtain the relative priority scales of definite numbers from individual responses. It can include normalizing the actual judgement to a relative form, which also belongs to a basic scale of numbers [17].

The recorded responses show the relative impact of one or more factors with respect to the other in a pairwise comparison, relative to the fundamental control factor [18].

By using a super matrix derived from the column priorities, the ANP integrates the outcome of interdependence and responses within the elements [17].

4.1.2. *Best-Worst Method*

Another method of solving multi-criteria decision-making problems is a Best-Worst Method (BWM), where a number of choices are evaluated relative to a number of elements, before the best one is chosen [19]. In this process, experts estimate the best and worst criteria as being either the most or least desirable [20].

Using pairwise comparison, all criteria are evaluated against each other and the maximin question is then derived to estimate the weights [19].

4.1.3. *Multi-attribute Utility Theory*

The Multi-attribute Utility Theory (MAUT) is another decision-making approach derived by Keeney and Raiffa [21, 22]. This tool transforms multiple criteria into a common scale and then compensates the poor ones with the higher scores of the other criteria [22]. Hence, it is also known as a compensatory approach [23, 24].

4.1.4. *Decision Expert*

The Decision Expert (DEX) is a 30-year-old qualitative decision support approach in the evaluation of decision choices, which has been used in scientific, technical and practical researches [25]. It combines the classical numerical MCDM with a rule-based experts judgment approach. This method guided many new algorithms and techniques for decision knowledge and evaluations with multiple criteria [25].

4.1.5. *Analytic Hierarchy Process*

The Analytic Hierarchy Process (AHP) mainly deals with decision-making in a complex world [26]. It includes identifying and arranging decision objectives, estimating criteria and constraints, and then evaluating alternatives using pairwise comparison for all elements in each hierarchy [26]. In addition, this model also estimates the relative importance of alternative courses of action [26].

The AHP is relatively simple, easy to use and is the most flexible of all the other multi-criteria decision models. Hence, it has been widely used and researched since 1980 in all applications containing multiple alternatives in decision-making [27].

There are thousands of AHP applications used in the complex decision-making process [28], which have delivered effective results in areas such as planning, resource allocation, deciding priority, and choosing the best alternative [29, 30]. Some of the application involved: selecting a type of nuclear reactor [31], decisions on global climate change [32], evaluating the quality of software systems [33], choosing faculty for the University of Pennsylvania [34], deciding where to locate offshore manufacturing plants [35], assessing risk in operating cross-country petroleum pipelines [36], selecting a third-party logistics (3PL) provider [37], and exploring the commercialization of future motor fuel technologies [38].

4.2. HDM overview

So far, the AHP and its variant, the HDM, are one of the most used methods for evaluating the alternatives in the area of decision-making [22]. Additionally, the AHP has been the most popular model in management science research and applications [22, 39].

The HDM is a MCDM method that represents the problem in a hierarchical disposition to help decision makers visualize the effectiveness of each decision criteria on the main model objective [40]. It is used to rank and evaluate the available alternatives based on quantitative and qualitative judgments [41].

The HDM converts all the judgments recorded in pairwise comparisons to relative weights on a ratio scale using mathematical operations on three matrices [42]. This ultimately quantifies all the responses stored by decision makers [42, 43]. In the case of a higher number of decision makers, the HDM becomes useful to identify consensus [42, 44]. The HDM links all the decision criteria at the operational level at each hierarchy to support the higher-level objectives of an organization [42], although it can be difficult to quantify the relationship between the benefits/mission at the top most level and all the other bottom levels with solely operational decisions, without having space for all intermediate hierarchies [42].

The total number of hierarchies in the HDM is derived using the logical sequence of the decisions involved. Additionally, this number does not need to be very high to avoid exceedingly large measurements nor too low to avoid excessive aggregation [42]. Hence, it is recommended to have an optimum number of five levels where each hierarchy can contain multi-dimensional decision factors. The top most level would be the Mission whereas the levels below are actionable Strategies [42]. Strategies can influence multiple Goals and the achievement of each Goal would result in delivering Objectives.

4.3. *The HDM model development*

The objective of this research is the highest level at the hierarchy of the model that is aimed to analyze different communication protocols for home automation and to find the best one for the home automation application. The second level is that of the criteria groups, the Perspectives. These groups will be compared to each other based on how important each one is in serving the Objective. The third level is the criteria, and this is where each group of criteria is compared against other criteria in the same group. The last level is communication protocols, the Alternatives, that are evaluated based on the criteria based on the upper level.

4.4. *Perspectives and criteria*

The literature review shows important points regarding the specifications for each communication protocol. Firstly, flexibility and expanding the ability of the network is important. The system architecture includes a personal computer to provide network management, remote access and wireless technologies, which can make the network complex and expensive [45].

Also, almost all systems require physical wiring and connections, which will make these intrusive installations costly [45].

Additionally, existing systems offering multiple approaches for user control and monitoring of the devices are generally limited to a single method that ultimately creates the problem of interface inflexibility. However, systems offering more than one interface can confuse users [45].

From these points, three perspectives were created to build the model: Flexibility, Network and Reliability. For more explanation about the model, this section describes the second and third levels of the hierarchy. There are four perspectives in this model: Flexibility, Network, Ease of Use, and Security. The model consists of 12 criteria (Figure 1).

4.4.1. *Flexibility*

This perspective represents the ease of building and expanding the network. This is constrained by the number of smart devices and brands that communicate using the protocol or using it as a language to communicate with central hubs. Cost is another factor to be considered in the initiation and expansion of the network. This perspective consists of three criteria:

Cost: The cost of using devices that are supported by the protocol. Some protocols seem to be less expensive due to the unnecessity of using special hubs to run the protocol as opposed to their counterparts, given that their technology is not proprietary. Thus, they do not command a premium price.

Interoperability: The support of the protocol by different brands, which gives the customer the freedom to choose from a range of certified devices by the protocol alliance, which generally lacks network interoperability for home networks as well as automation systems [45].

Adoption scale: Each protocol has a large selection of products, that is, the number of products supporting this protocol. A product that supports any given protocol mainly shows the

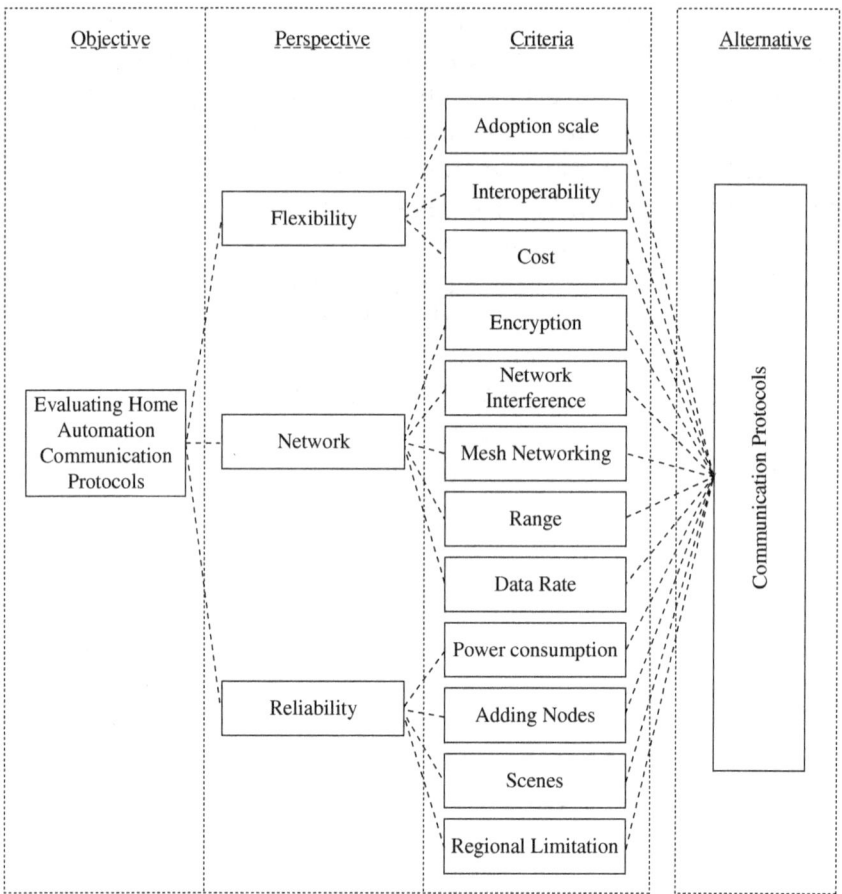

Figure 1. Communication protocols HDM.

communication protocol logo in its packaging. The date of this criterion is mainly collected from the communication protocol sponsor's website [46–49].

4.4.2. Network

This perspective represents the characteristics of the network. All protocols share the same characteristics, but each characteristic differs from one protocol to another. Different protocols transmit data for different ranges and different sizes. The radio frequency is different in all protocols, but if more than one protocol is being used in the same smart home, and those protocols have the same radio frequency, there could be a network interference. Mesh networking is another network feature that is recommended in a smart home in order to expand the network without affecting the signal strength. This perspective has five criteria:

Encryption: The process of converting information or data into a code, especially to prevent unauthorized access [50]. Encryption is generally used to transmit confidential digital data across networks. It can also be used to keep classified files stored on a desktop confidential.

The signals are transformed to readable and unreadable and vice versa using mathematical algorithms, also known as keys. Security experts can make more use of encryption in order to protect sensitive information [51].

Range: The maximum network coverage area from one point to another. This is measured in feet or meters.

Mesh networking: In some protocols, the signal transmits from device to device (nodes) until it reaches the hub and every device acts as a repeater.

Network interference: The radio frequency wave that the protocol runs on: the specific and different radio frequency wave of 908.42 MHz is used for Z-Wave than for ZigBee, WiFi and Bluetooth. This avoids a possible lag due to congestion on the WiFi/2.4 GHz band.

Data Rate: The data transmission rate that measures in Kbps (Z-Wave up to 100 Kbps, ZigBee up to 250 Kbps).

4.4.3. *Reliability*

This perspective focuses on how reliable it is to use and configure the network. It is also the ability to add more smart objects and nodes, and the usage of setting presets to control them simultaneously. Also, this perspective focuses on how easy it is to connect and run battery-operated smart objects. This perspective has four criteria:

Scenes: Scenes are nothing but definite settings set with a simple button. Instead of changing the state of temperature, lights, coffee pot, you can set them using a scene having all these preset settings.

Adding nodes: The ease of adding more smart devices and objects to the network.

Power consumption: Battery-operated smart home products will suffer due to the power consumption nature of the protocol.

Reginal Limitation: As devices run on radio frequencies that are specific for each country, a device will not work in another country because of the changed frequency.

4.5. *Data collection*

See Table 1.

5. Results and Analysis

5.1. *Perspectives ranking*

The pairwise comparison method was used to rank the perspectives based on their relative importance to the objective (Figure 2). The result of the ranking method shows that the network was ranked the highest (43%), followed by reliability (33%) and flexibility (25%).

The Network perspective ranked the highest (43%). This perspective includes five criteria: the Encryption method of the protocol, the Network Interference, the ability of Mesh Networking, the Range of the signal and the Data Rate

Table 1. Data collection—criteria vs. candidates.

Perspective	Criteria	WiFi	Bluetooth	Z-Wave	ZigBee
Flexibility	Adoption Scale	35,000	30,000	2,100	1,439
	Interoperability	700	31,000	600	281
	Cost	None	Hub	Hub	Hub
Network	Encryption	WPA2	AES-128	AES-128	AES-128
	Network Interference	2.4 GHz	2.4 GHz	915 MHz	2.4 GHz
	Mesh Networking	Star	Mesh	Mesh	Mesh
	Range	100 ft.	800 ft.	100 ft.	35 ft.
	Data Rate	100 Mbps	50 Mbps	100 Kbps	250 Kbps
Reliability	Power Consumption	High	Low	Low	Low
	Adding Nodes	256	1,000	232	256
	Adding Scenes	Limited	Limited	Yes	Yes
	Regional Limitation	No	No	Yes	Yes

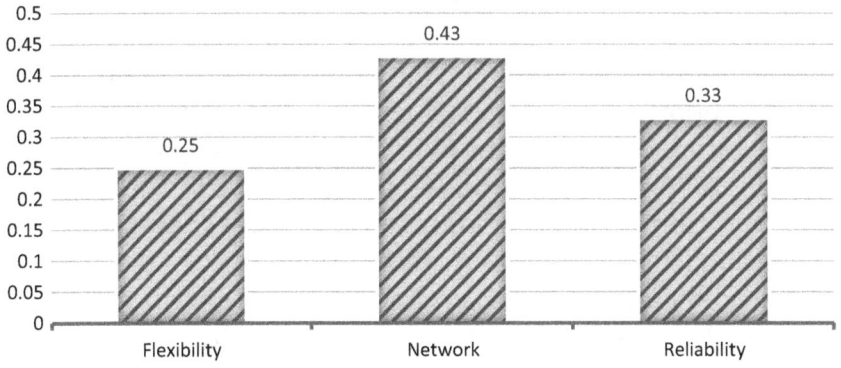

Figure 2. Ranking of perspectives.

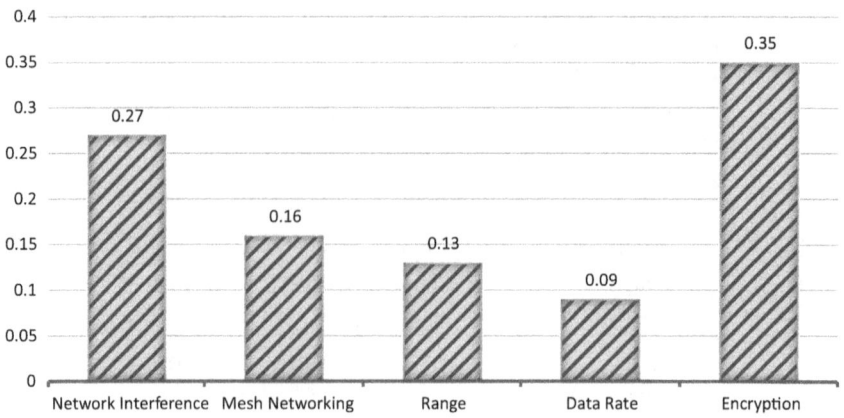

Figure 3. Criteria ranking—Network perspective.

(Figure 3). Among these criteria, Encryption was found to be the highest; encryption is the process of converting information or data into a code, especially to prevent unauthorized access. Then, followed by Network Interference, which depends on the radio frequency wave that the protocol runs on. As Z-Wave uses a different frequency (908.42 MHz in the US) from the others, it eliminates the lag due to congestion. Next came Mesh Networking, which is a feature in some protocols where the signal transmits from device to device (nodes) until it reaches the hub and then every device acts as a repeater. Network Range, which is the maximum network coverage area from one point to another, came next. Finally, the Data Rate that is measured in Kbps (Z-Wave up to 100 Kbps, ZigBee is up to 250 Kbps).

Reliability is the second perspective that includes Power Consumption, Adding Nodes, Adding Scenes, and Regional Limitation (Figure 4). Among these criteria, Power Consumption ranked the highest. This is an important criterion for battery-operated smart home products that will suffer due to the power consumption nature of some protocols. Adding Nodes, the second criterion in this perspective, is the ability to add more nodes in the network. It is measured by the maximum number of devices each protocol can manage. However, these

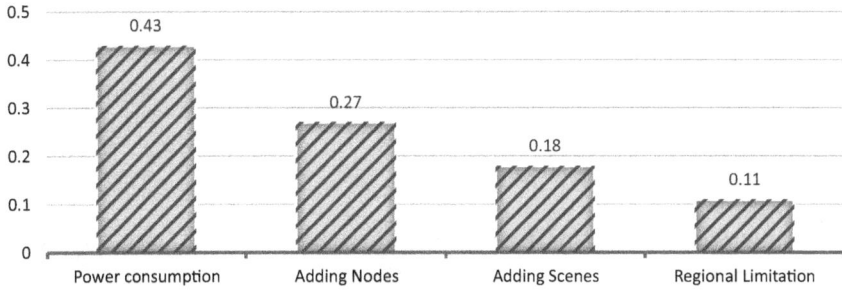

Figure 4. Criteria ranking—Reliability perspective.

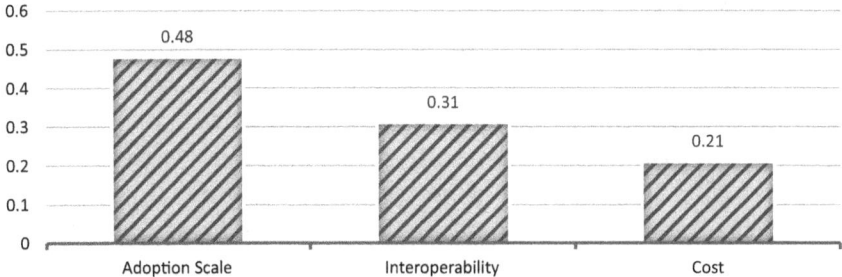

Figure 5. Criteria ranking—Flexibility perspective.

protocols support a higher number of devices than what is needed for a basic smart home network's need. Adding Scenes, the third criterion, allows the addition of different scenes with only one setting. Finally, in this perspective is Regional Limitation. As each country has its own frequency band, devices programmed in one country cannot run in another.

Lastly, the lowest ranked perspective is Flexibility (Figure 5). This perspective includes Adoption Scale, Interoperability, and the Cost of adopting the protocol. Among this perspective, Adoption Scale ranked the highest. Each protocol has a large selection of products, that is the number of products supporting this protocol. Interoperability, which is the support of the protocol by different brands that allows the customer the freedom to choose from a range of certified devices by the protocol alliance, was the second-ranked criterion. The last criterion in the ranking was the Cost of adopting the protocol. Some protocols are cheaper than their competitors because of their non-exclusive technology.

5.2. *Overall HDM ranking*

The alternatives were ranked based on their abilities in each criterion and their relative importance to the objective of the model. The decision model shows that the Bluetooth protocol and Z-Wave protocol ranked the highest, followed by ZigBee and then finally WiFi. The following section shows the criteria ranking for each alternative.

The overall model ranking shows that the best communication protocol for the home automation application is Z-Wave, the language that is designed for this application (Table 2). Z-Wave strengths are shown in the Network perspectives and its criteria. Also, it shares the same high ranking with ZigBee in the Reliability perspective. Z-Wave differentiates itself by transferring the signal in a lower radio frequency (900 MHz) as opposed to other communication protocols (2.4 GHz), which eliminates network interferences. It also featured in controlling low-energy smart devices and battery-operated devices that require a low amount of power. On the other side, the two main weaknesses in Z-Wave are its low data rate and short range. However, this protocol is not designed for these purposes. So, for the smart homeowner looking for a protocol to install security cameras or video applications, Z-Wave is not the best option.

The second-best communication protocol for home automation applications is Bluetooth, which is the best protocol in term of flexibility due to its high adoption scale with more than 30,000 devices and brands supported. Also, this is due to the low cost of Bluetooth devices.

Figure 6 shows the strengths and weaknesses of each protocol. WiFi is best when it comes to Data Rate, Adoption Scale, and Interoperability. Bluetooth also has an advantage from its ability to transmit signals for high ranges, up to 800 feet. The

Table 2. Overall HDM ranking.

Perspective	Criteria	WiFi	Bluetooth	Z-Wave	ZigBee
Flexibility	Adoption Scale	1.92	1.92	1.44	0.96
	Interoperability	0.93	1.24	0.93	0.62
	Cost	0.63	0.42	0.42	0.42
	Sum	3.48	3.58	2.79	2
Network	Encryption	1.4	1.05	1.05	1.05
	Network Interference	0.27	0.27	1.08	0.27
	Mesh Networking	0.16	0.48	0.64	0.64
	Range	0.39	0.52	0.39	0.26
	Data Rate	0.36	0.27	0.09	0.18
	Sum	2.58	2.59	3.25	2.4
Reliability	Power Consumption	0.43	1.29	1.72	1.72
	Adding Nodes	0.27	0.54	1.08	1.08
	Adding Scenes	0.36	0.36	0.72	0.72
	Regional Limitation	0.44	0.44	0.11	0.11
	Sum	1.5	2.63	3.63	3.63
Total sum		7.56	8.8	9.67	8.03

Z-Wave's strengths are cost and network interference as this protocol has a lower radio frequency than other protocols. ZigBee is the best when it comes to the lowest power consumption, followed by Z-Wave and Bluetooth.

6. Future Research

The intent of this paper was to develop a HDM to assess and evaluate the top communication protocols—WiFi, Bluetooth, Z-Wave and ZigBee—that can be used in smart homes for home automation purposes. However, the literature review still remains very limited on the technical details of these alternatives. Also, relatively new and available communication protocols, such as Insteon, Apple HomeKit and Thread (System, 2014),

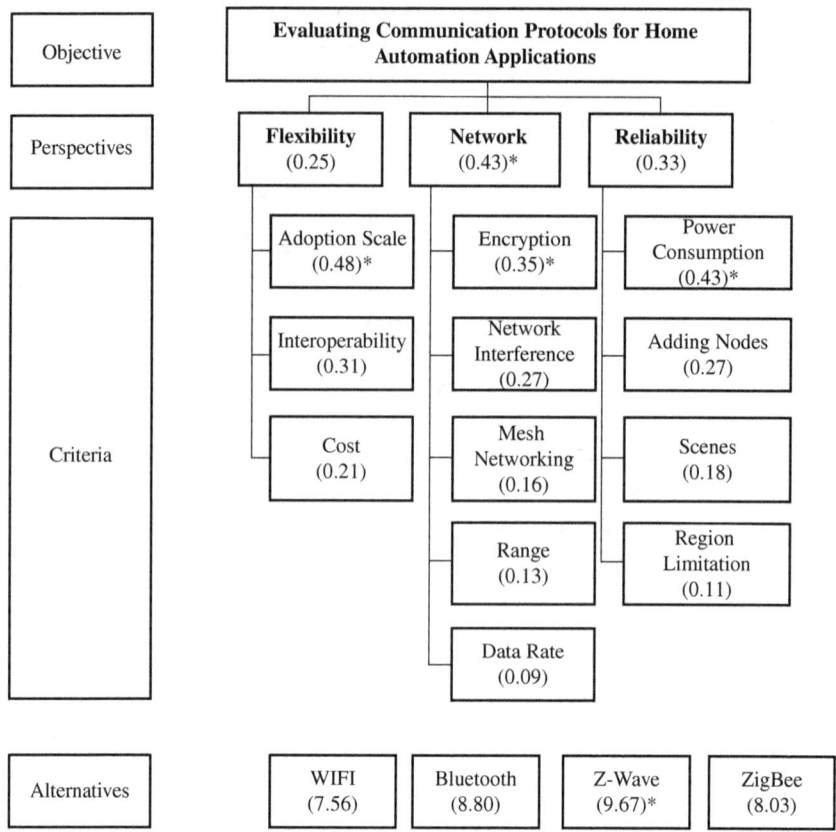

Figure 6. Overall HDM ranking.
Note: *Top-ranked criteria.

are not reviewed and synthesized well enough to be considered in this paper as they are used for a single line of products.

This model was assessed and evaluated based on an extensive literature review conducted by the authors. Therefore, another future research point is having expert panels to assess and evaluate the perspectives and criteria used in the model, and to evaluate the technology candidates accordingly. Using expert panels will give more reliable data and points of view.

References

1. N. Balta-Ozkan, R. Davidson, M. Bicket and L. Whitmarsh, "Social barriers to the adoption of smart homes", *Energy Policy,* **63** (2013) 363–374.
2. D. Weyns, H. V. D. Parunak, F. Michel, T. Holvoet and J. Ferber, "Environments for multiagent systems state-of-the-art and research challenges", *Lecture Notes in Computer Science,* **3374** (2005) 1–47.
3. C. D. Nugent, D. D. Finlay, P. Fiorini, Y. Tsumaki and E. Prassler, "Editorial: home automation as a means of independent living", IEEE Transactions on Automation Science and Engineering **5**, 1 (2008) 1–9.
4. A. G. Paetz, E. Dütschke and W. Fichtner, "Smart homes as a means to sustainable energy consumption: a study of consumer perceptions", *Journal of Consumer Policy,* **35**, 1 (2012) 23–41.
5. O. Vermesan and P. Friess, Internet of Things—Global Technological and Societal Trends (Aalborg: River Publishing, 2011).
6. I. Iskin and T. U. Daim, "Smart thermostats: are we ready?" *International Journal of Energy Sector Management,* **4**, 2 (2010) 146–151.
7. A. Talgeri and A. Kumar B. A., "Domotics—a cost effective smart home automation system using wi-fi as network infrastructure", *International Journal of Engineering Research and Applications,* **4**, 8 (2014) 52–55.
8. T. S. Cave, "Z-Wave vs ZigBee: which Is better for your smart home?" https://thesmartcave.com/z-wave-vs-zigbee-home-automation/.
9. K. Hong, S. Lee and K. Lee, "Performance improvement in ZigBee-based home networks with coexisting WLANs", *Pervasive and Mobile Computing,* **19** (2015) 156–166.
10. D. Capano, "Open standard wireless systems", *Control Engineering,* **63**, 1 (2016) 64.
11. C. Insteon and K. Alt, "Customize your smart home: Z-Wave vs ZigBee vs Insteon", 2017. http://www.asecurelife.com/z-wave-vs-zigbee-vs-insteon/. Accessed: 02 November 2017.
12. M. Walters, "Catch the Z-Wave", Security Dealer & Integrator, October 2014.
13. "Bluetooth SIG introduces Bluetooth 5", *Wireless News,* 2016.
14. R. Calem, "Battle of the networking stars: part one—Zigbee and Z-Wave wireless technologies fight for the home", *Digital Connect,* **1**, 1 (2005) 35.

15. I. B. Huang, J. Keisler and I. Linkov, "Multi-criteria decision analysis in environmental sciences: ten years of applications and trends", *Science of the Total Environment*, **409**, 19 (2011) 3578–3594.
16. A. Rolland, "Reference-based preferences aggregation procedures in multi-criteria decision making", *European Journal of Operational Research*, **225**, 3 (2013) 479–486.
17. T. L. Saaty, "Fundamentals of the analytic network process—dependence and feedback in decision-making with a single network", *Journal of Systems Science and Systems Engineering*, **13**, 2 (2004) 129–157.
18. S. H. Hashemi, A. Karimi and M. Tavana, "An integrated green supplier selection approach with analytic network process and improved Grey relational analysis", *International Journal of Production Economics*, **159** (2015) 178–191.
19. J. Rezaei, "Best-worst multi-criteria decision-making method", *Omega*, **53** (2015) 49–57.
20. G. van de Kaa, L. Kamp and J. Rezaei, "Selection of biomass thermochemical conversion technology in the Netherlands: a best worst method approach", *Journal of Cleaner Production*, **166** (2017) 32–39.
21. R. L. Keeney, H. Raiffa and D. W. Rajala, "Decisions with multiple objectives: preferences and value trade-offs", *IEEE Transactions on Systems, Man, and Cybernetics*, **9**, 7 (1979) 403–403.
22. J. Kim, Y. Suharto and T. U. Daim, "Evaluation of Electrical Energy Storage (EES) technologies for renewable energy: a case from the US Pacific Northwest", *Journal of Energy Storage*, **11** (2017) 25–54.
23. N. Sheikh, "Assessment of solar photovoltaic technologies using multiple perspectives and hierarchical decision modeling", PhD dissertation, Portland State University, 2013.
24. K. Gumasta, S. Kumar Gupta, L. Benyoucef and M. Tiwari, "Developing a reconfigurability index using multi-attribute utility theory", *International Journal of Production Research*, **49**, 6 (2011) 1669–1683.
25. M. Bohanec, M. Znidarsic, V. Rajkovic, I. Bratko and B. Zupan, "DEX methodology: three decades of qualitative multi-attribute", *Informatica*, **37**, 1 (2013) 49.
26. T. L. Saaty, Models, Methods, Concepts and Applications Of The Analytic Hierarchy Process (New York: Springer, 2012).

27. W. Ho and X. Ma, "The state-of-the-art integrations and applications of the analytic hierarchy process", *European Journal of Operational Research*, **43**, 2 (2009) 33–41.
28. J. E. de Steiguer, J. Duberstein and V. Lopes, "The Analytic Hierarchy Process as a means for integrated watershed management", in 1st Interagency Conference on Research in the Watersheds (ICRW), 2003, pp. 734–740.
29. N. Bhushan and K. Rai, Strategic Decision Making: Applying the Analytic Hierarchy Process (London, New York: Springer, 2004).
30. H. Martin and T. U. Daim, "Technology Roadmap Development Process (TRDP) for the service sector: a conceptual framework", *Technology in Society*, **34**, 1 (2012) 94–105.
31. G. Locatelli and M. Mancini, "A framework for the selection of the right nuclear power plant", *International Journal of Production Research*, **50**, 17 (2012) 4753–4766.
32. M. Berrittella, A. Certa, M. Enea and P. Zito, "An analytic hierarchy process for the evaluation of transport policies to reduce climate change impacts", Working Papers 2007.12, *Fondazione Eni Enrico Mattei*, 2007.
33. J. McCaffrey and A. Alghamdi, "Test run: the Analytic Hierarchy Process, evaluating defense architecture frameworks for C4I system using Analytic Hierarchy Process", *Journal of Computer Science*, **5**, 12 (2009) 1075.
34. J. Grandzol, "Improving the faculty selection process in higher education: a case for the Analytic Hierarchy Process", Educational Resources Information Center (US), *Association for Institutional Research*, **1** (2006) 103.
35. W. Atthirawong and B. McCarthy, "An application of the Analytical Hierarchy Process to international location decision-making", Proceedings of the 7th Annual Cambridge International Manufacturing Symopsium: Restructuring Global Manufacturing, University of Cambridge, Cambridge, 2002, pp. 1–18.
36. P. Kumar, "Analytic Hierarchy Process analyzes risk of operating cross-country petroleum pipelines in India", *Natural Hazards Review*, **4**, 4 (November 2003).
37. T. U. Daim, A. Udbye and A. Balasubramanian, "Use of Analytic Hierarchy Process (AHP) for selection of 3PL providers", *Journal of Manufacturing Technology Management*, 24, 1 (2012) 28–51.
38. T. U. Daim and M. R. Nava, "Evaluating alternative fuels in USA: a proposed forecasting framework using AHP and scenarios",

International Journal of Automotive Technology and Management, **7**, 4 (2007) 289–313.
39. J. S. Dyer, P. C. Fishburn, R. E. Steuer, S. Zionts, K. Deb and J. Wallenius, "Multiple criteria decision making, multiattribute utility theory: recent accomplishments and what lies ahead", Management Science, **54**, 7 (2008) 1336–1349.
40. D. F. Kocaoglu, "A participative approach to program evaluation", IEEE Transactions on Engineering Management EM-30, **3** (1983) 112–118.
41. T. U. Daim, W. Schweinfort, G. Kayakutlu and N. Third, "Identification of energy policy priorities from existing energy portfolios using hierarchical decision model and goal programming", International Journal of Energy Sector Management, **4**, 1 (2010) 24–43.
42. T. U. Daim, Hierarchical Decision Modeling, Essays in Honor of Dundar F. Kocaoglu (Portland: Springer International Publishing, 2016).
43. T. U. Daim and D. Fenwick, "Choosing a hybrid car using a hierarchical decision model", Inderscience Publishers, **3**, 3 (2011) 243–257.
44. B. Wang, D. F. Kocaoglu, T. U. Daim and J. Yang, "A decision model for energy resource selection in China", Energy Policy, **38**, 11 (2010) 7130–7141.
45. S.-H. Yang, F. Yao, X. Lu and K. Gill, "A Zigbee-based home automation system", IEEE Transactions on Consumer Electronics, **55**, 2 (2009) 422–430.
46. "Safer, smarter homes start with Z-Wave", Z-wave, 2 November 2017. http://www.z-wave.com/.
47. "Zigbee alliance", Zibee, 02 November 2017. www.zigbee.org/.
48. "Wi-fi alliance", 2 November 2017. https://www.wi-fi.org/.
49. "Bluetooth technology website", 2 November 2017. https://www.bluetooth.com/.
50. J. D. ,. I. Polley, "Encryption", Journal of the National District Attorneys Association, **32**, 2 (1998) 40.
51. L. Vangelova, "Encryption", Govexec.com, 1997.

Chapter 8

Personal Transformation: Smart House

Ahlam Alsuwiada*, Ahmed Al-Shareef*, Zuhair Alheayk*, You Hong Yong*, Wei Ming Jang*, Kenny Phan* and Tugrul Daim*,†,‡

*Portland State University, Portland, Oregon, USA
†Higher School of Economics, Moscow, Russia
‡Chaoyang University of Technology, Taiwan

Abstract

A "smart house" is one of the best solutions today in the market because it helps users to manage their activities by using special programs to automate electronic home devices [1]. The smart house concept began in the 20th century by the invention of home appliances. The vacuum cleaner was the first to be invented, followed by the sewing machine, food processors, refrigerator and washing machine. In 1939, popular magazines on mechanics introduced users to modern technology and got them to think about the future. In 1966, the first home computer ECHO IV was invented [2]. When people realized the importance of these modern appliances, demand for these appliances increased and this led to more research and development in smart houses. By 2000, smart houses began to develop quickly because different kinds of technology started to emerge that helped spread smart product technologies to users, who began to demand more of such products. In 2013, Microsoft showed in their marketing laboratories how they could help researchers to discover different kinds of smart house appliances [3]. Today, a smart house uses many applications such as remote mobile control, control lighting, video surveillance management, etc.

Keywords: Technology assessment, smart house, smart appliances.

1. Introduction

As the demand for smart appliances increased and with a variety of smart products in the market, users became spoilt for choice. However, it was difficult for them to choose the right products for their smart houses. The Hierarchical Decision Model (HDM) is one of the techniques that will help users make the right decision easily in choosing the features they need. So we conducted further research on each feature to gain more information and understand how it affected the smart house. We used pairwise comparisons to evaluate each feature from experts who answered our survey. Deciding the features of a smart house is a complicated process because experts have many ideas but different opinions.

In this paper, we reached the preference features through applying pairwise comparisons to decision-making. We used a simple tool that helped to recognize the features of a smart house—the HDM was used to simplify a large number of options.

2. Literature Review

Smart houses became very popular in recent years because demand for smart appliances has increased, but more information is needed to make them easy to use. Smart houses are considered to be one of the best standard systems because they connect with smart devices to give awareness and intelligence, so that they can supply the best service to users. A smart house makes life more convenient, comfortable and intelligent. It also increases energy efficiency and security [4].

Smart house systems consist of smart appliances, which have a smart card each. The frame of this smart card contains a sensor, digital meter, micro-controller and a LAN card. The sensor is the device that initiates the power consumption and determines its current status. A digital meter explains the consumed unit of electricity and cost. A micro-controller manages equipment that runs the electricity flow, whereas the LAN card links equipment that have ethernet ports, which are used to join the smart meter and smart card via a twisted pair cable [5].

Technology is needed to install smart home communication, and protocols like Z-Wave, X10, Insteon and ZigBee can be used to implement this technology. Z-Wave is the fastest way to send a message that comes from using a Source Routing Algorithm. X10 uses short radio frequency that enables activation between the transmitter and device. Insteon is similar to an electrical line, and the user may have an interface and at the same time use a wireless network to support more flexibility in placing the instrument. ZigBee explains the mesh-networking concept, since the signal that comes from the transmitter zigzags in a fashion similar to bees searching for the best way to the receiver. There are many examples for implementing these technologies in the home, such as cameras or video door phone which provides more than a doorbell, Doors can open by scanning one's fingerprint, audio systems distribute music, channel modulators take any video signal, and remote controls [6].

Many applications today are applicable in smart houses, and researchers try to make all these appliances part of the smart house by connecting them together into a network of appliances. Users today are more willing to make their appliances—both inside and outside their homes—part of a smart house and take the advantage of what a security system does in providing an immense amount of support in an emergency. Smart houses also support senior people who do not have ability to manage their activities without having to rely on other people. Such support can come in the form of daily reminders to take their medicine, alarm the hospital in times of emergencies, and track how much food they are eating. Smart houses can also help seniors in their day-to-day tasks at home, such as turning off stoves when food has been cooked. They can also assist senior people by using an intelligence appliance that helps to observe and learn from nurses' behaviors, activities and habits; alarms will be sent to medical facilities if the appliance observes strange behavior [6].

Users need to build smart houses but this will result in the demand of electricity becoming very high. As a result, power plants might not be able to supply all the electricity that is needed. Houses do not have a big effect on electricity consumption, but if we have millions of them, how are we to solve this problem? In addition, the cost of electricity has increased quickly in the last few decades. All these factors contribute to the use of a smart grid. Users began to use automated metering infrastructure in the house to supply real time information to help the end user monitor his/her electricity consumption. Smart house architectures, which depend on renewable energy such as wind turbines and solar panels, help to reduce carbon emission and electricity bills through a set of sensors and actuators that measure and control electricity consumption. Users thus need to make a distinction between a smart house and smart grid with the help of Information and Communications Technology (ICT), which can distribute electricity in a more efficient and cost effective way [7].

Smart houses are beneficial to people who have worked for a long time. With increasing difficulties in life the requirement to live comfortably can sometimes become very expensive.

Thus, people work hard to gain financial security and achieve personal goals. However, this leads to them becoming stressed and unwilling to do anything once they have returned home after work. Many researchers have found that people who work with stress ultimately succumb to health problems. A smart house could be the best solution to this problem; an automated light system could have a big effect on minimizing stress. The latter has the capacity to manage lighting conditions automatically. Many appliances are used today to automate electronic and light control, such as the Savant lighting control, Demo Eazy system, InelliSwitch and Green Room System. These appliances with some electronics companies like Panasonic and Samsung manage their home devices. Eye 2H, an intelligent system, is the solution that describes the process and facial expression to discover human expression. This manages the electronic and lighting equipment in smart homes [8].

However, despite all these benefits from the application of smart houses, many people do not have the means of owning one because it is very expensive as compared to smart products. Also, more work is needed to adapt a normal home to one's needs and the costs to reach that goal will not be realistic [9]. Also, smart houses do not have a wireless system that works perfectly, because the wireless signal sometimes interferes with other electronic products, leading to network failure, especially when the work becomes more complex when adding more appliances in the system [10].

Smart houses also have limitations; they do not have the ability to manage large equipment together in the system, and some instruments have to be left out of the system because there are only a specific number of sensors [11, 12]. Also, smart houses can lead to security issues, for example, thieves can compromise the safety of your home by gaining access to it by illegally accessing your appliances, i.e., iPhone, iPod, etc., and then remotely controlling it [9].

3. Methodology

HDM is a systematic approach or a conceptual tool that is used to help a decision maker apply a rational decision. It breaks a big complex decision problem down into simple sub problems, and then provides a way to consolidate all straightforward arrangements into a thorough result. It is also efficient for decreasing bias by constraining the decision maker to autonomously consider the diverse criteria involved in the choice. These autonomous estimations are made through pairwise comparisons; each element is compared with every other element to select its relative weight, and is then checked for consistency. A substantial set of examinations is exceptionally reliable when component weights remain relatively comparative when compared with the others.

Those procedures for creating an HDM need a couple of essential steps. Those that start will be selected as the main criteria. This set of criteria should be restricted to the practically impactful set that will make the choice from the important qualities acknowledged previously. Every component in the situated subcriteria can be broken down into constituent sets of subcriteria. The following model may be developed and the experts will address it by giving weights of relative importance to each of the qualitative and quantitative criteria, so that they can be contrasted with their companions. Components in each level are compared with different components in that same level [13].

4. Problem Definition

Deciding to buy a house is an essential financial decision people make during their lifetime. They take many factors into their considerations, such as the cost, neighborhood, nearby schools and the house's design. People take a long time to find the design they want for their own dream home. One option is buying a smart house. There are many benefits in having a smart house, but the greatest benefit is that of a contemporary lifestyle. However, buying a smart house with all its features could be very expensive, thus being out of reach for middle-class people [15]. Therefore, people are willing to give up some features in order to make smart houses more affordable.

The goal of this project is to determine the most favorable features that people desire in smart houses. In other words, this project proposes a typical smart house that is affordable for most middle-class individuals.

5. Identification of Objectives and Features

In this project, the mission is to determine the most favorable features in a smart house. To satisfy this mission, there are four objectives, which are Safety, Convenience, Environment and Intelligence. Under each objective, there are several goals to fulfill.

The first objective is Safety. Because of the rise in the number of crimes, for example, robbery and theft [16], people want a strong security system that will keep their families safe while at home. There are many devices or systems in the market today that can be used for safety purposes. This category contains the devices for Entrance Security, Fire/CO Gas Alarms, Children Monitors, and Surveillance. The second objective is Convenience. There are many devices or systems that will make things easier for people. These devices can be divided into four categories: Lighting, WiFi, Disability Assistance/Telecare and Entertainment. The third objective is Environment. In this objective, there are many devices that help to reduce environmental damages. These devices can be divided into three categories, which are Sprinkling, Solar Energy and Trash Recycling. The last one is Intelligence. This objective helps homeowners to take advantage of the current advances in technology. For example, Heating, Ventilation, and Air-Conditioning (HVAC), Motorized Drapes, Appliances and Smart Faucets are devices that fulfill the Intelligence objective [17].

Table 1 summarizes the features and their definitions [6,18].

Table 1. Device Types.

Safety	
Entrance	Devices such as camera and sensors are installed at entrances, to identify comers and protect homes from break-ins.
Fire/CO Gas Alarm	Devices are installed inside the home to detect the presence of fire or carbon monoxide.
Children Monitor	Devices are installed inside children's room to monitor them and notify the parents whenever something unusual is happening.
Surveillance	Devices are installed inside the house to observe unusual behavior.
Convenience	
Lighting	Sensors are installed inside the house to switch on/off lights, depending on the presence of people.
WiFi	This item helps to set up the equipment and can be controlled by either IOS or Android devices. This equipment can be controlled remotely too. It also becomes easy to control wall switches, lamps and garage doors.
Disability Assistance/ Telecare	This facilitates disabled seniors in obtaining implanted devices. These devices assist them in their daily activities, such as eating and drinking.
Entertainment	Distributes music from one location throughout your entire home with a whole-home distribution system, and also controls the speaker volume in each room, either individually or all at the same time.

(Continued)

Table 1. (*Continued*)

Environment	
Sprinkling	This smart sprinkler controller optimizes watering schedules based on soil moisture and weather predictions.
Solar Energy	Solar power is the best possible way to save energy. One can cover an especially large portion of your energy needs and gain greater independence, for example from rising energy costs.
Trash Recycling	A sensor recycling trash system helps keep garbage and recycling materials separate.
Intelligence	
HVAC	A HVAC system helps maintain good indoor air quality through adequate ventilation with filtration. It also provides thermal comfort.
Motorized Drape	Motorized curtains open smoothly and quietly via a press of a button on a wireless wall switch or hand-held remote control.
Appliances	A domestic device or piece of equipment, such as electrical and gas appliances, designed to perform a specific task.
Smart Faucet	The faucet is equipped with smart and intelligent features and is programmable through simple touches to preset the ideal temperature of water and to control water usage.

6. Implementation and Results

6.1. *Data gathering*

A questionnaire (refer to Appendix A) was used to collect the experts' opinions for the relative importance of each level in the HDM. The features of a smart house are defined literally with supplementary pictures as a consistent reference for the expert panel when they responded to the survey. There are four criteria

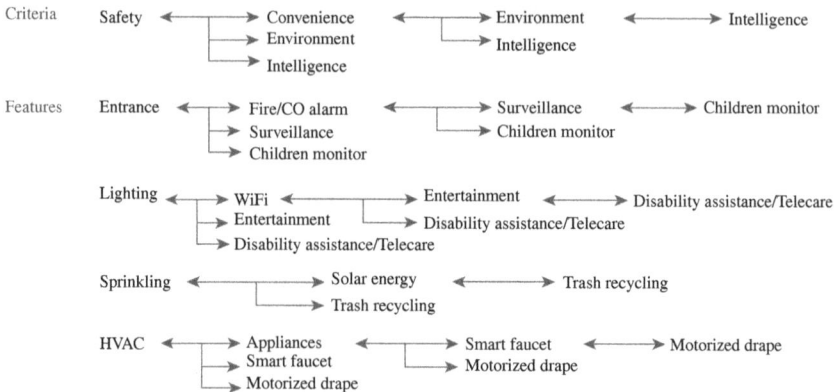

Figure 1. The mechanism of an expert questionnaire using the PCM.

that potential smart house buyers are most concerned about: Safety, Convenience, Environment and Intelligence. For every criterion, several representative features are listed. Though not all features of a smart house are included, this module provides a compressive assessment to demonstrate a feasible method in deciding the most favorable features that can be employed to accelerate the flourish of the smart house market.

These features were compared with a Pairwise Comparison Method (PCM) in a three-level HDM (see Figure 1). A questionnaire was developed for the survey. Its result was based on the HDM and input from the expert panel. The experts allocated a total of 100 points to the two elements in the proportion of their relative importance to the immediate upper level in the HDM diagram.

6.2. Outcome analysis

The relative importance rating (weight) of each feature contributes to the most favorable smart house being calculated. Though the experts on the panel may not have expertise in all the features listed in the questionnaire, they are treated as potential buyers of the smart house. Among 18 survey invitees, 15 experts had completed and submitted their responses.

The result of the comparison is quite concordant and the disagreement is only 0.04. Meanwhile, the individual inconsistencies of the members on the expert panel are distributed over an acceptable range of 0.00–0.08 (see Table 2).

Table 2. Overall view of the weights for the criteria and features.

Goal, Criteria, and Feature	Weight (Criterion to Goal)	Weight (Feature to Criterion)	Final Weight (Feature to Goal)
The most Favorable Feature			
Safety	0.40		
Entrance		0.21	0.08
Fire/CO gas alarm		0.34	0.14
Surveillance		0.19	0.08
Children monitor		0.25	0.10
Convenience	0.20		
Lighting		0.21	0.04
WiFi		0.34	0.07
Disability assistance/ Telecare		0.22	0.04
Entertainment		0.23	0.05
Environment	0.16		
Sprinkling		0.30	0.05
Solar energy		0.29	0.05
Trash recycling		0.41	0.06
Intelligence	0.24		
HVAC		0.27	0.07
Motorized drapes		0.20	0.05
Appliances		0.29	0.07
Smart faucet		0.24	0.06
Disagreement		0.04	

The outcome analysis demonstrates a human essential demand for the living safety beyond other criteria. As to the Feature level, there is no prominent item. The Fire/CO Gas Alarm and Children Monitor criteria have a slightly higher weight than the others. This could have been caused by safety concerns.

A numerical analysis of the preferences among different customer segments was also made based on the characteristics of the expert panel to gauge their influence on decision-making. These characteristics are:

- Gender—male or female;
- Matrimony——married or single;
- Children—with or without children (less than 18 years old);
- Estate—owns or leases a house;
- Age—20–30, 31–40, 41–50, over 50;
- Education level—high school, undergraduate studies, Masters, PhD;
- Nationality.

In the study, we analyze one characteristic at a time. Conjunctions among characteristics, such as a married man with kids, are not discussed in this paper. Neither are the psychological factors behind the results, i.e., why experts made their choices while doing the comparison.

Tables 3 and 4 analyze the input from experts, which shows there is no divergence in the demand for safety from people of different characteristics. The safety objective is either the first or second choice for all experts.

This result reveals the preference of potential consumers. People expect smart houses to provide them with a safer home and that the other fuzzy functions are secondary concerns. They are thus willing to pay more for safety features but within a limited budget. As its name implies, people are also looking forward to the intelligence that a smart house possesses. The safety and intelligence criteria represent almost 85% of the requirements from customers.

Table 3. The preference of the objective for the different characteristics of the experts.

Characteristic of Expert		Preference of Objective	
		#1 Choice	#2 Choice
Gender	Female	Safety	Intelligence
	Male	Safety	Intelligence
Matrimony	Single	Intelligence	Safety
	Married	Safety	Intelligence
Kid	w/o kid	Safety	Intelligence
	With kid	Safety	Intelligence
Estate	Lease	Safety	Intelligence
	Own	Intelligence	Safety
Age	21–30	Safety	Intelligence
	31–40	Safety	Convenience
	41–50	Safety	Intelligence
	51–	Safety	Convenience
Education	PhD	Safety	Convenience
	Master	Safety	Intelligence
	Bachelor	Safety	Intelligence
Nationality	China	Safety	Convenience
	Iraq	Safety	Convenience
	Saudi	Intelligence	Safety
	Taiwan	Safety	Intelligence

Table 4. The choice of objective made by experts.

	Preference of Objective	
	#1 Choice	#2 Choice
Safety	16	3
Intelligence	3	11
Convenience	0	5
Environment	0	0

Table 5. Feature preferences for the different characteristics of the experts.

Characteristic of Expert		Preference of Feature		
		#1 Choice	#2 Choice	#3 Choice
Gender	Female	Entrance	Fire/CO gas alarm	Surveillance
	Male	Fire/CO gas alarm	Children monitor	Entrance
Matrimony	Single	Fire/CO gas alarm	Appliances	WiFi
	Married	Fire/CO gas alarm	Entrance	Surveillance
Kid	w/o kid	Fire/CO gas alarm	Appliances	WiFi
	With kid	Fire/CO gas alarm	Entrance	Children monitor
Estate	Lease	Fire/CO gas alarm	Entrance	Surveillance
	Own	Appliances	Fire/CO gas alarm	WiFi
Age	21–30	Fire/CO gas alarm	Appliances	Children monitor
	31–40	Fire/CO gas alarm	HVAC	Solar energy
	41–50	Fire/CO gas alarm	Children monitor	Entrance
	50–	Fire/CO gas alarm	Entrance	Surveillance
Education	PhD	Fire/CO gas alarm	Surveillance	Children monitor
	Master	Fire/CO gas alarm	Children monitor	WiFi
	Bachelor	Fire/CO gas alarm	Entrance	Appliances
Nationality	China	Fire/CO gas alarm	Surveillance	Children monitor
	Iraq	Fire/CO gas alarm	Children monitor	Surveillance
	Saudi	Appliances	Fire/CO gas alarm	WiFi
	Taiwan	Entrance	Fire/CO gas alarm	Surveillance

In order to deeper understand the consumers' perspective from the feature aspect, we dissected the data and referred them to each characteristic of experts. As expected, features that fulfilled the safety objective were mostly named as the top choice. Experts of almost every characteristic regarded the fire/CO gas alarm as their highest priority. Other safety features also share a major portion, as shown in Tables 5 and 6 and Figure 2.

Nevertheless, if we segregate those safety features, the outcome presents a diverse preference in other features. Tables 7 and 8 and Figure 3 demonstrate a noticeable conclusion. Although the demand for WiFi is higher than the others, it does

Table 6. The experts' preference of feature.

	Preference of Feature		
	#1 Choice	#2 Choice	#3 Choice
Fire/CO gas alarm	15	4	0
Appliances	2	3	1
Entrance	2	5	2
Children monitor	0	4	4
Surveillance	0	2	6
HVAC	0	1	0
WiFi	0	0	5
Solar energy	0	0	1

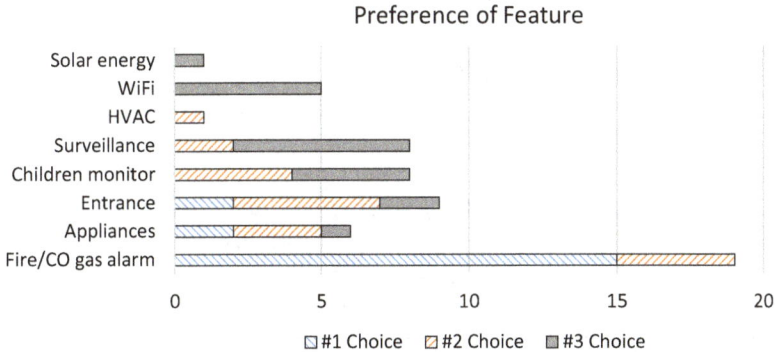

Figure 2. The choice of feature made by experts.

not stand out as the first choice. More experts selected appliances to be their item of most concern.

People expect a smart house to provide advanced protection, such as automatically detecting and reacting to a possible mishap. The fire/CO gas alarm system is already a standard feature in a modern house. However, by connecting it with the intelligent appliances, a smart alarm system can go one step further: once a fire alert is detected, the system can command

Table 7. The preference of feature (exclude safety objective) for the different characteristics of the experts.

Characteristic of Expert		Preference of Feature (Exclude Safety Objective)		
		#1 Choice	#2 Choice	#3 Choice
Gender	Female	Appliances	WiFi	Smart Faucet
	Male	Appliances	WiFi	Trash Recycling
Matrimony	Single	Appliances	WiFi	Motorized drapes
	Married	WiFi	Appliances	Trash Recycling
Kid	w/o kid	Appliances	WiFi	HVAC
	With kid	Trash Recycling	WiFi	Smart Faucet
Estate	Lease	Trash Recycling	WiFi	Smart Faucet
	Own	Appliances	WiFi	HVAC
Age	21–30	Appliances	WiFi	HVAC
	31–40	HVAC	Solar energy	WiFi
	41–50	Trash Recycling	Smart Faucet	Appliances
	51–	WiFi	Entertainment	Sprinkling
Education	PhD	Smart Faucet	Lighting	WiFi
	Master	WiFi	Trash Recycling	Motorize drapes
	Bachelor	Appliances	WiFi	HVAC
Nationality	China	Smart Faucet	Lighting	WiFi
	Iraq	Trash Recycling	Disability assistance/T	WiFi
	Saudi	Appliances	WiFi	HVAC
	Taiwan	WiFi	HVAC	Appliances

the source facility to take preventive actions, such as shutting down the furnace or turning down the stove.

The calculated features' weights for the different characteristics of the expert are listed in Appendix B.

7. Conclusion

The smart house is a highly customized merchandise. Identifying the favorite features of potential customers is always the top challenge for the service providers. A customer's preferences of

Table 8. The choice of feature (excluding the safety objective) made by the experts.

Preference of Feature (Exclude Safety)			
	#1 Choice	#2 Choice	#3 Choice
Appliances	8	1	2
HVAC	1	1	5
WiFi	4	10	4
Solar energy	0	1	0
Trash recycling	4	1	2
Smart faucet	2	1	3
Entertainment	0	1	0
Lighting	0	2	0
Disability assistance	0	1	0

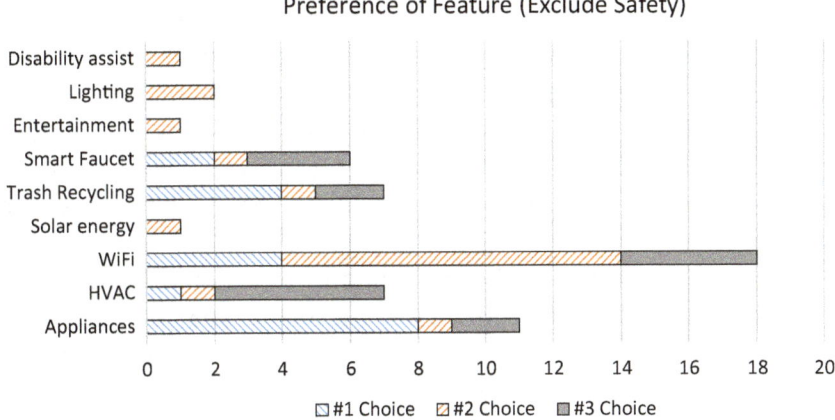

Figure 3. The choice of feature (excluding the safety objective) made by the experts.

his/her smart house have been summarized via an analysis in this research. Smart house builders, investors or component developers can make their business decisions based on this information:

- People expect smart house to provide them with a safer home;
- Intelligent facilities should be integrated to generate a more comfortable living space;
- WiFi is an essential demand for people living in a smart house.

To concentrate on a targeted market, the outcome of this report will be helpful in understanding the inclination of customers of specific characteristics. It will help developers to focus more on things that are really important rather than wasting precious resources on something that is perceived to be irrelevant by customers.

7.1 *Limitations*

When we made our survey, we selected people who were highly educated, such as degree, Masters and PhD holders, since we needed individuals with big backgrounds to help us receive more accurate answers. Since smart houses are new on the market, we wanted to avoid inaccurate information if we had sent this survey to random individuals.

As our project is not well known in the market, we sent clarifications for each feature with our survey. We also chose international students to obtain a variety in answers and opinions.

As we mentioned in the introduction the demand for smart products increased in 2000, which then rose year by year. As a result, users became unwilling to use older products. Also, as we collated our information from the survey, each one had their own preference in particular features.

The cost of a smart house differs from one to the other, so we did not put this into consideration because of the variety of smart products, which for the same feature had many different models and brands. These factors would have led to inaccurate information, so we ignored the cost for these reasons.

7.2 Future works

There are many features that can be considered when building a smart house, so we must add more features in the hierarchy to obtain more information and to prevent any bias that can happen. We divided the objective into four parts: Safety, Convenience, Environment and Intelligence, but in doing so, we ignored Saving Energy and categorized Solar Energy under Environment. However, we should have implemented Saving Energy as an objective instead. Under this objective, we would have included many features like Thermostat, Lighting, Shading, Ceiling Fan and Solar Energy.

Also, we see the final weight for safety is 0.4, 0.25 for intelligence, 0.20 for convenience, and 0.15 for environment. If we were to compare the result between safety and convenience, we will see a big difference. Hence, we need more convenience analysis to see whether there is any difference or change.

As we mentioned in the beginning, we collected our survey from international students and highly educated individuals who were all nearly of the same age. We then saw our survey focus on safety with convenience ranked as third. In the future, researchers who are creating a similar survey must focus on asking senior people whether they prefer convenience to safety, as we have noticed in existing literature review.

References

1. "Smart homes and X-10 home automation—introduction and overview", *Explainthatstuff.com*, 2015. http://www.explainthatstuff.com/smart-home-automation.html. Accessed: 16 March 2015.
2. "The history of smart homes", *M2mevolution.com*, 2014. http://www.m2mevolution.com/topics/m2mevolution/articles/376816-history-smart-homes.htm. Accessed: 16 March 2015.

3. "Home smart home: A history of connected household tech", *Mashable*, 2015. http://mashable.com/2015/01/08/smart-home-tech-ces/. Accessed: 16 March 2015.
4. Sining Ma, Xi Chen, Guicai Song, Juan Wang, Limin Sun and Jia Yan, *The Converged Service Oriented Architecture in Smart Home Service, 2012* International Conference on Cyber-Enabled Distributed Computing and Knowledge Discovery, IEEE.
5. Bilal shahid, Zubair Ahmed, Adnan Faroqi and Rao M. Navid-ur-Rehman, Implementation of Smart System based on Smart Grid Smart Meter and Smart Appliances, 2nd Iranian Conference on Smart Grid, May 23&24, 2012, Tehran, Iran (Icse 2012).
6. Rosslin John Robles and Tai-hoon Kim, Applications, Systems and Methods in Smart Home Technology: *A Review, International Journal of Advanced Science and Technology* Vol. 15, February, 2010 37.
7. A. M. Carreiro, C. H. Antunes and H. M. Jorge, Energy smart house architecture for a smart grid, 2012 IEEE International Symposium on Sustainable Systems and Technology (ISSST).
8. Lim Teck Boon, Mohd Heikal Husin, Zarul Fitri Zaaba and Mohd Azam Osman, Eye 2H: A proposed automated smart home control system for detecting human emotions through facial detection, The 5th International Conference on Information and Communication Technology for The Muslim World (ICT4M).
9. "Disadvantages", *Jacklyn-IntelligentHome*, 2015. https://sites.google.com/a/cortland.edu/jacklyn-intelligenthome/disadvantages. Accessed: 16 March 2015.
10. "What is a smart home? Advantages/disadvantages", *Drmerz.com*, 2015. http://drmerz.com/what-is-a-smart-home-advantages-disadvantages/. Accessed: 16 March 2015.
11. 2015. [Online]. Available: http://blog.gerhardsappliance.com/technology-news/advantages-and-disadvantages-of-smart-homes/. Accessed: 16 March 2015.
12. Nektarios Papadopoulos, Apostolos Meliones, Dimitrios Economou, Ioannis Karras, Ioannis Liverezas, A Connected Home Platform and Development Framework for Smart Home Control Applications, 2009 7th IEEE International Conference on Industrial Informatics, https://ieeexplore.ieee.org/xpl/conhome/5175248/ proceeding.

13. M. Leopoldo, V. Noppadon, B. Philip, D. Rodney and N. Thanaporn, "Use of PCM modeling in selection of an HDTV", Portland, 2015.
14. "Engineering and Technology Management—Portland State University", Portland State University, 2015. http://www.pdx.edu/engineering-technology-management/ Accessed: 16 March 2015.
15. "Smart home cost comparison", *Loxone*. http://www.loxone.com/enus/smart-home/cost-comparison.html. Accessed: 16 March 2015.
16. "Annual crime in the U.S. Report released", Federal Bureau of Investigation. http://www.fbi.gov/news/stories/2012/october/annual-crime-in-the-u.s.-report-released. Accessed: 16 March 2015.
17. A. Bee, "List of smart home features", *eHow*. http://www.ehow.com/list_7184897_list-smart-home-features.html. Accessed: 16 March 2015.
18. "A dozen good reasons why you need a smart home", *Dream Green House*. http://www.dreamgreenhouse.com/features/2014/reasons.php. Accessed: 16 March 2015.

Appendix A: The Questionnaire in the Survey Form Used to Gather Experts' Opinions

Compare these objectives with respect to their contribution to the mission

Objectives	Total 100 Points			Total 100 Points	
	Safety	[ ] Convenience		Convenience	[ ] Environment
	Safety	[ ] Environment		Convenience	[ ] Intelligence
	Safety	[ ] Intelligence		Environment	[ ] Intelligence

Compare these features with respect to their contribution to the objective—safety

Features of Safety	Entrance	[ ] Fire/CO alarm		Fire/CO alarm	[ ] Surveillance
	Entrance	[ ] Surveillance		Fire/CO alarm	[ ] Children monitor
	Entrance	[ ] Children monitor		Surveillance	[ ] Children monitor

Compare these features with respect to their contribution to the objective—convenience

Features of Convenience	Lighting	[ ] WiFi		WiFi	[ ] Disability assistance/Telecare
	Lighting	[ ] Disability assistance/Telecare		WiFi	[ ] Entertainment
	Lighting	[ ] Entertainment	Disability assistance/Telecare	[ ] Entertainment

Compare these features with respect to their contribution to the objective—environment

Features of Enviornment	Sprinkling	[ ] Solar energy		Solar energy	[ ] Trash recycling
	Sprinkling	[ ] Trash recycling			

Compare these features with respect to their contribution to the objective—intelligence

Features of intelligence	HVAC	[ ] Motorized drape		Motorized drape	[ ] Appliances
	HVAC	[ ] Appliances		Motorized drape	[ ] Smart faucet
	HVAC	[ ] Smart faucet		Appliances	[ ] Smart faucet

Appendix B. The Features' Weight for the Different Characteristics of the Experts

Mission, Objective, and Feature	Female	Male	Difference
The most Favorable Feature	Average		
Safety	0.3980	0.4000	0.0020
Entrance	0.1203	0.0784	0.0419
Fire/CO gas alarm	0.1076	0.1482	0.0406
Surveillance	0.0964	0.0774	0.0190
Children monitor	0.0736	0.0960	0.0224
Convenience	0.1860	0.2110	0.0250
Lighting	0.0360	0.0388	0.0028
WiFi	0.0834	0.0745	0.0089
Disability assistance/Telecare	0.0366	0.0416	0.0050
Entertainment	0.0300	0.0561	0.0261
Environment	0.1100	0.1600	0.0500
Sprinkling	0.0272	0.0419	0.0147
Solar energy	0.0465	0.0509	0.0044
Trash recycling	0.0363	0.0673	0.0310
Intelligence	0.3060	0.2290	0.0770
HVAC	0.0667	0.0621	0.0046
Motorized drapes	0.0744	0.0409	0.0336
Appliances	0.0868	0.0759	0.0109
Smart faucet	0.0781	0.0501	0.0280

Mission, Objective, and Feature	Single	Married	Difference
The most Favorable Feature	Average		
Safety	0.3250	0.4264	0.1014
Entrance	0.0441	0.1100	0.0659
Fire/CO gas alarm	0.1390	0.1331	0.0059
Surveillance	0.0439	0.0982	0.0543
Children monitor	0.0980	0.0851	0.0129
Convenience	0.2100	0.2000	0.0100
Lighting	0.0266	0.0419	0.0153
WiFi	0.1028	0.0683	0.0345
Disability assistance/Telecare	0.0350	0.0417	0.0068
Entertainment	0.0457	0.0481	0.0024
Environment	0.1000	0.1591	0.0591
Sprinkling	0.0300	0.0395	0.0095
Solar energy	0.0253	0.0582	0.0330
Trash recycling	0.0448	0.0614	0.0166
Intelligence	0.3650	0.2145	0.1505
HVAC	0.0852	0.0558	0.0294
Motorized drapes	0.0863	0.0396	0.0466
Appliances	0.1290	0.0615	0.0674
Smart faucet	0.0646	0.0575	0.0070

Mission, Objective, and Feature	w/o kid	with kid	Difference
The most Favorable Feature	Average		
Safety	0.3880	0.4050	0.0170
Entrance	0.0788	0.0992	0.0204
Fire/CO gas alarm	0.1496	0.1272	0.0224
Surveillance	0.0748	0.0882	0.0134
Children monitor	0.0848	0.0904	0.0056
Convenience	0.2000	0.2040	0.0040
Lighting	0.0238	0.0448	0.0210
WiFi	0.1008	0.0658	0.0350
Disability assistance/Telecare	0.0369	0.0414	0.0045
Entertainment	0.0384	0.0519	0.0135
Environment	0.0920	0.1690	0.0770
Sprinkling	0.0260	0.0424	0.0164
Solar energy	0.0272	0.0606	0.0334
Trash recycling	0.0388	0.0660	0.0272
Intelligence	0.3200	0.2220	0.0980
HVAC	0.0738	0.0586	0.0152
Motorized drapes	0.0718	0.0422	0.0296
Appliances	0.1205	0.0590	0.0615
Smart faucet	0.0539	0.0622	0.0083

Mission, Objective, and Feature	Lease	Own	Difference
The most Favorable Feature	Average		
Safety	0.4260	0.3460	0.0800
Entrance	0.0926	0.0920	0.0006
Fire/CO gas alarm	0.1503	0.1034	0.0469
Surveillance	0.0917	0.0677	0.0240
Children monitor	0.0914	0.0828	0.0086
Convenience	0.2010	0.2060	0.0050
Lighting	0.0414	0.0307	0.0108
WiFi	0.0678	0.0968	0.0290
Disability assistance/Telecare	0.0450	0.0298	0.0152
Entertainment	0.0468	0.0487	0.0020
Environment	0.1660	0.0980	0.0680
Sprinkling	0.0365	0.0378	0.0013
Solar energy	0.0571	0.0341	0.0230
Trash recycling	0.0724	0.0261	0.0463
Intelligence	0.2070	0.3500	0.1430
HVAC	0.0510	0.0890	0.0381
Motorized drapes	0.0407	0.0747	0.0340
Appliances	0.0569	0.1248	0.0679
Smart faucet	0.0584	0.0615	0.0031

Mission, Objective, and Feature	21–30	31–40	41–50	51–	Difference
The most Favorable Feature		Average			
Safety	0.3900	0.3300	0.4100	0.4600	0.1300
Entrance	0.0825	0.0825	0.0986	0.1326	0.0501
Fire/CO gas alarm	0.1430	0.0990	0.1134	0.1470	0.0480
Surveillance	0.0781	0.0660	0.0944	0.1017	0.0357
Children monitor	0.0863	0.0825	0.1037	0.0787	0.0250
Convenience	0.2000	0.2500	0.1667	0.2450	0.0833
Lighting	0.0256	0.0700	0.0575	0.0473	0.0444
WiFi	0.0828	0.0500	0.0552	0.0856	0.0304
Disability assistance/ Telecare	0.0438	0.0350	0.0228	0.0507	0.0279
Entertainment	0.0478	0.0650	0.0311	0.0614	0.0339
Environment	0.1244	0.2000	0.1800	0.1450	0.0756
Sprinkling	0.0276	0.0660	0.0395	0.0609	0.0384
Solar energy	0.0429	0.0040	0.0586	0.0479	0.0411
Trash recycling	0.0540	0.0500	0.0819	0.0363	0.0457
Intelligence	0.2856	0.2200	0.2433	0.1500	0.1356
HVAC	0.0661	0.0924	0.0555	0.0504	0.0420
Motorized drapes	0.0607	0.0374	0.0483	0.0261	0.0346
Appliances	0.0959	0.0374	0.0686	0.0432	0.0585
Smart faucet	0.0628	0.0528	0.0709	0.0303	0.0406

Mission, Objective, and Feature	PhD	Master	Bachelor	Difference
The most Favorable Feature		Average		
Safety	0.3900	0.3957	0.4067	0.0167
Entrance	0.0788	0.0757	0.1164	0.0406
Fire/CO gas alarm	0.1170	0.1554	0.1164	0.0389
Surveillance	0.1053	0.0790	0.0821	0.0263
Children monitor	0.0889	0.0856	0.0918	0.0062
Convenience	0.2200	0.2057	0.1933	0.0267
Lighting	0.0671	0.0321	0.0348	0.0350
WiFi	0.0652	0.0822	0.0761	0.0170
Disability assistance/ Telecare	0.0303	0.0468	0.0352	0.0165
Entertainment	0.0575	0.0447	0.0473	0.0128
Environment	0.1750	0.1643	0.1083	0.0667
Sprinkling	0.0546	0.0399	0.0277	0.0270
Solar energy	0.0620	0.0600	0.0330	0.0290
Trash recycling	0.0585	0.0644	0.0477	0.0167
Intelligence	0.2150	0.2343	0.2917	0.0767
HVAC	0.0488	0.0611	0.0716	0.0228
Motorized drapes	0.0418	0.0617	0.0443	0.0199
Appliances	0.0493	0.0599	0.1125	0.0632
Smart faucet	0.0752	0.0515	0.0634	0.0237

Mission, Objective, and Feature	C	1	S	T	Difference
The most Favorable Feature		Average			
Safety	0.3900	0.4680	0.3317	0.4400	0.1363
Entrance	0.0788	0.0861	0.0837	0.1478	0.0690
Fire/CO gas alarm	0.1170	0.1825	0.1050	0.1217	0.0775
Surveillance	0.1053	0.0941	0.0637	0.0963	0.0417
Children monitor	0.0889	0.1052	0.0793	0.0743	0.0309
Convenience	0.2200	0.1920	0.2083	0.1950	0.0280
Lighting	0.0671	0.0292	0.0369	0.0329	0.0379
WiFi	0.0652	0.0612	0.0935	0.0825	0.0323
Disability assistance/ Telecare	0.0303	0.0625	0.0229	0.0443	0.0397
Entertainment	0.0575	0.0391	0.0551	0.0354	0.0222
Environment	0.1750	0.1720	0.1150	0.1250	0.0600
Sprinkling	0.0546	0.0347	0.0299	0.0460	0.0247
Solar energy	0.0620	0.0587	0.0410	0.0393	0.0227
Trash recycling	0.0585	0.0786	0.0441	0.0398	0.0388
Intelligence	0.2150	0.1680	0.3450	0.2400	0.1700
HVAC	0.0488	0.0405	0.0842	0.0747	0.0437
Motorized drapes	0.0418	0.0434	0.0660	0.0423	0.0242
Appliances	0.0493	0.0482	0.1179	0.0729	0.0697
Smart faucet	0.0752	0.0359	0.0769	0.0501	0.0410

Chapter 9

Personal Transformation: Wearable GPS Device for Children

Bhawinee Banchongraksa*, Jessie Truong*, Lu Chuan Chieh*, Mufeed Yacoub*, Papit Meteekotchadet* and Tugrul Daim*,†,‡

Portland State University, Portland, Oregon, USA
†*Higher School of Economics, Moscow, Russia*
‡*Chaoyang University of Technology, Taiwan*

Abstract

A missing child is a parent's worst nightmare. When a child goes missing, even for a very short time, it is hard for a parent not to entertain a variety of horrifying scenarios and outcomes. In recent years, GPS technology has offered a practical solution to this dilemma in the form of smartphones and wearable tracking devices. However, many schools do not permit children who are 10 years and below to bring phones to school. So, wearable tracking devices have become the default option for parents wanting to keep track of their young children. Without missing a beat, the wearable tracking device industry has come to the same conclusion and has introduced many wearable devices to the market. The quantity and variety of these wearable devices may certainly be for the consumer's benefit as manufacturers compete, but it puts them (the consumers) in a moderately overwhelming position as they attempt to browse through the options of unfamiliar technology.

This paper will address the decision-making process of selecting a wearable tracking device by using the Hierarchical Decision Model (HDM). Through a literature review, we will identify relevant features that we will use to construct the model, and then identify the top options for the model to choose from.

Keywords: Technology assessment, wearable technologies, safety.

1. Introduction

From the variety of smartwatches and fitness trackers out in the market today, there are more wearable devices out there than ever before. Also, since wearable devices are used in a range of applications, including healthcare, sports and fitness, and there are continuous innovations in the industry, we can safely presume that wearable devices will be around for quite a while. One of the trends that can be observed in this growing industry is the level of integration of these devices with people of all ages, including children.

All parents want to keep their children safe, but differing parenting styles is still a heated topic. Every parent defines safety differently, for example, free-range parents would never leash their child while helicopter parents would never let their child out of sight. However, when it comes to crowded events and theme parks or giving their children the independence to venture out into the world alone, will this latest trend—wearable technology devices with GPS tracking capabilities—in child safety be common ground for these two sets of parents? This device allows parents to pinpoint the exact position of the child via their smartphones or tablets. The devices come in many colors with different designs and different features; choosing one could be a hair pulling decision for some parents. After reviewing articles regarding wearable devices for children, we conducted a study to help parents decide which wearable device is the most suitable based on their needs and financial status.

1.1. *Technology definition*

Wearable technology is a category of technology that keeps track of personal information mostly relating to an individual's health, fitness, location and biofeedback, via a device worn on that individual. These devices have communication capabilities, which allow the user to access data in real time through another connected device [1]. A wearable Global Positioning System (GPS) tracker is a wearable device used to track a person. Its main function is a GPS tracking system that provides the exact location and information of the person being tracked. This data can either be stored within the GPS tracking device or sent to a central database [2]. Thus the devices can display one's location either in real time or during analysis later.

1.2. *What do they do? What are they used for?*

The wearable GPS tracking device is used to monitor a child's location and activities. The device allows parents to track their children whenever they are out of sight, thus giving them greater peace of mind. The device can also connect the child to his/her parents via an application on a smartphone or internet-enabled computer, which can show in real time a child's route and location.

The features—GPS range, battery life, real-time tracking and monthly service fee—of each tracking device brand may be different, but there are some criteria that parents have to take into account because a child may choose a product solely on physical attributes such as weight, material and color. The wearable tracking device can have several additional features besides a tracking system. For example, some devices come with a 1- or/and 2-way communication such as calling and texting, while others can send emergency alerts.

2. Problem Definition

According to the National Center for Missing and Exploited Children, roughly 800,000 children are reported missing each year in the United States [3]. The causes vary; some were kidnapped by strangers or family members, some drowned, while others may have wandered off and gotten lost.

Statistics from the FBI's National Crime Information Center show that ~203,900 children were abducted by family members, ~58,200 children by family acquaintances, and ~115 children by strangers [4]. The National Center for Missing Children reports that 20% of the children abducted by nonfamily members are found dead [5].

Other situations, such as drowning and getting lost typically involve young children when they wander away from their parents in public. Also, children with mental disabilities or behavioral issues such as Attention Deficit Disorder and autism tend to wander off from their guardians whenever they are distracted [6]. Hence, smartphones might not be the best option since its functions might be too complicated for these children. Moreover, there are a number of studies that postulate that children should not be allowed to use smartphones for reasons related to learning and behavior development. Currently, the average age for a child getting his/her first smartphone is 10.3 years [7]. Consequently, taking into account the concerns listed above, a wearable tracking device may be the next best alternative for parents of young children. However, there are many brands of wearable track devices in the market today, each providing a variety of different features. Therefore, we adopt the Hierarchical Decision Model (HDM) in this paper to assist parents in selecting the wearable tracking device best suited for them and their children.

3. Literature Review

Currently, there are many GPS tracking devices out there on the market [8]. By understanding their limitations and focusing on what features are most important to parents, we can select a tracking device that will work best for them. The goal in this segment is to analyze the top ten GPS tracking devices for children in the market through reviewing online articles.

We start out with a list of the top 15 devices that were reviewed by the articles. The list also includes all the features each device offers. To meet our goal, we had to first cut down our devices to ten. In helping us to make the cut, we interviewed technology experts to help define technical features. Also, we interviewed existing and users and non-users for a list of features from a GPS tracking device that would be important to them. These lists are then combined and curated into a list of devices that match the features highlighted by the participants. We then compared these devices with each other to highlight their similarities and differences for parents to use as a quick reference. We also rated these devices, as a means to help parents make their final decision.

From the articles [8], we felt that as a kid-friendly product the device needed to be of a compact size, which would enable them to fit neatly in backpacks or on small wrists. The device would also require capabilities such as a real-time GPS signal that would allow the parent to know where his/her child is, and a panic alert to notify parents whenever the device is taken off the child's wrist. Another capability worthy of consideration is the device being able to allow a child to send out SOS text messages during emergencies. Below are the criteria that we firmly feel parents need to consider when purchasing such a device.

3.1. *Price*

One not only pays for the GPS tracking device, but its monthly service fees as well. This monthly fee could determine whether you sign up for the service contract. On the other hand, different devices offer attractive marketing deals to convince potential customers to sign up, such as free usage for the first couple of months [2].

3.2. *Battery life*

Battery life may be a critical feature for any technology product, but it is even more critical in a tracking device. It affects the performance of the product, for example, it could cause bad signals to the receiver device. We believe parents should only consider devices that have batteries that last at least 24 hours.

3.3. Waterproof

Kids can be clumsy. Thus, owning a waterproof device is recommended for them.

3.4. Service range

Since this product is a tracking device, GPS should be a must have. The range of this signal should be strong. Additional features such as 3G and WiFi can be considered.

3.5. Communication

For children in kindergarten or First Grade, this might not be a necessary feature to consider. However, we feel a messaging feature is necessary for older children. This is because in events where children are in a dangerous situation and cannot talk, messaging would be the safest means to call for help.

3.6. Compatibility

This is important as parents may have different network providers. A device that is compatible with more than one platform is a winner in this competitive market.

3.7. Real time

This is the most important feature in the device. Being able to know where one's child is at any given moment is as good as being there in real life.

3.8. *Distraction-free*

This feature is a good-to-have because it does not distract the children whenever they are in school or class.

3.9. *Monthly fee*

The majority of these tracking devices require a monthly service fee. This might be a drawback for parents who are on a tight budget.

3.10. *Panic alert*

This feature is good for children with special needs. In the event of the device being removed from a child's wrist, an alert gets sent to the child's parents. We recommend this feature to parents who have children with autism.

3.11. *Ease of use*

The device needs to be simple enough for a child to navigate, as well as for the parents to set it up and then use it. We can categorize this into two sub categories—hardware and software. We will define the former as a subcriterion under this criterion, while the latter will get defined as a subcriterion under the Technical Performance criterion. More details will be shown in the model section.

3.12. *Appeal*

A child's product needs to be appealing, so these devices may come in different colors and attractive shapes.

Tables 1 and 2 offer a summary of the devices that we filtered and collected from the articles we reviewed. These devices are highly recommended to parents who are planning to purchase such a device.

Table 1. Part 1 of our review's final list of children GPS tracking devices.

	HereO	AmbyGear	Filip 2	Caref GPS	Wherecom K3
Price	$199	$99	$149.99	$67.95	$129.99
Battery Life	Up to 60 hours	7 Days	Up to 2 Days	1 Day	Up to 2 Days
Water Proof	Splash-proof	Yes	Water-resistant (IP 63)	Yes	—
Range of Service	GSM, GPS, WiFi	GPS, WiFi, bluetooth	GPS, 3G/GSM, WiFi	GPS	3G, GPS, WiFi
Screen	Yes/EPD	Yes/4 colors LED	Yes	Yes	Yes/LCD
Communication	Text	Text	Text/Call	Text/Call	Call
Compatibility	iOS, Android	iOS, Android	iOS, Android	iOS, Android	iOS, Android, Windows
Real Time	Yes	Yes	Yes	Yes	Yes
Distraction-free	N/A	Yes	—	—	—
Monthly Fee	$4.95/month	No	$10/month	$9.99/month	—
Panic Alert	Yes	Yes	Yes	Yes	Yes
Others		1. Send text/alert without service 2. Games 3. Find me apps			
Amazon Rating	2.5/5	—	2/5	2.5/5	—

Table 2. Part 2 of our review's final list of children GPS tracking devices [19–23].

	PocketFinder	Tinitell	My Buddy Tag	Lineable	Paxie Band
Price	$129.95	$149.00	$40.00	$5.00	$175.00
Battery Life	2~3 Days	1~3 Days	1 Year	1 Year	Up to 36 hours
Water Proof	Yes	Yes	Yes	Yes	Yes
Range of Service	GPS, 3G/GSM Cell ID, WiFi	GSM, GPS	Bluetooth/Limit Range	Bluetooth/Limit Range	GPS
Screen	No	No	No	No	No
Communication	Email	Call	No	No	No
Compatibility	iOS, Android	iOS, Android	iOS, Android	iOS, Android	N/A
Real Time	Yes	Yes	—	—	Yes
Distraction-free	N/A	N/A	No	No	—
Monthly Fee	$12.95/month	$12/month	No	No	$9.99/month
Panic Alert	Yes	N/A	Yes	—	Yes
Others	For Pet/ Elderly User	—	out of range alert	Group package	—
Amazon Rating	3/5	—	2.5/5	3/5	—

Tracking devices are meant to help keep children safe while their parents are busy keeping up with their daily chores. Our goal is to help parents feel at ease when deciding which device fits their needs and financial status. We recognize that the list of devices in Tables 1 and 2 can be overwhelming, so we have limited the scope to just five devices. In addition, we will use the HDM model to narrow down these devices to just one—the best device based on the experts' evaluations.

4. Relevant Brands

In our process of selecting the best wearable tracking device, we decided to select relevant brands and consider a select few to be our alternatives. First, we conducted a literature review on the top brands of the devices and collected some suggestions from users. Even though we focused specifically on wearable devices, we found up to ten brands that had the minimum features we desired. We then picked five out of these ten brands based on their comparable level of technologies used and competencies. With these five alternatives, the analysis would not be too complex but still varied enough to proceed with the HDM. The five alternatives we chose were: HereO GPS Watch, AmbyGear Smartwatch, Filip 2, Caref GPS, and Omate Wherecom K3. None of these brands is well known in the high-tech industry. So, branding is not considered in our analysis.

4.1. *HereO GPS watch*

HereO was founded in 2011. Its GPS watch was officially unveiled in 2015. HereO claims its GPS watch is small enough to fit a child's wrist [9]. The watches have colorful designs which are attractive to kids. It is even available in popular cartoon characters such as Hello Kitty. It is important to note that HereO GPS watches are intentionally built with simple features to make it easy for children to use. However, its cost price is the highest among the selected alternatives.

4.2. *AmbyGear*

The AmbyGear Smartwatch was developed under Ambit Networks, which was founded in 2013. The AmbyGear Smartwatch was released in 2015 [10]. As it focuses mostly on children 6 years and above, it possesses more complex features and functions, which include games and apps. Also, its cost price is much lower than the HereO GPS Watch, even though it provides more services. The AmbyGear Smartwatch is clearly marketed as the "internet for kids". However, some customers view the many features provided as unnecessary distractions for their children.

4.3. *Filip 2*

The Filip 2 was released in 2014 by Filip, which was founded in 2012 [11]. The Filip 2 typically focuses on tracking and communication functions similar to the HereO GPS Watch. Unlike the AmberGear Smartwatch, it does not provide additional services. The Filip 2 was inspired by an actual event in its founder's life. The founder's 3-year-old son, Filip, once wandered out of sight in a crowd and father and son were separated for 30 minutes [11]. This inspired the founder to come up with the Filip 2; its objective is to keep parents and kids connected. The Filip 2 can also make and receive calls and its cost price is only a little above average.

4.4. *Caref GPS*

The Caref GPS watch was released in 2014 and the company that created it was founded in 2012 [12]. Its features and functions are very similar to the Filip 2. Interestingly, Caref, the founder, was similarly inspired as Filip. However, the cost price of the Caref GPS watch is much lower than that of the Filip 2. In fact, it is the cheapest smartwatch among our alternatives.

4.5. *Omate Wherecom K3*

The Omate Wherecom K3 was released in 2016. The company, Omate, was founded in 2013 [13]. Focusing on children 6 years and above, the Omate Wherecom K3's features and functions are more complex than the other alternatives. Its additional functions are mostly about entertainment. The K3 can also make and receive calls, and its cost price is about average.

5. Methodology—HDM

5.1. *The HDM*

The HDM is a decision-making method for analyzing complex and multi-criteria decisions [14, 15]. The HDM was originally conceptualized by Dundar Kocaoglu in 1979 with the same fundamental concepts as the Analytical Hierarchy Process (AHP), but using a different pairwise comparison scale and judgment quantification technique [16]. Two years later, Kocaoglu developed the general form of the HDM together with David Cleland. This model would go on to consist of five levels of decision elements, namely, Mission, Objectives, Goals, Strategies and Actions (MOGSA) [17]. These levels are flexible enough to undergo appropriate changes to accommodate a vast variety of cases and structures. The number of levels may also increase or decrease depending on the complexity or simplicity of the subject.

The fundamental advantage of the HDM is its ability to break down problems into a hierarchy of more easily comprehensible sub-problems to be evaluated independently. Using the pairwise comparison, subject experts provide values for the priorities of the items in the model at each level. In simpler terms, experts choose in a series of one-to-one comparisons which item is more important than the other. The HDM gives decision makers the ability to organize feelings, intuition and logical thinking in the decision-making process [18].

5.2. Criteria selection and model building

To address the research objective, a HDM was developed as shown in Figure 1. The model's structural content is derived from a literature review and a team brainstorming session that came up with different factors for consideration in children wearable tracking devices. The model is composed of four levels.

Level 1—Mission: This level represents the mission we set out to achieve, which is to assist parents in selecting a GPS tracking device for children from ages 5 to 10.

Level 2—Criteria: This level includes the main categories for consideration. As all team members have technical backgrounds, we acted as the technical experts and selected these criteria. For any technology, we feel there are at least three important criteria a buyer should consider before making a purchase. These important criteria are: the technical

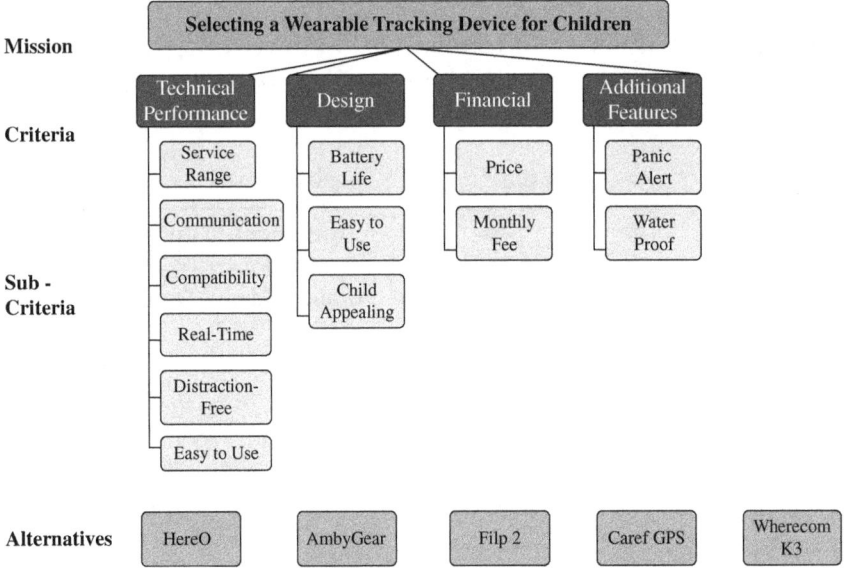

Figure 1. The HDM model.

performance of the device, the design of the device, and the financial capability of the buyer. Anything else is additional.

Level 3—Subcriteria: This level includes a list of more specific features for consideration. We sat down and interviewed several existing product owners to ask them what are the features/functionalities important to them in a wearable tracking device. The final list was generated in combining this list to the list that we collected from our literature/articles reviewed.

Level 4—Alternatives: This level presents the options available. The five products mentioned earlier will be used as our alternatives. The details of our research model can be seen in Appendix B.

In this model, we had seven experts from our social circle doing pairwise comparisons for our project. Table 3 shows information about our experts, for example, age, gender, jobs, marital status, and the number of children they have. Essentially, we can also consider whether they owned the products or are potential buyers.

Before our experts did their pairwise comparisons, we sent them information and specifications of the five alternatives we chose. Doing this would familiarize them with the alternatives and the type of technology they used. Moreover, we also made

Table 3. List of experts.

	Gender	Age	Occupation	Number of Children/ Grandchildren	Current Product Owner	Potential Buyer
Expert 1	Female	31	Retailed Manager	1	Yes	—
Expert 2	Female	35	Dental Assistant	2	No	Yes
Expert 3	Female	60	Interpreter	2	Yes	—
Expert 4	Male	55	Teacher	1	No	Yes
Expert 5	Female	48	Business Owner	0	No	Yes
Expert 6	Female	32	Piano Teacher	0	Yes	—
Expert 7	Female	46	Business Owner	2	Yes	—

sure that our experts were able to understand the model and knew how to carry out the pairwise comparison. We had earlier sent them the link to the model and provided them with instructions. We even followed up with them—we let the ones who understood the process proceed, while we met and assisted those who were still not clear on how to do pairwise comparisons in the model. This ensured that we got correct results. Table 3 shows a list of our experts.

6. Data Analysis and Results

After choosing the right experts, we decided to collate the pairwise comparison scores by sitting with and interviewing the experts to ensure they understood the scoring process. Prior to our meeting with them, we provided them with an information sheet containing the specifications and features of each device. The aim of this action was to bring our experts up to par and for them to have a leveled knowledge of the basics.

After the scores were entered, we received our results from the HDM software. Table 4 shows the criteria level results. The numbers highlighted in yellow show the top rated criteria for each expert. According to Table 4, we see that most of our experts felt that Technical Performance was really important for them. On the other hand, we can also see that Expert 2

Table 4. Criteria importance—consolidated.

	Expert 1	Expert 2	Expert 3	Expert 4	Expert 5	Expert 6	Expert 7	Mean
Technical Performance	0.43	0.10	0.33	0.43	0.42	0.32	0.42	0.35
Design	0.17	0.40	0.27	0.17	0.15	0.22	0.13	0.22
Financial	0.12	0.30	0.22	0.12	0.38	0.32	0.37	0.26
Additional Features	0.28	0.20	0.18	0.28	0.05	0.14	0.08	0.17
Inconsistency	0	0	0	0	0.01	0	0.01	

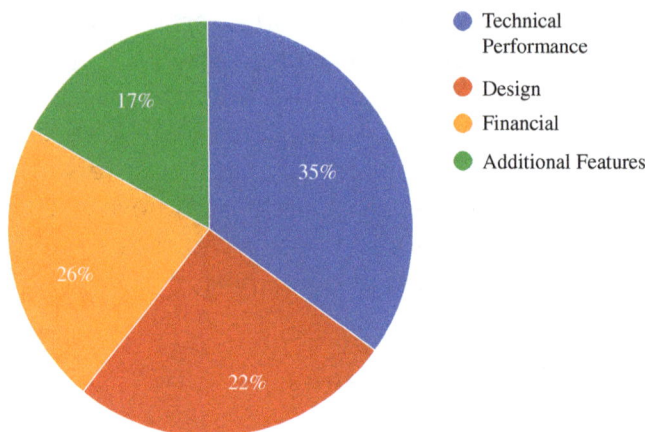

Figure 2. Criteria importance consolidated chart.

valued Design over Technical Performance. Financial was her second priority; there was no point in buying an expensive device for her sons since they are prone to misplacing items.

Figure 2 shows the experts' rating percentages at the criteria level. As we can see, Technical Performance has the highest rate at this criteria level. Due to the safety concerns that are tied to the Selection objective, it is not surprising that Technical Performance is considered to be the experts' most important aspect at 35%. The Financial aspect falls in second place, as it received 26%. Next, Design received 22% because Easy to Use is also a criterion that parents are understandably concerned with. Meanwhile, we can see that Additional Features is the lowest criterion in this level because our expert probably felt additional features were the least necessary of all the criteria. At this second level of the HDM, the inconsistency is well within the allowed range of 0–0.01.

After getting the result at the criteria level, we analyzed the results at the subcriteria level of our HDM model, which is represented in Table 5. All our experts had compared each subcriterion relative to its particular criteria group. Table 5 shows the consolidated data with regard to the subcriteria level. Also, the

Personal Transformation: Wearable GPS Device for Children 315

Table 5. Subcriteria evaluation data—consolidated.

	Expert 1	Expert 2	Expert 3	Expert 4	Expert 5	Expert 6	Expert 7	Mean	Criteria Importance	Subcriteria Weight
Price	0.5	0.42	0.52	0.45	0.8	0.5	0.6	0.54	0.26	0.14
Battery Life	0.33	0.44	0.27	0.3	0.6	0.24	0.63	0.40	0.22	0.09
Water Proof	0.5	0.73	0.48	0.65	0.3	0.48	0.3	0.49	0.17	0.08
Service Range	0.18	0.16	0.2	0.22	0.26	0.2	0.36	0.22	0.35	0.08
Communication	0.16	0.13	0.2	0.27	0.26	0.2	0.22	0.20	0.35	0.07
Compatibility	0.18	0.12	0.05	0.07	0.07	0.05	0.14	0.09	0.35	0.03
Real Time	0.21	0.19	0.26	0.24	0.25	0.23	0.17	0.22	0.35	0.08
Distraction-free	0.08	0.1	0.06	0.09	0.08	0.06	0.05	0.07	0.35	0.03
Monthly Fee	0.5	0.58	0.48	0.55	0.2	0.5	0.4	0.45	0.26	0.12
Panic Alert	0.5	0.27	0.52	0.35	0.7	0.52	0.7	0.51	0.17	0.09
Easy to Use(T)	0.18	0.3	0.23	0.11	0.09	0.26	0.07	0.18	0.35	0.06
Easy to Use(D)	0.33	0.31	0.56	0.44	0.2	0.6	0.21	0.38	0.22	0.08
Child Appealing	0.33	0.25	0.17	0.26	0.2	0.16	0.16	0.22	0.22	0.05

results for each expert concerning the subcriteria level are found in Appendix A. The numbers highlighted in yellow in Table 5 show the top rated subcriterion for each expert. From the results we can see that each expert's priorities are different. Also, the Criteria Importance column in Table 5 contains the values in the criteria evaluation. For example, price is the subcriterion under Financial, so the Criteria Importance of Price will be 0.26.

Next, it is important to note that we have two Easy to Use subcriteria: one under Technical Performance and the other under Design. Another important explanation to make is that the values under the Subcriterion Weight come from multiplying the Mean and Criteria Importance values. For example, the price's original weight of 0.54 when multiplied by 0.26 gives us 0.14 for the Subcriterion Weight.

Table 6 shows our subcriteria weights ranging from high to low.

Table 6. Weights of each subcriterion.

Subcriteria	Weight
Price	14%
Monthly Fee	12%
Panic Alert	9%
Battery Life	9%
Easy to Use (Design)	8%
Waterproof	8%
Service Range	8%
Real Time	8%
Communication	7%
Easy to Use (Technical Performance)	6%
Appeal	5%
Compatibility	3%
Distraction-free	3%

We can see from Table 6 that the top four subcriteria are Price, Monthly Fee, Panic Alert and Battery Life. We notice that Financial is critical for our experts because Price and Monthly Fee are the subcriteria for Financial. This result is particularly important: as children between 5 to 10 years may easily lose their wearable devices, parents may understandably be reluctant to buy a device that is too expensive. Distraction-free is of the lowest weight because our experts feel that this subcriterion does not matter when considering the safety of children. Compatibility may also be a least weighted subcriterion, but experts have placed it above Distraction-free since these devices can be paired with Android and iOS systems. A pie chart of our subcriteria can be seen in Figure 3.

After the subcriteria level, we will move to the last level, which is Alternatives, shown in Table 7 and Figure 4. At this level, we will score our devices relative to the subcriteria.

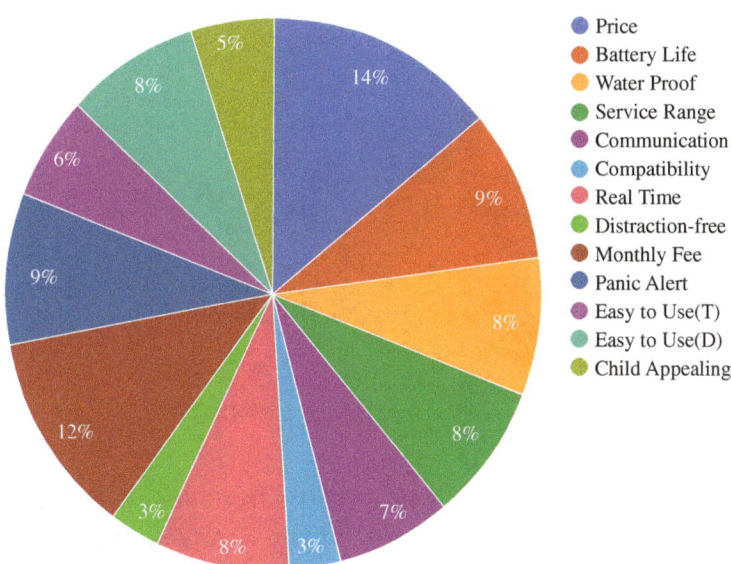

Figure 3. Subcriteria evaluation data—consolidated chart.

318 Digital Transformation

Table 7. Overall results.

	HereO GPS Watch	AmbyGear Smartwatch	Filip 2	Caref GPS	Wherecom K3	Inconsistency
Expert 1	0.19	0.21	0.2	0.2	0.2	0.01
Expert 2	0.25	0.3	0.15	0.17	0.13	0.03
Expert 3	0.22	0.22	0.19	0.21	0.16	0
Expert 4	0.22	0.2	0.18	0.19	0.21	0.01
Expert 5	0.15	0.22	0.15	0.3	0.18	0.01
Expert 6	0.17	0.27	0.18	0.23	0.15	0
Expert 7	0.19	0.22	0.18	0.2	0.21	0.01
Mean	0.2	0.23	0.18	0.21	0.18	
Minimum	0.15	0.2	0.15	0.17	0.13	
Maximum	0.25	0.3	0.2	0.3	0.21	
Std. Deviation	0.03	0.03	0.02	0.04	0.03	
Disagreement						0.028

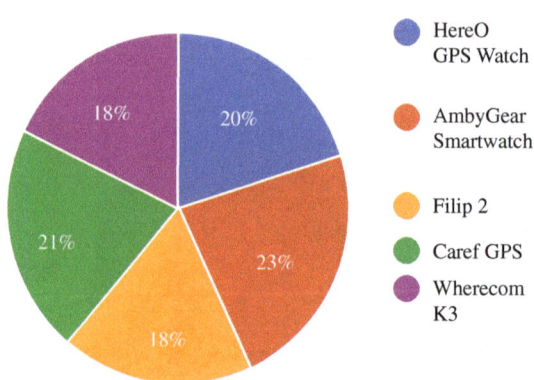

Figure 4. Overall results chart.

Table 7 is the result of the Alternatives level. We can see from the pie chart in Figure 4 that the AmbyGear Smartwatch is the best choice for a wearable tracking device for children.

The numbers highlighted in yellow in Table 7 show the top-rated device for each expert. We notice that Experts 1, 3, 6 and 7, who own these types of devices, are in agreement in choosing the AmbyGear Smartwatch as their top choice. On the other hand, the scores of Experts 2, 4 and 5, who do not own these types of devices, were not as consistent. It is safe to conclude that these results correlate to the experts' level of experience with these products.

The last critical area of the HDM model analysis is inconsistency. The importance of low inconsistency is a gauge of how well an expert can move through the pair-wise comparisons and not become lost and hence inconsistent.

Figure 5 shows the overall results of the HDM, of which all the figures are mentioned level by level, so we can get the value for the whole model.

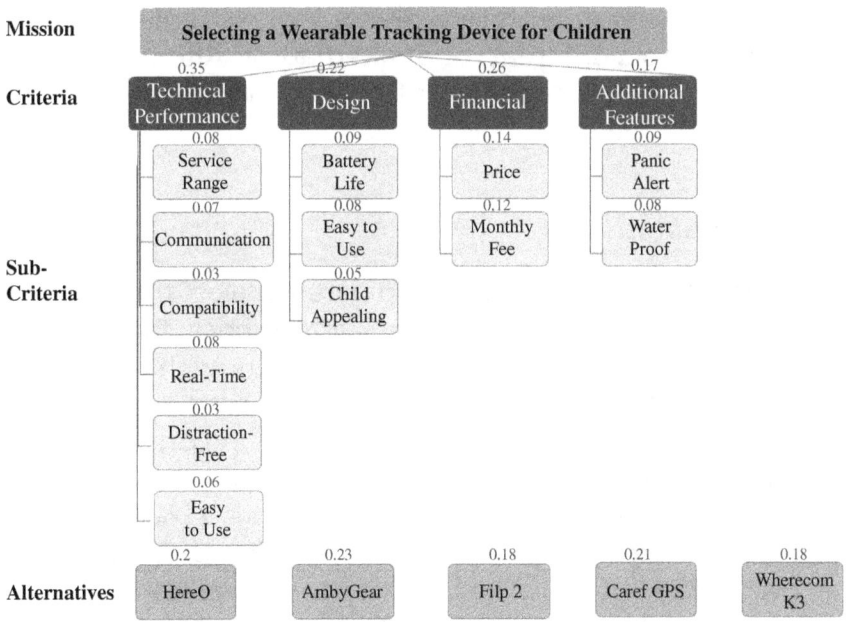

Figure 5. Overall results of the HDM.

7. Limitations and Future Research

One of the limitations in our project was the selection of the experts. Based on their current jobs, the majority of them were not professionals in technical areas or specialized in wearable devices. In future, we intend to invite experts specialized in specific areas, such as electronics, design and marketing. This will be for three steps: Selection of Criteria, Selection of Subcriteria, and Pairwise Comparison. We can even have them do evaluations in the areas they have expertise in. For example, experts who have background in electronics will be professionals in technical features. Experts who have background in design or marketing will be professionals in design on hardware features. Then, we can see whether there will be significant changes in the result.

We selected a variety of product models available on the market, but there are still a considerable number of different options that could have been chosen (perhaps yielding different results). Therefore, future research works could apply the same model to a different set of alternatives. Alternatively, we could still use the same set of alternatives but add more criteria.

8. Conclusion

The goal of this study was to help parents select the best children GPS tracking device at an affordable price. By using the HDM that was built with criteria collected through a literature review and interviews with experts, we achieved our goal. The result from the HDM pairwise comparison indicated that the AmbyGear Smartwatch was the winner with a 23% overall score. The runner-up was the Caref GPS with a 21% overall rating, and our second runner-up was the HereO GPS Watch with an overall score of 20%. The last two alternatives were tied with an 18% overall score. The margin between these devices is very small. The overall inconsistency and disagreement rates were very low, thus we did not perform any validation or sensitivity test for our model.

We feel our result might have been personally biased because some of our experts are product owners and they might favor one model over the other. Another area that might have affected our model is the Additional Features criteria. The features from this criteria were not available for all products. This shows our lack of expertise in selecting criteria for the model. Also, our naming convention for the subcriteria Easy to Use was not clear enough. We used the same name as the child node for the criteria Performance and Design. This caused our experts some confusion when they made the pairwise comparisons.

In conclusion, we feel our model could be improved in a number of ways. Perhaps we could start by improving the selection of criteria for evaluating the alternatives. Our HDM's experts should be restricted to people who have never owned such products, in order to avoid bias opinions. Lastly, we will make the naming convention clear in future models, to avoid confusing our experts.

References

1. "Wearable technology: patent landscape analysis", *LexInnova*. http://www.wipo.int/export/sites/www/patentscope/en/programs/patent_landscapes/reports/documents/lexinnova_wearable.pdf. Accessed: 16 March 2017.
2. "How does a GPS tracking system work?" *EETimes*. http://www.eetimes.com/document.asp?doc_id=1278363. Accessed: 16 March 2017.
3. "Missing children in America: unsolved cases", *ABC News*. http://abcnews.go.com/US/missing-children-america-unsolved-cases/story?id=19126967. Accessed: 16 March 2017.
4. "Missing children fast facts", *CNN*. http://www.cnn.com/2013/10/22/us/missing-children-fast-facts/. Accessed: 16 March 2017.

5. "Child abduction facts", *Parents*, 11 June 2015. http://www.parents.com/kids/safety/stranger-safety/child-abduction-facts/. Accessed: 16 March 2017.
6. "Should parents use GPS tracking on their kids?" *GPS Tracking*. https://www.liveviewgps.com/should parents use gps tracking on their kids.html. Accessed: 16 March 2017.
7. J. Donovan, "The average age for a child getting their first smartphone is now 10.3 years", *TechCrunch*, 19 May 2016. https://techcrunch.com/2016/05/19/the-average-age-for-a-child-getting-their-first-smartphone-is-now-10-3-years/. Accessed: 16 March 2017.
8. E. Malinowski, "Best GPS trackers for kids 2017", *Tom's Guide*, 10 March 2017. http://www.tomsguide.com/us/best-gps-child-trackers,review-2884.html. Accessed: 16 March 2017.
9. "Meet hereO", *The HereO Family*. https://www.hereofamily.com/. Accessed: 16-Mar-2017.
10. A. R. Menon, "AmbyGear—the smartwatch that makes kids smarter and keeps them safe", *TechStory*, 25 December 2016. http://techstory.in/ambygear/. Accessed: 16 March 2017.
11. "The world's first smart locator and phone for kids", *FiLIP*. http://www.myfilip.com/. Accessed: 16 March 2017.
12. "Precise innovation for employees", *NPP*. https://mynpp.com/offer/precise-innovation. Accessed: 16 March 2017.
13. "Omate Wherecom k3 kids'smartwatch offers standalone 3G support, Android 5.1", *Owler*. https://www.owler.com/reports/omate/omate-wherecom-k3-kids--smartwatch-offers-standalo/1455925200434. Accessed: 16 March 2017.
14. B. F. Baird, *Managerial Decisions Under Uncertainty: An Introduction to the Analysis of Decision Making* (New York: Wiley, 1989).
15. T. U. Daim, *Hierarchical Decision Modeling Essays in Honor of Dundar F. Kocaoglu* (Cham: Springer International Publishing, 2016).
16. H. Chen and J. Li, "A sensitivity analysis algorithm for the constant sum pairwise comparison judgments in hierarchical decision models", 2011.
17. D. Kocaoglu and M. G. Iyigun, "Strategic R&D program selection and resource allocation with a decision support system application", *Proceedings of the 1994 IEEE International Engineering*

Management Conference (IEMC'94), Dayton North, Ohio, 17 to 19 October 1994.

18. T. Pandejpong, "Strategic decision: process for technology selection in the petrochemical industry", in *System Science: Engineering Management* (Portland: Portland State University, 2002).
19. "PocketFinder 3G GPS trackers for children, pets, seniors and vehicles", *PocketFinder*. http://pocketfinder.com/. Accessed: 17 March 2017.
20. "A wearable mobile phone for kids", *Tinitell*. http://tinitell.com/. Accessed: 17 March 2017.
21. "My buddy tag", *My Buddy Tag*. https://www.mybuddytag.com/. Accessed: 17 March 2017.
22. "Let's protect our kids together", *Lineable*. http://lineable.net/. Accessed: 17 March 2017.
23. "Always there, even when you can't be", *SAFE Family Wearables— GPS Tracking, Heart Rate, Activity Tracking for Kids*. http://www.oursafefamily.com/. Accessed: 17 March 2017.
24. "10 best wearable tracking devices to keep kids safe", *InventorSpot.com*, 2 August 2015. http://inventorspot.com/articles/10-wearable-tracking-devices-keep-children-safe. Accessed: 16 March 2017.

Appendix A. Experts Individual Results of Subcriteria

Expert 1	Technical Performance	Design	Financial	Additional Features
Price			0.5	
Battery Life		0.33		
Water Proof				0.5
Service Range	0.18			
Communication	0.16			
Compatibility	0.18			
Real Time	0.21			
Distraction-free	0.08			
Monthly Fee			0.5	
Panic Alert				0.5
Easy to Use	0.18	0.33		
Child Appealing		0.33		
Inconsistency	0.01	0	0	0

Expert 2	Technical Performance	Design	Financial	Additional Features
Price			0.42	
Battery Life		0.44		
Water Proof				0.73
Service Range	0.16			
Communication	0.13			
Compatibility	0.12			
Real Time	0.19			
Distraction-free	0.1			
Monthly Fee			0.58	
Panic Alert				0.27
Easy to Use	0.3	0.31		
Child Appealing		0.25		
Inconsistency	0.01	0	0	0

Expert 3	Technical Performance	Design	Financial	Additional Features
Price			0.52	
Battery Life		0.27		
Water Proof				0.48
Service Range	0.2			
Communication	0.2			
Compatibility	0.05			
Real Time	0.26			
Distraction-free	0.06			
Monthly Fee			0.48	
Panic Alert				0.52
Easy to Use	0.23	0.56		
Child Appealing		0.17		
Inconsistency	0	0	0	0

Expert 4	Technical Performance	Design	Financial	Additional Features
Price			0.45	
Battery Life		0.3		
Water Proof				0.65
Service Range	0.22			
Communication	0.27			
Compatibility	0.07			
Real Time	0.24			
Distraction-free	0.09			
Monthly Fee			0.55	
Panic Alert				0.35
Easy to Use	0.11	0.44		
Child Appealing		0.26		
Inconsistency	0.03	0.02	0	0

Expert 5	Technical Performance	Design	Financial	Additional Features
Price			0.8	
Battery Life		0.6		
Water Proof				0.3
Service Range	0.26			
Communication	0.26			
Compatibility	0.07			
Real Time	0.25			
Distraction-free	0.08			
Monthly Fee			0.2	
Panic Alert				0.7
Easy to Use	0.09	0.2		
Child Appealing		0.2		
Inconsistency	0	0	0	0

Expert 6	Technical Performance	Design	Financial	Additional Features
Price			0.5	
Battery Life		0.24		
Water Proof				0.48
Service Range	0.2			
Communication	0.2			
Compatibility	0.05			
Real Time	0.23			
Distraction-free	0.06			
Monthly Fee			0.5	
Panic Alert				0.52
Easy to Use	0.26	0.6		
Child Appealing		0.16		
Inconsistency	0	0	0	0

Expert 7	Technical Performance	Design	Financial	Additional Features
Price			0.6	
Battery Life		0.63		
Water Proof				0.3
Service Range	0.36			
Communication	0.22			
Compatibility	0.14			
Real Time	0.17			
Distraction-free	0.05			
Monthly Fee			0.4	
Panic Alert				0.7
Easy to Use	0.07	0.21		
Child Appealing		0.16		
Inconsistency	0	0	0	0

Appendix B. Research Model

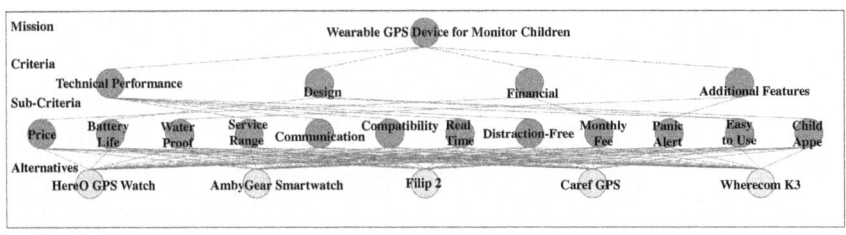

Chapter 10

Personal Transformation: Smartwatches

Alexander Blank*, João Ricardo Lavoie*, Felix Maier*, Kenny Phan* and Tugrul Daim*,†,‡

*Portland State University, Portland, Oregon, USA
†Higher School of Economics, Moscow, Russia
‡Chaoyang University of Technology, Taiwan

Abstract

Wearable devices are starting to invade our daily lives at a rapid pace. Smartwatches are the most popular among these devices, but because of their novelty and the market, manufacturers are trying to understand which is the best way to approach this market and the kind of characteristics these devices might or might not have.

Students are usually technology enthusiasts and often fall into the "early adopters" category when it comes to new technologies. However, there is not much research work in the literature concerning smartwatches and the type of characteristics that should be considered when buying one of these products.

The present study aims to assess different smartwatch models to determine which is the most suitable for Portland State University (PSU) students. The methodology used was the Hierarchical Decision Model (HDM), a Multi-Criteria Decision Making (MCDM) method that decomposes complex problems and situations in a hierarchical fashion, dividing it into smaller fractions to make the problem easier to approach and assess. The method relies on experts' judgments in order to assess the components of the model. After the experts have done their assessments, mathematical routines are applied and the relationships between components, as well as the best alternative to solve the problem, are presented.

The criteria contained in the model were based on a survey and on a literature review. The alternatives were chosen to represent the diversity of the market, and the experts were students who are technology-oriented and well informed about the products and the market. After the application of the model, an analysis is made on the main results (alternatives scores) and also the importance of each criteria and subcriteria.

Keywords: Technology assessment, smartwatches, students.

1. Introduction into Smartwatches

1.1. *History*

The stationary watch was first invented in the 15th century, followed by pocket watches in the 16th century. The first wristwatch was produced in the middle of the 19th century. Inspired by movies from the middle of the 20th century, people started thinking about other functionalities that a watch could possess (besides just displaying the time). Thus, the desire for functional variety was born. Watches that could fulfill this desire soon followed in the late 20th century when the first digital watch, called Pulsar, was released. Ten years later Intel produced the first LCD watch while Casio and Seiko entered the market with watches featuring a built-in database and blood pressure sensors (or touch screens), although those features were not widely accepted by customers yet [1]. In addition, the Seiko Receptor from 1990 was wirelessly connected to the Internet for the first time, an important aspect when defining what a smartwatch is made of. After that, companies like AT&T, Samsung and IBM entered the market through the 1990s, whereas Microsoft came into play in 2004 with a platform called Smart Personal Object Technology (SPOT). However, all these different concepts were either not good enough or the time was not ripe for them. A breakthrough was achieved in 2012 by the Kickstarter campaign of Pebble, which raised over US$10 million and sold over 70,000 units of smartwatches within one month, instantly elevating the then unknown brand into the smartwatch market. Based on this incarnation of smartwatches, other companies like Samsung, Motorola and Apple turned to their Research and Development (R&D) departments and came out with their own versions of new smartwatches [2]. Nevertheless, all the smartwatches that are out on the market today are only slightly different from each other, which underlines the fact that the purpose of today's smartwatches are not entirely defined yet.

1.2. *Definition*

Given the fact that smartwatches are very much a new technology where even the producers are not in agreement about what characteristics a smartwatch should exhibit, it is not surprising there is no clear and correct definition out there so far. Some say that a smartwatch "is a multipurpose device, usually worn on the wrist, that runs computing applications" [3], whereas others define it as "a computer-based wristwatch that provides an extension to a smartphone via Bluetooth" [4]. The Smartwatch Group states three attributes that are responsible for defining a smartwatch: it is (1) worn on the wrist, (2) able to indicate time, and (3) able to wirelessly connect to the internet [5]. That a smartwatch is worn on the wrist is thereby the only equal attribute that all three definitions state. Surprisingly, only one definition comes up with the attribute of indicating time, although one might argue that this should have been the first criterion. Two of the definitions furthermore mention the possibility of connecting the smartwatch to the smartphone (and by that connecting it to the internet). One argument is of the usage of apps. While the first definition states this clearly as a crucial aspect, the Smartwatch Group thinks that apps are not a compulsive criterion. However, there is no right or wrong when defining a smartwatch, and overall, every attribute mentioned above can be included when defining a smartwatch.

1.3. *What do they do and what are they useful for?*

The most modern smartwatches basically do what a smartphone does, but in a downscaled format. The majority of smartwatches are armed with a touchscreen, which makes them, except for one main button or the typical clock wheel, buttonless and by that fashionable. The material used varies from smartwatch to smartwatch, from plastic to aluminum, which is reflected in the cost price most of the time. Save for a few exceptions, all smartwatches are connected via Bluetooth to another mobile device in order to be connected to the internet. However, some smartwatches are only compatible with devices within the same brand. This also means that smartwatches serve only as a complement to other devices, otherwise they are mostly unusable [6].

The first group of usability can be declared as messaging. As the complement to a mobile phone, one can answer calls, read and reply to text messages and emails with a smartwatch. Some smartwatches today even have internal speakers so that a call can be conducted without taking one's mobile phone out of one's pocket. For text input the user can usually choose voice input while some smartwatches also offer a manual keyboard input. Furthermore, the smartwatch can display notifications from almost every kind of app that is installed on the mobile phone, such as Facebook or Google Now. Although it is not possible to reply to every kind of notification one can push a specific button on the screen so that the respective app opens on the smartphone.

Talking about internal speakers leads to the second group, called media. The combination of these speakers and a high-resolution screen makes it possible to watch videos, pictures, or listen to music. Only a few smartwatches right now have a camera that can shoot pictures and videos. Due to the permanent connection with another mobile device it is also possible to operate, for example, the music player on the other device or even a television.

A big market trend that smartwatches also try to cover is health and fitness. For this purpose, most smartwatches are equipped with a preinstalled chronograph for running, a timer, a workout app with different modes, as well as a heart rate monitor. For the heart rate monitor there are sensors installed at the back of the watch that touch the skin. Furthermore, third-party supplier apps, such as Runkeeper, can be installed to keep track of running miles. For exact measurements, all smartwatches are endued with GPS [7].

A fourth group of usage can be denoted as additional features. For example, there are certain smartwatches that are resistant to water and dust, which will allow one to listen to music while in the shower. Other convenient aspects are mobile payments, video chats, child monitoring, or identification and authentication [8].

1.4. *Market numbers*

Figure 1 shows the smartwatch market measured by volumes of shipment in 2014. It displays an unsuspected constellation when looking at the shipments in 2014, because everybody would expect the market to be dominated by the major technology companies. However, there are two relatively unknown brands among the top three, namely Pebble and Fitbit. Samsung though, is indisputably in first place with 1.2 million shipments, which is almost as much as the sum of shipments in the second and third place. It is also noteworthy that Apple is not included in this list even though they launched their smartwatch in 2015. However, one can expect them to be close to or even ahead of Samsung in the next year(s) [9].

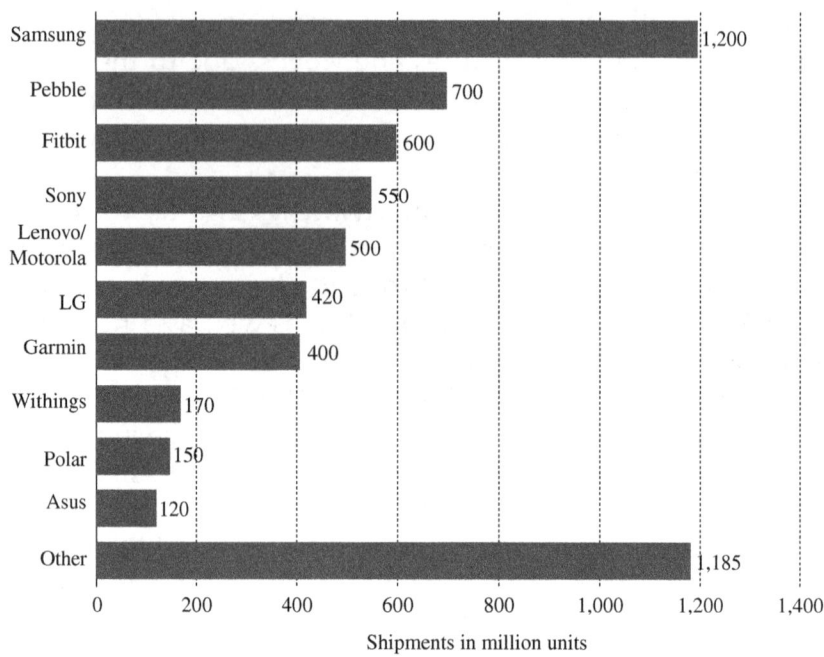

Figure 1. Smartwatch unit shipments worldwide by company in 2014 (in millions).

Furthermore, it is interesting to look at the growth rate of worldwide shipments between 2013 and 2015. The total of 1.23 million smartwatches from 2013 in relation to US$5.5 million in 2014 means a growth rate of 447%, whereas the growth between 2014 and 2015, which is expected to have a shipment of US$24.92 million, would imply a growth of another 453% [10]. In 2020, the amount of shipped smart smartwatches per year is expected to exceed US$100 million [11]. Those numbers expressed in terms of market volume cater for a sum of US$711 million in 2013, which is ten times more than it was in 2012 [12]. In 2014, the market volume climbed further and summed up to US$1.29 billion and it is forecasted to reach US$8.7 billion by 2015 [13]. Following this rapid increase experts estimate the market volume to be at US$32.9 billion in 2020 [14].

2. Problem Definition

The previous section proves that smartwatches have been adopted by the masses and that further growth will occur in the future. This means that the number of smartwatch producers will grow further as well, triggered by the basic concept of market supply and demand. If the smartwatch market follows the same rules and mechanisms as most markets, then this growth of producers will last until the market is saturated, which will probably take at least another ten years. Naturally, the bigger the number of providers, the more difficult the decision will be for the customer in terms of finding a smartwatch that will best suit his or her requirements. The customer would first have to weigh the importance of different aspects, such as price, design, functionality, compatibility, and many others, before deciding which smartwatch best fulfills these criteria. Given the large number of providers today and assuming that a customer has, for example, five important criteria, one can imagine the decision process to become pretty complex. In addition, there is always a certain degree of uncertainty, such as unknown brands or the appropriate price, when assessing and evaluating new technologies.

3. Relevant Providers

Before starting with the Hierarchical Decision Model (HDM) the team had to figure out who and how many relevant providers there should be. The team decided that five providers should be considered in the HDM, so that the model and the subsequent analysis do not become too complex. It also allows for enough variety between the respective alternatives. However, instead of taking the five most popular or famous smartwatch providers, the team decided to cover a variety of different companies. Hence, the relevant providers are Samsung, Apple, Motorola, Pebble and MyKronoz. In this way there are three major companies, of which two—Samsung and Motorola—are already successful in the smartwatch market. The remaining two are relatively unknown brands, though Pebble is already established in the market and MyKronoz is more of an unknown quantity. Moreover, by including very different providers in the analysis the team opens up the opportunity for surprising results, which would not have been the case if only the five top providers had been considered. If there was a provider who offers more than one model of smartwatch, the team will select the best performing model.

Samsung's Gear 2 was released in April 2014. Within a year, it helped Samsung become the leader in the smartwatch market. Similar to Samsung's smartphones and tablets, the Gear 2 combines highest technologic standards with a solid design and a slightly above average price. One downside is its lack of compatibility, since the watch can be only connected with Samsung devices.

The Apple Watch was released exactly a year after the Samsung Gear 2. Apple's pricing of the product, which is US$50 higher than the Samsung Gear 2, is justified by its inimitable look. Apple also depends on the brand loyalty of its customers and this may be the reason why it should not be worried even though the Apple Watch lags behind in some technical aspects.

The Moto360 by Motorola was released in September 2014 and it is the only watch among these alternatives that looks like a regular watch with a round face. This look is indeed a decisive criterion for many customers and so Motorola finds itself in 5th place in shipments in 2014. Looking at the technical details, one realizes that the Moto360 is in no way inferior to its competitors.

The Pebble Steel is the "wonder child" among the smartwatches. As stated in the introduction, Pebble sparked the popularity of smartwatches and this is a good reason why it is in 2nd place for shipments in 2014. Although the watch's technical capabilities are significantly lower compared to the other alternatives, so too is its price. The unique selling point of the

Pebble Steel, however, is its simplicity in function—basic functions are efficiently performed and paired with the hip image of an emerging start-up company.

The MyKronoz Ze Watch[2] was chosen to be the unknown alternative that comes with a low price. The Swiss watch that came out in the 3rd quarter of 2014 shows the least technical capabilities, but it can score with regard to compatibility. MyKronoz was intentionally chosen because many rankings and articles focus on its famous and highly technical products.

4. Methodology—the Hierarchical Decision Model

4.1. *The Hierarchical Decision Model*

The HDM is a Multi-Criteria Decision Making (MCDM) method and was developed in the 1980s by Dundar Kocaoglu [15]. The basic idea of a HDM (as well as with any other MCDM method) is that it represents the problem in a hierarchical disposition, so that the decision makers can visualize which items (criteria and subcriteria) affect the objective/mission. In Ref. [16], "[the] HDM helps the decision maker by presenting the decision problem as a cascade of problems that are simpler to handle". Reference [17] mentions, "This model breaks down the various elements of the problem into smaller sub-problems such that the decision problem is represented as a hierarchy". For Ref. [16], "[the] HDM is a tool used in a decision making to rank and evaluate the available choices that you have and then determine the best among them". Lastly, in Ref. [18], "it is a tool that helps decision makers to quantify and incorporate quantitative and qualitative judgments into a complex problem".

The HDM has been used in a variety of cases and for several purposes, one of which is a technology assessment that evaluates and tells which alternative is the best option in a particular setting, given the criteria established to evaluate the

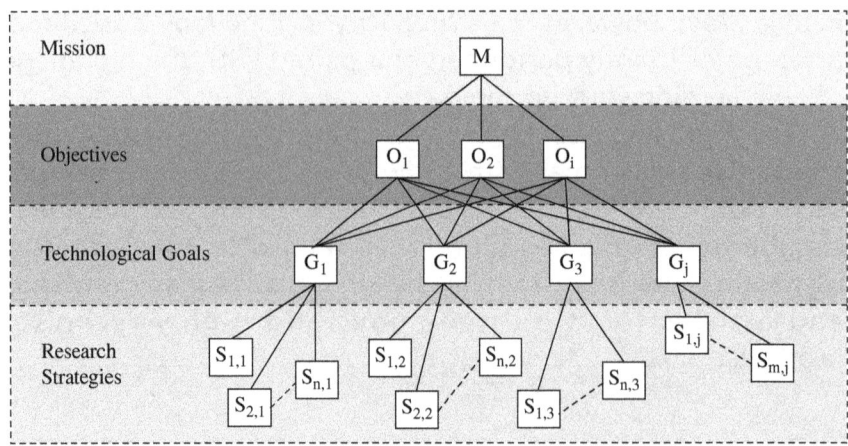

Figure 2. The HDM structure [20].

alternatives. According to Ref. [16], HDMs and systems alike can help decision makers in various ways, because these models give decision makers the ability to assess all the alternatives that are in place in a systematic way, by taking into account the weights or importance of each criteria involved. It also gives them the ability to, at the end of the process, realize what is the best way to solve the problem in question.

The basic structure of the HDM can vary depending on each application's needs. The most traditional structure is the MOGSA (see Figure 2), a five-level structure containing Mission, Objectives, Goals, Strategies and Actions. However, simpler structures can be used, such as a three-level structure containing Mission, Criteria and Alternatives, or a four-level structure containing Mission, Criteria, Subcriteria and Alternatives. According to Ref. [19], by using the HDM, one can easily regard a problem through diverse perspectives, and all the criteria under these perspectives can be compared to each other to form a ranking. In the end, a vital question can be posed: "In the judgment of the decision makers and experts, which perspective or criteria are more important than others?"

In order to apply the HDM, it is necessary to select experts (in the specific studied fields) who will help create the model and evaluate the relationships between the objective, criteria, subcriteria and alternatives. The experts will make pairwise comparisons among the items in the model (criteria, subcriteria and alternatives) to determine its weights and relationships using the constant-sum method (dividing a total of 100 points between the items being evaluated). The results of the comparisons are then extracted into matrices, which in turn will have their values normalized and processed in order to rank the alternatives. In the end, it is possible to determine which alternative is the best, considering the criteria and evaluations made by the experts involved. As Ref. [21] states, the pairwise comparisons within the HDM can help decision makers identify the relevance of each and every component of the model with regard to each other. After this step, the partial results (of each pairwise comparison) are extracted into a matrix, after normalizing the results through arithmetic means. Also, according to Ref. [21], "[the] HDM process also incorporates redundancy, which helps to reduce the measurement error and brings consistency to the results".

As stated earlier, the HDM has been applied in several different settings and fields, proving that it is indeed an effective method. Fields and areas that were explored using the HDM are (but not limited to) computer selection [17], agriculture [20], university housing [16], selection of graduate school [21], transportation options [22], solar photovoltaic technologies [19], health technology assessment [23], the semiconductor industry [18], energy [24], etc.

Engineering and research managers are frequently faced with multi-level decisions under conflicting objectives and criteria. They develop technical strategies to fulfill multiple goals, allocate resources to implement multiple strategies, and evaluate their projects and programs in terms of time, cost and performance characteristics [15]. In Ref. [25] the authors say that decision makers are nowadays facing ever complex and difficult situations, one of the reasons being that the world is becoming more and more complex. In order to help deal with these complex situations and problems, a variety of methods has been developed to try portraying complex problems in simpler ways, such as decomposing them into hierarchical levels, as the HDM does.

As Ref. [17] states, the decision process is as important as the decision itself. Thus, choosing the right method to aid in the decision process can be the difference between success and failure—"the best decision model to use when subjective judgment is needed to evaluate and select a solution with many criteria is the HDM". After selecting the research topic, our team sought to choose a well-established methodology that could analyze and consider different aspects of a problem. The relevance of the HDM to our team project was extremely high, since it gave us the opportunity to regard the problem and all its components in a systemic setting, taking into consideration complex interactions between several criteria and subcriteria involved in the topic. Furthermore, the method is not mathematically complex, resulting in a very agile and easy application. Analyzing the results achieved in the project, the group is sure that this was the right method to use and that the project's development would have been much harder had we chosen another method.

Obviously, every method has its strengths and weaknesses, and advantages and limitations. Taking this discussion to the HDM method, one of its weaknesses is that it is mandatory to find and have access to experts, in order for it to be applied. If, for one reason or another, it is not possible to rely on good experts, the outcome may be seriously compromised. Moreover, because the analyses are based on pairwise comparisons, and as the number of considered alternatives rises, the method's effectiveness decreases. This is due to the fact that the experts may eventually feel tired and lose concentration as time goes by, especially if they use time-consuming reasoning to get to a final conclusion regarding a comparison. Thus, the more comparisons one expert has to do, the more likely he/she will lose concentration and the less accurate his comparisons will be. Another weak point is that the method is not very flexible. All the analyses are based on a singular scenario (with all the pre-established criteria and subcriteria), therefore, should any criteria be changed, added or removed, the model should be rebuilt and the process restarted, disregarding all the pairwise comparisons already done. According to Ref. [16], the model's reliability and strength is extremely dependent on how the experts make their comparisons and judgments. Therefore, it is very important for decision makers to select the right expert pool (in terms of individual expertise and also in terms of the number of experts that should be invited) to aid in the decision-making process. The authors in Ref. [16] go on to dwell on one of the weaknesses of the method: "Since the expert's judgment differs from one problem to another, it is necessary to develop a new model when a change occurs. This makes the model become less interchangeable. Thus, as the circumstances change, the decision model has to be modified accordingly".

As opposed to its weaknesses and limitations, the HDM method has several advantages and strengths. For instance, if the decision maker finds and has access to the correct type of experts, all the reasoning and analyses that these experts make will result in the outcome being highly trustworthy. Furthermore, one of the biggest barriers in the decision-making process is to translate qualitative data into quantitative information (which is much easier to assess), and the HDM is a great way of doing this—it has the ability to interpret subjective aspects of a problem in order to analyze it objectively. Also, there are a number of sensitivity analysis algorithms, such as in Ref. [25], developed for this methodology, so that the decision maker can test the robustness of the model and make changes and adjustments whenever necessary.

4.2. *Criteria selection and model building*

The criteria selection is one of the most crucial parts of the HDM application. In order for the model to be reliable and robust, the criteria have to represent important aspects and characteristics involved in the situation that one is trying to assess or solve.

Conversely, the model builder cannot flood the model with too many criteria, or else the high number of pairwise comparisons may jeopardize the consistency and ultimately render useless results. Therefore, non-relevant aspects must not be included. Even among the pool of important criteria, only the most important ones should be considered to be a part of the model.

Aiming to find the most important criteria for this mission (which is defining the best smartwatch for students in Portland), the team opted for two different and complimentary approaches: a survey and a literature review. A small survey was conducted among the classmates of the Decision-Making Class of Spring 2015, and a small literature review was made in some of the academic databases—the team looked for analogous works that could serve as a starting point for the criteria definition.

The survey has yielded responses with a broad array of criteria that (to the eyes of the respondents) would be important during the process of selecting a smartwatch (the survey instrument can be seen in Appendix A). Figure 3 presents the criteria mentioned in the survey.

From this comprehensive list of possible criteria, the team computed the number of times each criterion was mentioned and came up with a narrowed list of the ten most cited ones:

- Look
- Compatibility

Accessories / Add-ons
Answer Calls with Voice/Text Messages
Apps
Battery Life
Call Answering
Camera
Color
Comfort in Use
Compatibility
Customer Support
Display Resolution
Ease of Synchronization
Ease of Use
Easy Access to Apps
Friendly User Interface
Look
Material of Case
Music Playing
Number of Available Functions
Operating System
Price
Processor
Security
Size
Speaker
Storage
Text Messages
Time Showing Styles
Usability of Apps
Video Calls
Video Playing
Voice Input
Warranty
Water Resistance
Weather Resistance
Weight

Figure 3. Criteria from the survey.

- Price
- Apps
- Display resolution
- Weight
- Processor
- Storage
- Water resistance
- Battery life

Considering only the narrowed list, the graph in Figure 4 shows the percentage of occurrences of each criterion.

After finalizing the survey, the team proceeded to search in the literature for past research works that could be used as a form of leverage to validate the criteria from the survey and to find other missing criteria.

The first impression from the literature was that a smartwatch must be a multi-functional product. Following Ref. [26],

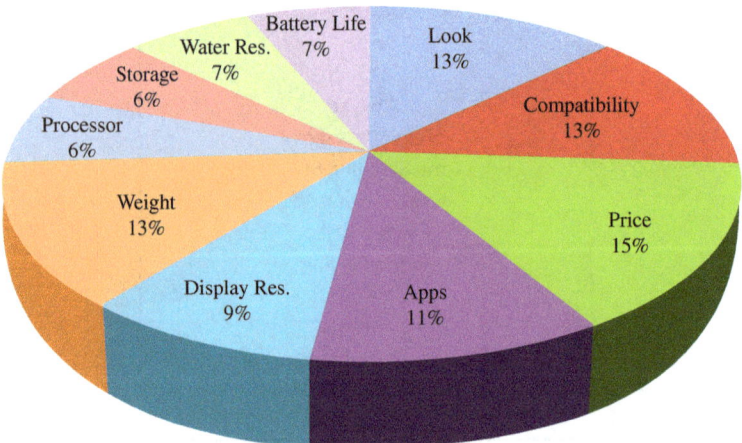

Figure 4. The percentages of the most important criteria from the survey.

smartwatches will certainly increase user satisfaction when offering and delivering many different functionalities. However, according to Ref. [26], "the main advantage of such devices is the fact that they are worn constantly and so do not require extra effort by the user to hold the device".

As Ref. [27] states, companies want to dodge or at least be prepared for market saturation on regular devices (like smartphones), therefore they are investing in wearable devices such as smartwatches. However, these new products' features must be comparable to those of which they are about to replace/complement.

The criteria gathered from the literature has validated the survey results, as most of the consulted works mention cost as a vital component [16, 28, 29] and also technical performance and physical attributes [27–29]. In addition to the validation of previous criteria found through the survey, the literature review also introduced a new criterion—brand. According to Ref. [28], the brand can have a strong influence in the decision process and can tell a lot about what customers are expecting from a particular product. Moreover, Ref. [28] mentions that the "brand can indicate the usability, quality, warranty, service and reliability".

Table 1 shows a summary with information about the alternatives with regard to each criterion [30–40].

After selecting the criteria, the team went on to create the model. The criteria identified in the previous step were called "subcriteria" and were grouped into four different "criteria" groups—Technical Performance, Physical Attributes, Financial, and Additional Features. The final model contains four levels, as seen in Table 2.

The subcriteria were allocated within the criteria groups as in Table 3.

Figure 5 presents the finalized model.

Table 1. Summary of alternatives/criteria.

Criteria \ Model	Samsung Gear 2	Apple Watch	Moto360	Pebble Steel	MyKronoz Ze Watch[2]
Display Resolution	102,400 Pixels	92,480 Pixels	92,800 Pixels	24,192 Pixels	4,096 Pixels
Processor	1.0 GHZ	1.0 GHZ	1 GHZ	80 MHZ	80 MHZ
Storage	4GB	8GB	4GB	0	0
Battery Life	3 days	18 hours	1 day	7 days	3 days
Weight	68 g	62 g	49 g	56 g	35 g
Look	—	—	—	—	—
Price	$299	$349	~$250	$149	$78
Weather Resistance	Dust and Water	Water	Dust and Water	Water 50 meter	None
Compatibility	Samsung Devices	iOS Devices	Android > 4.3 version	iOS and Android	All Bluetooth Devices
Apps	Play Store	App Store	Android Apps	Pebble App Store	None
Brand	—	—	—	—	—

Table 2. Model structure.

Level	Name
Level 1	Mission
Level 2	Criteria
Level 3	Subcriteria
Level 4	Alternatives

Table 3. Distribution of subcriteria.

Criteria	Subcriteria
Technical Performance	Display Resolution
	Processor
	Storage
	Battery Life
	Weight
	Look
Physical Attributes	Battery Life
Financial	Price
Additional Features	Water Resistance
	Compatibility
	Apps
	Brand

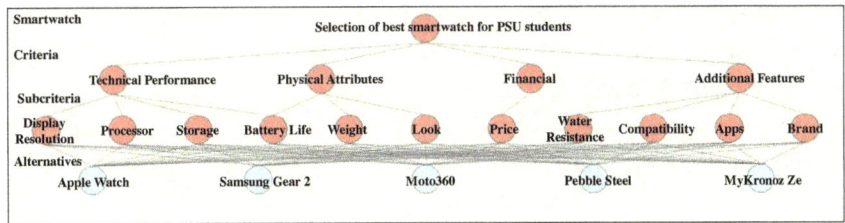

Figure 5. Finalized model.

5. Data Analysis and Results

Choosing the right experts to assess the elements of the HDM is an essential part of the method—the better informed and qualified the experts are, the more reliable the results will be. Evidently, decision makers should look for experts with the same background as that of the topic itself. Usually, for very specific and narrowed topics, the most proper experts would be professionals and researchers with strong technical backgrounds and experience in the particular industry field the decision is about. Nonetheless, when it comes to more general and broader topics, appropriate experts can be found more easily. In the case of this project, for example, the mission was to "select the best smartwatch for PSU students". Given the characteristics of the topic, appropriate experts would have to understand the needs and daily lives of students, and preferably be technology-oriented and well informed about wearable electronic devices and technologies. As three of the authors of this project are currently PSU students, technology-oriented, well informed (web-based and literature-based research on the topic was conducted prior and during the project) and have a background in technology management (ETM Department students), they were considered to be suitable experts. In addition, three other experts with the required pre-requisites were selected, forming a pool of six experts.

As aforementioned, the model is assessing five different technology alternatives, as follows:

- Apple Watch
- Samsung Gear 2
- Motorola Moto360
- Pebble Steel
- MyKronoz ZE Watch[2]

The overall results can be seen in Table 4.

Expert 1 ranked the Pebble Steel as the top choice, followed by the Samsung Gear 2, Motorola Moto360, Apple

Table 4. Overall results.

Selection of the best Smartwatch for PSU students						
Expert / Alternative	Apple Watch	Samsung Gear 2	Moto 360	Pebble Steel	MyKronoz Ze	Inconsistency
Expert 1	0.19	0.21	0.20	0.22	0.18	0.00
Expert 2	0.20	0.23	0.18	0.19	0.20	0.01
Expert 3	0.22	0.24	0.18	0.18	0.19	0.00
Expert 4	0.19	0.18	0.21	0.21	0.21	0.03
Expert 5	0.19	0.21	0.21	0.21	0.18	0.00
Expert 6	0.15	0.17	0.19	0.20	0.29	0.02
Mean	0.19	0.21	0.19	0.20	0.21	
Minimum	0.15	0.17	0.18	0.18	0.18	
Maximum	0.22	0.24	0.21	0.22	0.29	
Std. Deviation	0.02	0.02	0.01	0.01	0.04	
					Disagreement	0.03

Watch and MyKronoz ZE Watch[2]. Expert 2 ranked Samsung as the first choice, followed by Apple and MyKronoz, then Pebble and then Motorola. Expert 3's analysis concluded that Samsung was the top choice, followed by Apple, then MyKronoz and then Pebble and Motorola. Expert 4 considered Motorola, Pebble and MyKronoz as the top choices, then Apple and lastly Samsung. Expert 5 attached the same importance to Samsung, Motorola and Pebble, then Apple and then MyKronoz. Expert 6 considered MyKronoz to be the top choice, followed by Pebble, then Motorola and then Apple. The consolidated results (represented by the arithmetic mean, see the bold row in Table 4) point out that MyKronoz and Samsung are tied as the top choice, with 21% of the contribution to the Mission, followed by Pebble (20%), then Apple and Motorola (19% each). The chart in Figure 6 graphically represents the results:

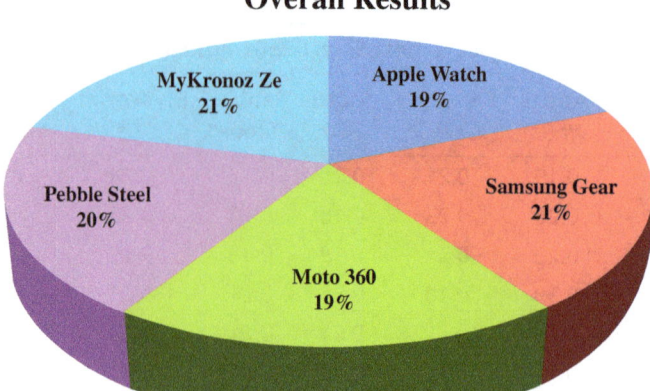

Figure 6. Overall results chart.

The inconsistency level of the experts was extremely low. Experts 1, 3 and 5 had no inconsistency, while the inconsistency percentages from Expert 2, 4 and 6 were 1%, 3% and 2%, respectively.

Another noticeable aspect of the results is the disagreement level among the experts—3%, which is a very low and acceptable rate. When the first three experts (the authors of this project) had submitted their results, the rate was even lower at 0%. This was due to the fact that before performing the pairwise comparisons, these experts have thoroughly studied and discussed all the alternatives with regard to the criteria contained in the model. Such knowledge acquired during the research process yielded similar evaluations, especially with regard to the technical specifications of the products.

As relevant and impactful as the final results themselves were the importance and ranking of the criteria and subcriteria. By means of analyzing these criteria, one is able to infer how much each criterion influences the decision-making process and how important each criterion is to the overall mission.

Table 5 shows the results concerning the second level of the model—the criteria level. Please refer to Appendix B to see the results from each expert concerning the criteria level. The same figure depicts the consolidated importance of the criteria, considering the six experts combined:

Combining the six experts, the pie chart in Figure 7 shows that the most important criterion is Technical Performance

Table 5. Criteria evaluation data—consolidated.

Criteria	Expert 1	Expert 2	Expert 3	Expert 4	Expert 5	Expert 6	Mean
Technical Performance	0.57	0.37	0.28	0.18	0.24	0.32	0.33
Physical Attributes	0.12	0.25	0.22	0.27	0.24	0.19	0.22
Financial	0.24	0.27	0.3	0.23	0.15	0.24	0.24
Additional Features	0.06	0.11	0.2	0.32	0.38	0.25	0.22

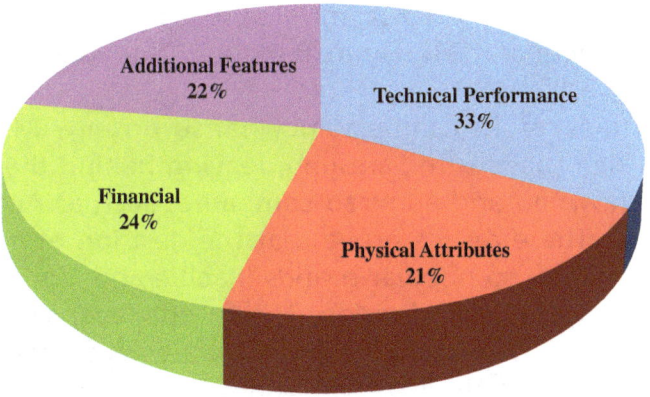

Figure 7. Criteria importance—consolidated.

(33%), way ahead of Financial (24%), and Additional Features and Physical Attributes (both with 22%).

After analyzing the criteria level, the subcriteria level is now put into perspective. Each expert compared each subcriteria within its particular criteria group. Table 6 shows the consolidated results from the six experts. Please also refer to Appendix B for the results from each expert concerning the subcriteria level. The Criteria Importance column contains the values gathered in the criteria evaluation (see Table 5). It is important to notice that the Criteria Importance for Battery Life is the summation of the Technical Performance and Physical Attributes criteria, once that subcriteria belong to these two criteria groups simultaneously. Another important explanation is that the values from Subcriteria Importance come from the multiplication of the Mean value (arithmetic mean of the three experts' evaluations) with the Criteria Importance value.

This combined analysis gives the overall importance/contribution of each subcriterion towards the Mission. The most important subcriterion is Price, with a 24% contribution to the Mission. The second place belongs to Battery Life (18%), way ahead of Display Resolution (9%) and Processor, Look and Apps (all 8% each). With 7% of the contribution, Compatibility occupies 7th place, then comes Weight (6%), Storage (5%), Brand (4%) and finally the least important subcriterion is Water Resistance with only 3%. Figure 8 represents the same data in a graphical fashion.

The final results can be somewhat surprising—MyKronoz sharing first place with Samsung, leaving behind the notoriously renowned and admired companies such as Apple and Motorola. These results prove that the decision of including smaller and not so popular brands (Pebble and MyKronoz) in this study was a good decision. The criteria level also yielded interesting results. Financial aspects are always a strong factor in every purchase decision, especially when it comes to students. Nevertheless, in this case, the Technical Performance criterion outcompeted the Financial criteria by a large margin

Table 6. Subcriteria evaluation data—consolidated.

Subcriteria	Expert 1	Expert 2	Expert 3	Expert 4	Expert 5	Expert 6	Mean	Criteria Importance	Subcriteria Importance
Display Resolution	0.28	0.22	0.29	0.21	0.35	0.22	0.26	0.33	0.09
Processor	0.23	0.23	0.27	0.29	0.21	0.27	0.25	0.33	0.08
Storage	0.14	0.22	0.20	0.14	0.09	0.18	0.16	0.33	0.05
Battery Life	0.49	0.39	0.28	0.28	0.27	0.33	0.34	0.54	0.18
Weight	0.25	0.29	0.31	0.14	0.41	0.22	0.27	0.22	0.06
Look	0.13	0.27	0.38	0.66	0.41	0.45	0.38	0.22	0.08
Price	1.00	1.00	1.00	1.00	1.00	1.00	1.00	0.24	0.24
Water Resistance	0.16	0.20	0.12	0.13	0.13	0.15	0.15	0.22	0.03
Compatibility	0.22	0.28	0.36	0.41	0.26	0.36	0.32	0.22	0.07
Apps	0.50	0.32	0.29	0.25	0.48	0.28	0.35	0.22	0.08
Brand	0.12	0.20	0.24	0.21	0.13	0.22	0.19	0.22	0.04

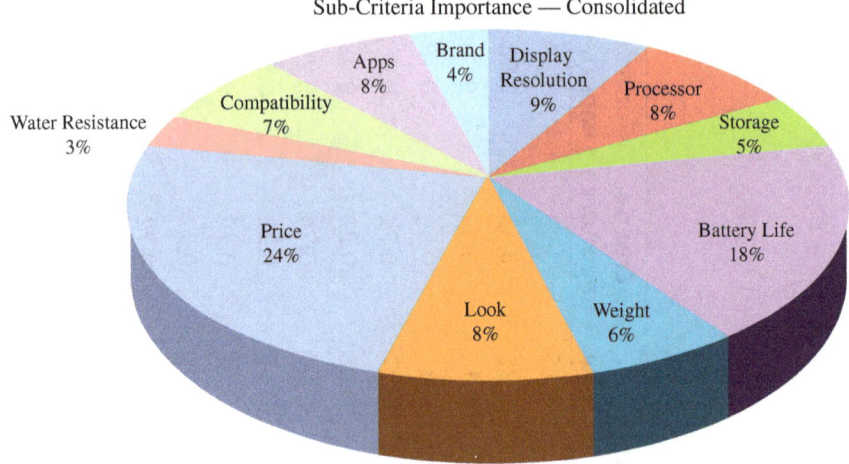

Figure 8. Subcriteria importance—consolidated.

(33% against 24%), demonstrating that students are willing to pay a higher price if the product delivers better technical performance. Also, the Additional Features and Physical Attributes criteria were close to the Financial criteria, meaning that in order to get the attention of the customer, the product has to present a whole package of advantages; it does not work to be extremely strong in a particular aspect but weak in another. Notwithstanding the fact that Financial is not the most important criteria, Price is the most important subcriterion. When one analyzes each subcriterion isolated, Price has the biggest importance, while Battery Life has the second-biggest importance (18%). Way behind are the others, starting with 9% (Display Resolution). It is relevant to point out, however, that although Price is the most important subcriterion, all the other subcriteria under the Technical Performance criteria group, when added up, are more important than Price, which leads us again to the conclusion that being strong in a single aspect (being cheaper, for instance) does not guarantee that customers will purchase the product. Rather, customers are looking for products that offer a more thorough solution.

6. Limitations and Future Research

Evidently, this research has its limitations and constraints. The selection of the experts, although justified, can also be a limitation. Future studies can be conducted using experts with a technical background and experience in the electronics industry (preferably with experience in wearable devices), in order to check whether this leads to significant differences in the outcomes.

With regard to the selection of experts, an interesting initiative would be to "slice up" the HDM model and send some parts to a certain pool of experts and other parts to other pools of experts. For instance, experts in the marketing of such products could evaluate subcriteria such as Look and Brand. Conversely, engineers would be more suitable to assess subcriteria such as Processor and Storage. The number of alternatives is also a limitation of this study. The group sought to select a diverse and representative sample among those models that are available in the market, but still there are a considerable number of different options that could have been chosen (perhaps yielding different results). Therefore, future research works could apply the same model to a different set of alternatives, or to the same ones but with more alternatives.

Lastly, the scope of the study could have been too broad. The focus of this work was to select the best smartwatch, regardless of the fact that these products can be used for a huge array of different applications. For example, any given smartwatch might be best suited for healthcare purposes, but not good enough for fitness purposes. Thus, it is recommended to repeat the study that only considers a single type of application, e.g., healthcare, fitness, entertainment, etc.

7. Conclusion

The goal of this study was to help in the selection of the best smartwatch for PSU students. The goal was achieved by using a MCDM method called a HDM. The model was built using criteria gathered through a survey and through literature review. The alternatives considered represent the diversity of smartwatches currently on the market, belonging to both renowned companies and small innovators.

After the application of the HDM, the Samsung Gear 2, along with the MyKronoz ZE Watch^2, was selected to be the best smartwatch for PSU students, taking into consideration the judgment of the experts and the criteria contained in the HDM. Although the products manufactured by Samsung and MyKronoz were the top choices, the rest of the alternatives did not trail by a big margin. On the contrary, the results were remarkably tight—the first alternative had 21% of the total score, while the last alternative had 19%. Such a scenario with tight differences between competitors shows that there is not much differentiation within the available products yet. The smartwatch market and the wearable device market, as a whole, are still very new, unexplored and unknown. Hence, in the near future, when the market is more mature, it should be clearer which products are performing better or worse. It will also become clearer who will be the leader, quick followers or laggards. Consequently, whichever brand that faster understands the needs and characteristics of this newborn market will surely be in a better position to lead the way.

References

1. "History smartwatches", *Smartwatch Group Blog*. http://www.smartwatchgroup.com/history-smartwatches/. Accessed: 29 May 2015.
2. *TechRadar*. http://www.techradar.com/us/news/wearables/before-iwatch-the-timely-history-of-the-smartwatch-1176685. Accessed: 29 May 2015.

3. E. J. Horn, "What is a smartwatch?" *Tom's Guide*. http://www.tomsguide.com/us/what-is-a-smartwatch,news-17560.html. Accessed: 29 May 2015.
4. "Definition of smartwatch", *PCMag.com*. http://www.pcmag.com/encyclopedia/term/65348/smartwatch. Accessed: 29 May 2015.
5. "What is a smartwatch definition", *Smartwatch Group Blog*. http://www.smartwatchgroup.com/blog/2013/10/28/what-is-a-smartwatch-definition/. Accessed: 29 May 2015.
6. R. Valdes and N. Chandler, "How smart watches work", *How stuff works*? http://electronics.howstuffworks.com/gadgets/clocks-watches/smart-watch.htm. Accessed: 29 May 2015.
7. "How to buy a smartwatch or fitness tracker", *Cnet.com*. http://www.cnet.com/topics/wearable-tech/buying-guide/. Accessed: 29 May 2015.
8. M. Sullivan, "These 5 smartwatch uses will make you stop loving your smartphone", *Venture Beat*. http://venturebeat.com/2014/11/21/these-5-smartwatch-functions-will-start-the-fall-of-the-smartphone/. Accessed: 29 May 2015.
9. "Smartwatch unit shipments worldwide by vendor from 2Q'15 to 4Q'17", *Statista*. http://www.statista.com/statistics/422196/smartwatch-unit-shipments-by-company-worldwide/. Accessed: 29 May 2015.
10. "Shipments of smart watches worldwide from 2013 to 2015", *Statista*. http://www.statista.com/statistics/302722/smart-watches-shipments-worldwide/. Accessed: 29 May 2015.
11. "The wearables report: growth trends, consumer attitudes, and why smartwatches will dominate", *Business Insider*. http://www.businessinsider.com/the-wearable-computing-market-report-2014-10. Accessed: 29 May 2015.
12. A. Adams, "The size of the smartwatch market and its key players", *Forbes*. http://www.forbes.com/sites/arieladams/2014/03/07/the-size-of-the-smartwatch-market-its-key-players/. Accessed: 29 May 2015.
13. N. Hughes, "Just 6.8m smartwatches sold in 2014 at an average price of $189", *Apple Insider*. http://appleinsider.com/articles/15/02/25/just-68m-smartwatches-sold-in-2014-at-an-average-price-of-189. Accessed: 29 May 2015.

14. "Global smartwatch market size, global trends, company profiles, segmentation and forecasts 2013–2020", *Globe News Wire*. http://globenewswire.com/news-release/2015/02/10/704752/0/en/Global-Smartwatch-Market-Size-Global-Trends-Company-Profiles-Segmentation-and-Forecasts-2013-2020.html. Accessed: 29 May 2015.
15. D. F. Kocaoglu, "A participative approach to program evaluation", *IEEE Transactions on Engineering Management* **3** (1983) 112–118.
16. S. Munkongsujarit, W. Schweinfort, I. Iskin, R. Colon, N. Tanatammatorn, N. Phopoonsak and A. Almobarak, "Decision model for a place to live at PSU: the case of international graduate students", in *Portland International Conference on Management of Engineering and Technology (PICMET)*, August 2009, pp. 513–534.
17. R. A. Taha, B. C. Choi, P. Chuengparsitporn, A. Cutar, Q. Gu and K. Phan, "Application of Hierarchical Decision Modeling for selection of laptop", in *Portland International Conference on Management of Engineering and Technology (PICMET)*, 5 to 9 August 2007, pp. 1160–1175.
18. K. Phan and D. F. Kocaoglu, "Innovation measurement framework to determine innovativeness of a company: case of semiconductor industry", in *Portland International Conference on Management of Engineering and Technology (PICMET); Infrastructure and Service Integration,* 27 to 31 July 2014, pp. 747–757.
19. N. J. Sheikh, Y. Park, and D. F. Kocaoglu, "Assessment of solar photovoltaic technologies using multiple perspectives and Hierarchical Decision Modeling : Manufacturers Worldview", in *Portland International Conference on Management of Engineering and Technology (PICMET); Infrastructure and Service Integration,* 27 to 31 July 2014, pp. 491–497.
20. P. Gerdsri and D. F. Kocaoglu, "HDM for developing national emerging technology strategy and policy supporting sustainable economy: a case study of nanotechnology for Thailand's agriculture", in *Portland International Conference on Management of Engineering and Technology (PICMET)*, 27 to 31 July 2008, pp. 27–31.
21. T. Turan, M. Amer, P. Tibbot, M. Almasri, F. Al Fayez, and S. Graham, "Use of Hierarchal Decision Modeling (HDM) for selection of graduate school for master of science degree program in engineering", in

Portland International Conference on Management of Engineering and Technology (PICMET), 27 to 31 July 2008, pp. 535–549.
22. J. Thompson, B. Barnwell, T. Calderwood, A. Kumar and S. Vang, "Decision model for Portland metro bike commuters", in *Portland International Conference on Management of Engineering and Technology (PICMET); Technology Management in the Energy Smart World*, 31 July to 4 August 2011.
23. L. Hogaboam, B. Ragel and T. Daim, "Development of a Hierarchical Decision Model (HDM) for Health Technology Assessment (HTA) to design and implement a new patient care database for low back pain", in *Portland International Conference on Management of Engineering and Technology (PICMET); Infrastructure and Service Integration*, 27 to 31 July 2014, pp. 3511–3517.
24. B. Wang, D. F. Kocaoglu, T. U. Daim, and J. Yang, "A decision model for energy resource selection in China", *Energy Policy* **38**, 11 (2010) 7130–7141.
25. H. Chen and J. Li, "A sensitivity analysis algorithm for the constant sum pair-wise comparison judgments in Hierarchical Decision Models", 2011.
26. V. K. Seetharamu, J. Bose, S. Sunkara, and N. Tigga, "TV remote control via wearable smartwatch device", *Annual IEEE India Conference*, Pune, India, 11 to 13 December 2014.
27. C.-H. Wang, "A market-oriented approach to accomplish product positioning and product recommendation for smart phones and wearable devices", *International Journal of Production Research* **53**, 8 (2015) 1–12.
28. J. Belding, E. Loanzon, H. Millward, L. Seboni, D. Sibanda, and T. Torgeson, "A decision model for purchasing the highest value printer for home use for the least cost", in *Portland International Conference on Management of Engineering and Technology (PICMET)*, 2 to 6 August 2009, pp. 494–512.
29. B. Saatchi, L. Pham, H. Pham, C. F. Pai and Y. Tran, "Decision model for selecting a sedan car", in *Portland International Conference on Management of Engineering and Technology (PICMET); Technology Management in the IT-driven services*, 28 July to 1 August 2013, pp. 393–400,
30. "Apple", *Apple Inc.* www.apple.com. Accessed: 29 May 2015.

31. "Samsung", *Samsung*. www.samsung.com. Accessed: 29 May 2015.
32. "Motorola", *Motorola*. www.motorola.com. Accessed: 29 May 2015.
33. "Pebble", *Pebble*. www.getpeble.com. Accessed: 29 May 2015.
34. "MyKronoz", *MyKronoz*. www.mykronoz.com. Accessed: 29 May 2015.
35. W. Shanklin, "Comparing the six Samsung Gear smartwatches", *GizMag*. http://www.gizmag.com/samsung-gear-s-vs-gear-live-gear-2-neo-gear-fit-galaxy-gear/33602/. Accessed: 29 May 2015.
36. *TechRadar*. http://www.techradar.com/reviews/phones/moto-360-1264259/review/5. Accessed: 29 May 2015.
37. "Pebble technical specs", *Reddit*. http://www.reddit.com/r/pebble/wiki/tech_specs. Accessed: 29 May 2015.
38. A. Kingsley, "Moto 360 smartwatch powered by a four-year-old processor", *ZDNet*. http://www.zdnet.com/article/moto-360-smartwatch-powered-by-a-four-year-old-processor. Accessed: 29 May 2015.
39. B. Detwiler, "Pebble Steel teardown reveals easy-open case, removable components, and more storage", *Cnet.com*. http://www.cnet.com/news/pebble-steel-teardown-reveals-easy-open-case-removable-components-and-more-storage/. Accessed: 29 May 2015.
40. *Smartwatches.specout*. http://smartwatches.specout.com/compare/118-136/MyKronoz-ZeWatch-2-vs-MyKronoz-ZeSplash. Accessed: 29 May 2015.

Appendix A. Survey Instrument

Criteria for Smart Watches

What is most important for YOU?

What is important for you when considering to buy a Smart Watch?
Please write down your criteria, preferably in bullet-points

Appendix B. Experts Individual Results

Criteria	Expert 1
Technical Performance	0.57
Physical Attributes	0.12
Financial	0.24
Additional Features	0.06
Inconsistency	0.02

Criteria	Expert 2
Technical Performance	0.37
Physical Attributes	0.25
Financial	0.27
Additional Features	0.11
Inconsistency	0.01

Criteria	Expert 3
Technical Performance	0.28
Physical Attributes	0.22
Financial	0.30
Additional Features	0.20
Inconsistency	0.01

Criteria	Expert 4
Technical Performance	0.18
Physical Attributes	0.27
Financial	0.23
Additional Features	0.32
Inconsistency	0.00

Criteria	Expert 5
Technical Performance	0.24
Physical Attributes	0.24
Financial	0.15
Additional Features	0.38
Inconsistency	0.00

Criteria	Expert 6
Technical Performance	0.32
Physical Attributes	0.19
Financial	0.24
Additional Features	0.25
Inconsistency	0.00

	Expert 1			
Subcriteria	Technical Performance	Physical Attributes	Financial	Additional Features
Display Resolution	0.28	—	—	—
Processor	0.23	—	—	—
Storage	0.14	—	—	—
Battery Life	0.35	0.62	—	—
Weight	—	0.25	—	—
Look	—	0.13	—	—
Price	—	—	1.00	—
Water Resistance	—	—	—	0.16
Compatibility	—	—	—	0.22
Apps	—	—	—	0.50
Brand	—	—	—	0.12
Inconsistency	0.00	0.03	0.00	0.00

Personal Transformation: Smartwatches

Subcriteria	Expert 2			
	Technical Performance	Physical Attributes	Financial	Additional Features
Display Resolution	0.22	—	—	—
Processor	0.23	—	—	—
Storage	0.22	—	—	—
Battery Life	0.33	0.43	—	—
Weight	—	0.29	—	—
Look	—	0.27	—	—
Price	—	—	1.00	—
Water Resistance	—	—	—	0.20
Compatibility	—	—	—	0.28
Apps	—	—	—	0.32
Brand	—	—	—	0.20
Inconsistency	0.01	0.01	0.00	0.01

Subcriteria	Expert 3			
	Technical Performance	Physical Attributes	Financial	Additional Features
Display Resolution	0.29	—	—	—
Processor	0.27	—	—	—
Storage	0.20	—	—	—
Battery Life	0.24	0.31	—	—
Weight	—	0.31	—	—
Look	—	0.38	—	—
Price	—	—	1.00	—
Water Resistance	—	—	—	0.12
Compatibility	—	—	—	0.36
Apps	—	—	—	0.29
Brand	—	—	—	0.24
Inconsistency	0.02	0.00	0.00	0.01

	Expert 4			
Subcriteria	Technical Performance	Physical Attributes	Financial	Additional Features
Display Resolution	0.21	—	—	—
Processor	0.29	—	—	—
Storage	0.14	—	—	—
Battery Life	0.36	0.19	—	—
Weight	—	0.14	—	—
Look	—	0.66	—	—
Price	—	—	1.00	—
Water Resistance	—	—	—	0.13
Compatibility	—	—	—	0.41
Apps	—	—	—	0.25
Brand	—	—	—	0.21
Inconsistency	0.01	0.01	0.00	0.01

	Expert 5			
Subcriteria	Technical Performance	Physical Attributes	Financial	Additional Features
Display Resolution	0.35	—	—	—
Processor	0.21	—	—	—
Storage	0.09	—	—	—
Battery Life	0.35	0.18	—	—
Weight	—	0.41	—	—
Look	—	0.41	—	—
Price	—	—	1.00	—
Water Resistance	—	—	—	0.13
Compatibility	—	—	—	0.26
Apps	—	—	—	0.48
Brand	—	—	—	0.13
Inconsistency	0.00	0.00	0.00	0.00

	Expert 6			
Subcriteria	Technical Performance	Physical Attributes	Financial	Additional Features
Display Resolution	0.22	—	—	—
Processor	0.27	—	—	—
Storage	0.18	—	—	—
Battery Life	0.33	0.33	—	—
Weight	—	0.22	—	—
Look	—	0.45	—	—
Price	—	—	1.00	—
Water Resistance	—	—	—	0.15
Compatibility	—	—	—	0.36
Apps	—	—	—	0.28
Brand	—	—	—	0.22
Inconsistency	0.00	0.00	0.00	0.00

Chapter 11
Personal Transformation: Drones

Donavon Nigg*, Sarah Alobaidi*, Rushikesh Jirage*,
Tejas Deshpande*, Haitham Alkharboosh* and Tugrul Daim*,[†],[‡]

*Portland State University, Portland, Oregon, USA
[†]Higher School of Economics, Moscow, Russia
[‡]Chaoyang University of Technology, Taiwan

Abstract

2016 was known to be the year of the drones. The drone technology and market had grown rapidly since 2012, with more than US$1.9 billion in sales in 2015 [1]. Recently, almost all electronic components, including those which make up a drone, are driven by cheaper cost but with high compatibilities to different technologies such as GPS and accelerometers, amongst which the "Follow-Me Drones Camera System" has been a revolutionary product. This autonomous system can be useful for professional photographers and hobbyists, and for commercial purposes such as surveys, construction, agriculture, as well as for capturing video footage without the owner having to actually handle the device.

Each innovative idea that changes our world starts with making a right decision. This study aims to assess different drones by different manufacturers that provide Follow-Me technology, and to choose the best one by using the Hierarchical Decision Model (HDM). The HDM will focus on a niche consumer base—active sports enthusiasts—to evaluate a total of six autonomous drone systems in the market based on a set of evaluation criteria. These criteria are constructed based on interviews with experts in the subject areas of photography and remote sensing, as well as on consumer reports and literature reviews.

The HDM uses two expert panels, of which the second panel (known as Expert Panel 2) consists of subject matter experts who have a good working knowledge about the products and can thus compare one against the other based on the given criteria. Consumers who know which drones are best suited to their preferences fill Levels 2 and 3 of the HDM. After recording all the responses and evaluating all alternatives against each other, the highest score will be considered the winner— the best drone system amongst all.

Keywords: Technology assessment, drones, sports.

1. Introduction

According to International Drone Racing Association's CEO Charles Zablan in an interview with *The New York Times*, drone technology was so expensive and unattainable three years ago that only professional cinematographers were able to use it [2].

Now, full drone racing kits with cameras are available on Google for about US$1,000. Drones are Unmanned Aerial Vehicles (UAVs) that are pilotless, non-crewed aircrafts that rely on remote control or on-board computers while in-flight [3]. Initially, drones were commonly used for military purposes, but are now widely used for search and rescue operations, and civil applications such as policing, firefighting and so forth. The technology is also attracting hobbyists and enthusiasts who operate drones on relatively smaller scales [3]. In only a few years, drones have evolved quickly into tools of leisure. According to Juniper Research, as a result of the rapidly increasing popularity and usefulness of drones, experts expect demand for commercial drones to reach almost 4 million, rising up to 16 million a year by 2020. Like several other technologies that have become commonly available over the time, these flying robots are expected to be useful and entertaining as well. They not only bring eye-catching aerial video perceptives to life but also inspire people to invent games and create arts that did not exist before. They have become flying extensions of the human desire to innovate, help people and have fun [2].

1.1. *History of drones*

The concept of UAVs is not new. The idea of flying the first UAV was conceived on 22 August 1849 when Austria attacked the Italian city of Venice. The first pilotless aircraft, "Aerial Target", was developed in 1916 during the First World War but never used. In November 1917, the Automatic Airplane was demonstrated for the US Army. Upon the success of this demonstration, several UAVs were developed during the First and Second World Wars. During the Second World War, drones were used to train anti-aircraft gunners and to carry out aerial attacks. Not long after, the US Air Force (USAF), concerned about losing pilots over hostile territory, began planning for unmanned flights using UAVs. By 1973, the US adopted UAVs during the Vietnam War, with more than 3,435 UAV missions carried out, of which 554 were lost in combat.

UAV technology started growing fast in the 1980s and 1990s, especially during the Persian Gulf War in 1991, when technology became cheaper and more capable. While UAVs had mainly been used by the military until then, the Central Intelligence Agency (CIA) also used them after the September 11 attacks for intelligence gathering operations over Afghanistan, Pakistan, Yemen and Somalia in 2004. As of 2008, The USAF has employed 5,331 UAVs, which is twice the number of manned planes it has. By 2013, it was stated that UAVs were used by at least 50 countries and that many of them, including Iran, China and Israel, made their own UAVs. As mentioned earlier, UAVs are not limited to military and government purposes but have become popular with hobbyists and outdoor enthusiasts for recreational use. The use of UAVs has also become increasingly popular in commercial and private markets. Amazon, the largest retailer in the world, has also started to develop its own drones for fast deliveries [3].

Rapidly increasing innovation in drone technology, coupled with administrative support by regulatory bodies like the Federal Aviation Administration (FAA) have motivated many established companies and new entrants to develop a wide variety of drones during the past three years. The technology has already created large market segments in private and commercial sectors such as photography, sports, agriculture, insurance and safety. Due to the ease of access to drone technology and low entry barriers, many manufacturers have since entered the drone market and have become established with a wide range of drone products for different market segments. Companies like DJI Innovations, Parrot, Yuneec, Hubsan and Ehang are some of the major players in the drone market giving strong competition to each other within their product lines [7, 8].

1.2. *"Follow-me drones" technology and benefits*

With an increasing popularity and technological innovations in drones, there are many different features that users can pick from within the technology. An example is a Follow-Me Drones System, which is an advanced UAV technology system with the ability to fixate on a moving object and requires no special effort for navigation. This provides users with much more creative freedom in filming an object of interest, be it a person, pet or plane. A drone simply follows this object [4].

1.3. *How does the technology work?*

Follow-me is an intelligent fight mode that turns the drone into a hands-free aerial camera crew. Initial versions of follow-me technology didn't really "follow" the object; they used the GPS signal provided by the phones or control device [4]. Currently, there are two types of this technology: one with follow-me transmitter technology known as a Ground Station Controller (GSC), and the other uses recognition software such as DJI ActiveTrack. Most of the follow-me drones use a GPS enabled devices. These include smartphones, tablets or a GSC with a transmitter in the form of either a wearable transmitter or a mobile phone. The drone is encoded to follow the object using this transmitter, which is supposed to capture video footage at all times. The location of the device is then sent to the drone. Whenever the subject moves the system/drone is programmed to track its coordinates. Ultimately, a virtual connection is created between the drone and a GPS-equipped device, which allows the drone to trace the moving object by turning and moving along with it. Most of such UAVs can remain still and yet still trace the object by simply rotating, even though the purpose of follow-me technology is to "follow" the object and capture it [4].

Follow-me technology using GPS tracking is still used by many companies like Yuneec and 3DR. It is not meant to truly see an object, but to follow it using the map coordinates that it was provided with. However, recent softwares like "Active Track" from DJI can really see the object; its tracking algorithm is smart enough to recognize a human shape and keep it in the center-frame of the camera [6]. An average speed of an auto follow drone is around 25 mph. In a few follow-me systems, the speed is clocked; other units vary slightly [4].

1.4. *Benefits*

For filming and videography professionals and hobbyists, having a drone camera that can follow objects is a special feature where the pilot can lock the drone's direction onto a specific moving target [2]. The drone will then automatically follow the object via the use of sensors and tracking software. This allows the user freedom to be creative and focus on capturing photos and videos without having to continuously worrying whether the device is following the object or not. Other advantages of self-following drones are that they can be used for filming action sports such as mountain biking and skateboarding, where each movement the sportsman makes is captured.

1.5. *Market numbers*

According to a report by the NPD Group's Retail Tracking Service, the demand and sale for drones has increased 224% from April 2015 to April 2016. Additionally, US officials have estimated that almost 1 million consumer drones were sold in the United States during the 2015 holiday season [1]. This was a 445% hike in drone unit sales from the prior holiday season in 2014.

According to the new research by Market and Markets, the global drone market is expected to grow at a Compound Annual Growth Rate (CAGR) of 32% from 2015 to 2020, making it a 5.6-billion-dollar industry. Amongst all the applications, precision agricultural drones are expected to have the highest demand with a CAGR of 42%. Other applications with higher demand can be in public commercial sectors such as law enforcement, media production, retail, inspection and safety, mapping services and education [5].

In the case of geographical growth, the Asia-Pacific region expects to see the highest demand—a CAGR of 38% from 2015 to 2020, where the top players in the market can be DJI Innovations, Parrot, 3D Robotics and PrecisionHawk [5].

Table 1 shows the US FAA's forecast with a potential figure of 1.9 million annual sales in 2016, rising to 4.3 million units sale by 2020.

Table 1. US FAA sales forecast summary from 2016 to 2020.

	Sales Forecast Summary Million sUAS Units				
	2016	2017	2018	2019	2020
Hobbyist (model aircraft)	1.9	2.3	2.9	3.5	4.3
Commercial (non-model aircraft)	0.6	2.5	2.6	2.6	2.7
	2.5	4.8	5.5	6.1	7.0

Note: Numbers may not add due to rounding.

2. Problem Definition

What is the best drone camera system on the market today? This can only be answered by understanding what the camera system will be used for, what the customer needs are, and lastly, which product matches most closely to those needs. Market numbers show that drone cameras are flying off the shelves in the United States at an aggressive pace and this trend looks set to continue into the future. This means that the already high number of drone camera system manufacturers will continue to grow; this will result in more differentiated products of many different features. As a result, consumers will face more difficult decisions when trying to choose the right drone camera system that best suits their needs.

Consumer will first need to identify their needs, followed by the features that will best support their needs. They then should weigh the importance of these different needs (i.e., criteria such as price, speed, control, compatibility and many other factors) in order to purchase the best product. For example, a customer looking to invest between $500 and $1,500 for a sports action flying camera system could use a decision-making process to help him make an informed decision for the best choice. Our model was built to address such scenarios.

3. Alternative Products

Before building the Hierarchical Decision Model (HDM), the team sought five to six products for comparison, which were close competitors in terms of price, value, features, etc. But after searching for alternative products on the market, we realized that this wasn't as a simple a task as we envisioned. We found out that there were multiple products available in a wide range of prices, some companies had wide product lines of different products providing similar features at lower prices, and some providers allowed enough variety between respective alternatives, such as DJI, 3DR, Hubsan, etc. Hence, instead of selecting only five to six close competitors, we considered 11 products for comparison and then screened out a few according to our selection criteria. In the end, DJI, Hubsan, Yuneec, 3DR, AirDog, Ehang and Lily were the providers we considered for comparison. In addition, some of them provided more than one product in the "Follow-Me" Drone Camera System category [4, 8].

3.1. *DJI Phantom 3*

The DJI Phantom 3 is currently one of the most popular and successful drones on the market that uses follow-me technology. Its inbuilt camera is of 30 fps with a 4K (or 2.7K) video capacity and 12 MP for photos. It is placed on a 3-axis gimbal that can be controlled remotely. The camera has a professional f/2.8 lens with a field view of 94°. It has an amazing set of features including GPS-assisted Hover, a vision positioning system, an altitude set height, return-to-home technology, automatic flight logs, a following mode, auto pilot and a first-person-view. The drone can be controlled using a DJI Devo remote control or with a tablet on either Android or iOS operating systems. Along with great quality features, the flight time of the DJI Phantom 3 is 23 minutes. It is powered by a small battery and has a control range of 2 to 5 km [8].

3.2. *DJI Phantom 4*

The DJI Phantom 4 is a globally successful follow-me drone because of its advanced technology and ease of use. It is a fully autonomous quadcopter with a high-precision camera having 12 MP (photo) and 4K & HD (video), and a 94° FOV fitted on a 3-axis gimbal.

Due to its great photographic and video graphic abilities, it is fully capable of competing with the advanced technology of the GoPro Hero 4. The Phantom 4 also has obstacle sensors that help it avoid obstacles autonomously, and features such as Tap Fly (which enables the drone to fly towards the required direction with a single tap on a phone/tablet screen), Visual Tracking, and Smart Return Home (which enables the drone to return to the pilot once it runs out of battery). It also has a Dual Satellite Positioning System, Vision Positioning System and Auto-Pilot System. It also has foldable arms, thus making it easy to carry and fit into small spaces. All the hardware and software used for the Phantom 4 are of high quality and updated constantly [8].

3.3. *DJI Phantom 4 Pro*

Along with all the features and design of the Phantom 4, the Phantom 4 Pro provides longer flight time, better camera quality and a greater number of sensors and specifications, thus making it a solid upgrade of the standard version. The major difference between both models is the extra 3 minutes a Pro can last in the air as opposed to the standard version. The camera of the Pro has also been improved to 20 MPX and has better sensors and features than its predecessors [8].

3.4. *DJI Mavic Pro*

The Mavic Pro is the first premium selfie drone available on the market. It is incredibly small and folds easily, making it mobile and easy to carry. It is able to capture silk smooth aerial footage with no jitters, which seems to be the case with some of its competition. Its 4K camera is equipped with awesome features such as Active Track, Tap Fly, and more [8].

3.5. *Hubsan 501S*

The Hubsan 501S is an interesting drone with a lower price tag of $219. It uses follow-me technology and has a long flight time of 20 minutes. Some of its rich features include 1080 P, 5.8 GHz FPV, headless mode, One Key Automatic Return, GPS Hold, and Altitude Hold. The H501S was an instant hit as soon as it was announced due to its lower price. Brushless motors and an integrated 4.3 FPV screen are two main added features in the Hubsan 501S [10].

3.6. *3DR Solo*

The 3DR Solo, manufactured by 3DR Robotics, is one of the most powerful drones on the market. It has a top speed of 89 km/h and when effectively combined with its follow-me feature can lock and record fast-moving objects like cars, people and speedboats. Its advanced photography and videography features involve the Selfie, Orbit, Follow Me, and Pixhawk 2 Auto-Pilot modes. The 3DR Solo does not include a camera, but is compatible with most of the GoPro cameras, including HERO 3, 3+ and 4. The drone includes a 3-axis stabilized gimbal and supports a flight time of 20 to 25 minutes with a control distance of 800 meters [8].

3.7. *3DR IRIS+*

The 3DR IRIS+ is an all-in-one stylish and powerful aerial vehicle with a self-directed, compact and durable design. It runs on the new Pixhawk autopilot system and features Copter autonomous capabilities such as automatic take-off and landing, custom mission planning supported by GPS waypoint navigation, stabilized loitering, return to launch, circling function and more. This quadcopter features a 5,100-mAh 3S battery capable of providing 15 to 20 minutes of flight time [10, 11].

3.8. *Yuneec Typhoon H*

The Yuneec Typhoon H is a powerful hexcopter with six motors that was specially built for high-quality photography and videography. It has a 4K UHD 30 fps, HD 1,080 p, 120 fps video / 12.4 MP still camera with a 3-axis anti-vibration gimbal with

360° rotation supported by follow-me technology. Some of its functionalities involve an integrated autonomous Flight mode, Team mode, ST16 all-in-one-controller, Orbit Me, Point of Interest mode, Journey/Selfie mode, Dynamic Home return, and powerful collision avoidance. The Typhoon H has a long control range of 1,600 meters and a flight time of 25 minutes [8].

3.9. *AirDog*

The AirDog was specially designed for autonomous follow-me mode flights for all sporting needs. It is a quadcopter with the ability to follow an AirLeash device automatically to capture high-quality footage using a GoPro camera. Due to this ability, it can be used to record sports like skating, surfing and soccer. The AirDog does not come with a remote controller; instead it follows and can be controlled by an AirLeash, a standalone product that can be attached to the user's arm or wrist. An AirDog is also very easy to carry, due to its retractable arms and it being lightweight. However it does not come with a camera. The AirDog supports some features like Lap or Track recording that can be configured and uploaded using smartphone applications. It uses a 14.8 V 5,600 mAh lithium polymer battery to power up and has a flight time of 10 to 18 minutes with a maximum control range of 250 meters [8].

3.10. *Ehang GhostDrone 2.0*

The Ehang Ghost Drone 2.0 is known as the easiest drone to fly due to its reliance on an easy to use Android or iOS application. The GhostDrone 2.0 consists of a 4K camera, a 3-axis gimbal, VR Google camera control, a smartphone tilt control and some advanced boilerplate autopilot algorithms. It supports a flying range of 25 meters on a single charge with more than half a mile distance range. Some of its cool modes include the Waypoint Mode, Companion Mode, Avatar Mode, flight planning mode, as well as the "single tap" free Ehang Play app. Most importantly, the GhostDrone 2.0's warranty coverage includes free repair or total replacement protection for up to three events and covers the shipping cost for both ways [9].

3.11. *Lily*

Lily has been one of the best drone camera systems advertised in the market because of its fully autonomous (follow-me) mode. Lily is extremely portable and lightweight. Its waterproof function makes it a good fit for outdoor events and water sports. Its built-in camera of HD 1,080 p recording at 60 fps and its 720 p, 120 fps slow motion recording makes it interesting. Unfortunately, this drone is not available on the market today [8].

4. Methodology

4.1. *HDM*

4.1.1. *HDM history*

Engineers and managers frequently face the problem of multilevel decision-making under conflicting goals and criteria, or making decisions without complete information. They develop strategies and allocate resources, but when it comes to evaluating the results and strategies, it becomes difficult to measure the efficiency of managerial decisions. The HDM is an approach to provide solutions to such complex analysis and evaluations. The HDM is a Multi-Criteria Decision Making (MCDM) method and was developed in the 1980s by Dundar Kocaoglu. The HDM's approach is to make complex problems simpler by breaking them down into various sub-problems. This tool is used to evaluate and rank alternatives/choices and then chooses the best one. It can be categorized into a decision made under certainty, a decision made under uncertainty, and a decision made under risk. The HDM refines the Analytic Hierarchy Process (AHP) developed in 1970 by Thomas Saaty, which was to elicit and evaluate judgments from two or more elements, including generating criteria, classifying criteria and screening decision alternatives [12–14].

4.1.2. *HDM structure*

The basic structure of a HDM varies according to its applications and uses. As shown in Figure 1, a HDM is a tree consisting of nodes that starts with a Mission, then Objectives, followed by Goals (or Criteria), and lastly, Alternatives. Each criterion is evaluated according to its importance to the Objective. The Alternatives are evaluated according to their preference with respect to each criterion or goal. The Objectives and the Criteria are factors in the decision-making process, and the lines which connect an Objective to its Criteria mean that each criterion should be compared to the alternative based on the preference for specific objectives. Likewise, the greater the number of alternatives lines, the greater the preference for that particular criterion.

Based on Bruce Baird's *Managerial Decisions Under Uncertainty*, the HDM is a software that has options to create three main

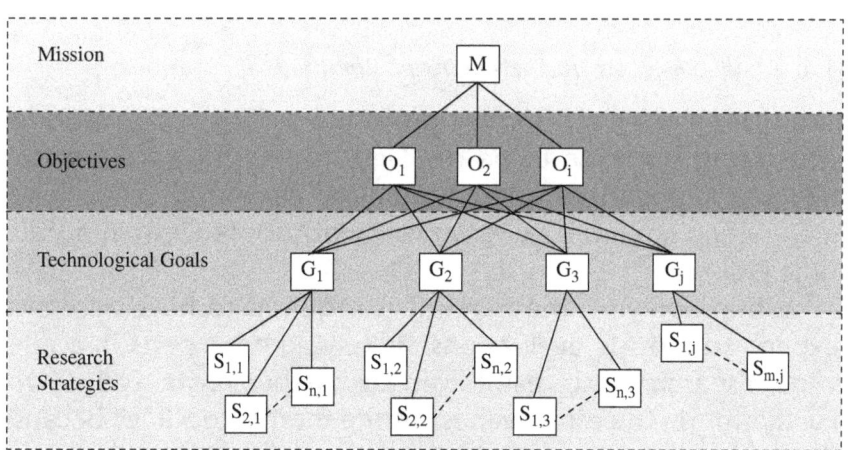

Figure 1. HDM basic structure [27].

elements, which are the Objectives, the Criteria and the Alternatives, which will allow a decision maker to enter his inputs and then send it to the experts for scoring in order to make a final decision [18]. Hence, the steps any decision maker should be involved in while implementing the HDM are: problem definition, information gathering, data collection, development and weighing of choices, selecting the best alternative, and taking follow-up actions with effective planning and execution steps [15].

4.1.3. *Application areas*

The HDM is used in several areas in public and private sectors [16, 18]: relative priorities for police calls, allocation of patrol resources to all precincts, high school selection, evaluation of R&D programs, personnel allocation, medical care evaluation, higher education scenarios, transportation planning, evaluating and scheduling alternatives for Material Requirements Planning (MRP), choosing software components from different software retailers, estimating the quality of research or investment proposals, and energy policies. Also, it is used to evaluate price, counts, or subjective judgment that inputs into a numerical matrix.

4.1.4. *Weaknesses and strengths of the HDM*

The HDM is criticized for not providing sufficient direction for structuring the issue to be resolved, forming the scale of the hierarchy for options and criteria, and gathering group judgment when a team is geographically distributed within a short time [16].

When applying the HDM, it is required to have access to experts to provide evaluations. By relying on experts in a relevant research field, results can be further narrowed down. Furthermore, the effectiveness of the method declines because the analysis depends on pairwise comparisons. This is due to the fact that as it is time-consuming for experts to give their judgments according to the comparison, they might lose focus, grow tired and not complete the process. The more comparisons an expert does, the more likely this person will lose concentration, thus possibly affecting the accuracy of the comparison. On the other hand, the HDM has several strengths and advantages. The result will be trustworthy because the analysis relies on experts' opinions. Because it is challenging to translate the decision-making process from qualitative to quantitative data, which is easier to evaluate, the HDM is a great method to do so. This methodology can be developed to examine the robustness of an HDM by using some sensitivity analysis algorithms, thus allowing decision makers to change or adjust variables whenever necessary [17].

4.2. *Criteria selection and model building*

In order to evaluate the differences between the follow-me camera drones that are on the market today, a full list of 29 criteria was created (see Table 2). As can been seen, there are many similarities in the products available. However, there are some differences, both slight and very distinctive.

We conducted market research and found 11 different products on the market that claimed to do follow-me aerial photography. From the list of criteria we found the following: four items to use as screening criteria, seven items that were irrelevant to the study, and five items that were merged with other criteria. We ended up with a total of 11 criteria that we used to evaluate the products.

Our screening criteria was used in the form of a "Yes" or "No" answer, and it consisted of the following questions:

1. Does the product have altitude tracking? (For both the 3DR products the answer was "No", so these products were removed from our list.)
2. Is the device compatible with a cell phone app? (For the Hubsan 501S the answer was "No", so it was removed too.)

Table 2. Criteria definition.

	Name	Definition	Used as
1	Battery life	How long the unit can fly	Evaluation Criteria = Performance
2	Battery Charging time	Down time between flights	Irrelevant due to all units have swappable batteries
3	Camera Resolutions	Scored by mega pixels	Evaluation Criteria = Performance
4	Durability	Will it withstand a hard landing	Evaluation Criteria = Value
5	Altitude Tracking	Can the unit adjust altitude with the object being filled	Screening criteria (Must have It)
6	Anti-Collision Technology	Can the unit avoid trees and obstacles while flying	Evaluation Criteria = Feature
7	Down Time for charging	Down time between flights	Irrelevant due to all units have swappable batteries
8	Night flight	Can the drone be flown and film at night	Irrelevant due to FAA regulations and line of sight
9	Price	Cost of the unit in US dollars	Evaluation Criteria = Value
10	Flight Time	Number of minutes the unit can fly and film	Evaluation Criteria = Performance
11	360 Degree camera angle	Can the unit film in 360 degrees w/o rotating the drone	Evaluation Criteria = Feature
12	Control Range	Measured in feet from the person controlling the unit	Evaluation Criteria = Performance
13	Water Proof	Can the unit be submersed in water	Merged with Water resistant
14	Speed of Travel	How fat the unit will travel up, down and forward	Evaluation Criteria = Performance

Personal Transformation: Drones 383

#			
15	Altitude / position hold	Can the drone hold position and continue filming	Irrelevant all the units can do this
16	Water Resistant	Is the unit able to resist water such as light rain	Evaluation Criteria = Feature
17	Weight of unit	The lighter the unit the longer it can fly	Merged with flight time.
18	Global Positioning System	Does the unit have built in GPS	Irrelevant all units have GPS
19	Built in Camera	Is the camera a add on purchase	Screening criteria (Must have it)
20	Compactness	Easy to store and carry to sporting event	Merged with satisfaction = Value
21	Ascend time	Speed in which the unit will ascend	Merged with speed = Performance
22	Descend time	Speed in which the unit will descend	Merged with speed = Performance
23	Availability	Is the unit for sale today and available	Screening criteria (Must have It)
24	Maintenance / Warranty	Does the unit come with maintenance / warranty	Evaluation criteria Satisfaction = Value
25	Different filming actions	Can the unit do multiple types of sports	Irrelevant all drones will film all sports
26	Application compatibility	Can the unit connect to smart phones	Screening Criteria (Must have it)
27	Ease of use	Set up and learning to operate the unit	Merged with user satisfaction = Value

3. Was the product actually a drone and camera system? (For the AirDog and the 3DR the answer was "No" since they both did not come with a camera, so those products were removed.)
4. Is the product currently available on the market? (The Lily is currently not available on the market, so it was removed from future research.)

If we had left these five products on the list as well as these four criteria, then a customer would probably never reach the decision of purchasing one of these five since the HDM would not connect a feature to a product and therefore not end up selecting the product. Also, if we had left all 11 products and 15 criteria on the HDM, too much time would be required for our panel of experts to get a good response. It is likely that the HDM would end up being disliked and not used.

Figure 2 shows how we screened out 5 of the 11 products.

Other items on our long list of criteria were removed from evaluation because they either had the same features, or none

Products / criteria	Screening Criteria			
	Altitude tracking Y/N	App Compatibility W/ smartphone	Camera included Y/N	Available today Y/N
DJI phantom 4 pro	YES	YES	YES	YES
DJI mavic	YES	YES	YES	YES
DJI phantom 4	YES	YES	YES	YES
Yuneec Typhoon H	YES	YES	YES	YES
DJI phantom 3	YES	YES	YES	YES
Ehang Ghostdrone 2.0	YES	YES	YES	YES
Airdog auto follow	YES	NO	NO	YES
3DR IRIS	NO	YES	NO	YES
HUBSAN H 501S	YES	NO	YES	YES
3DR solo	NO	YES	NO	YES
Lily				NO

Figure 2. Screen criteria of alternatives.

of the units had this feature and were therefore irrelevant. This list is as follows:

1. "Down time" and "Battery Charging time" were not needed since all remaining units had a spare battery option, thus allowing the user to have a charged battery ready.
2. "Night flight" was removed due to federal regulations on flying drones at night and not having a clear line of sight with the drone.
3. "Altitude/position hold" and "Built in GPS" were both found on all remaining drones and removed.
4. Lastly we found all to have built-in GPS systems, so we removed that criteria from the list.

Another issue with our long list of criteria was that some items needed to be merged. We used our second expert panel, Expert Panel 2, to help review this list and merged the following:

1. "Weight of the Unit" was merged with "Flight Time" since weight has a direct effect on the time the drone could fly.
2. "Compactness" and "Ease of Use" were merged and used to evaluate customer satisfaction.
3. "Ascend Time" and "Descend Time" were both merged with speed since all three criteria address the speeds at which the drone can fly in different direction.
4. We found that in all the drones camera systems that were touted to be water resistant, only one was actually waterproof. We originally kept this as an evaluation criterion, but we later realized that a "Yes/No" answer to only one device was problematic to our model. In the end, we still combined "Water Resistant" with "Water Proof", but gave the product that was waterproof a higher rating in comparison to the other products that were water resistant.

The next step was to look at the remaining criteria and determine the best way to use them as evaluation criteria, in order to compare and select the best product. We divided our evaluation criteria down to three objectives—Value, Performance and Features. In order to weigh the following criteria we used Expert Panel 2, a group of five EMT students, to do a deep dive research on all the remaining products.

4.2.1. *Value*

Three items—Price, Durability and Satisfaction—fell under the evaluation objective for Value.

- Price was easy because it was based on the price of the product from online vendors listed.
- Durability was given a rating of between 1 and 100 and these ratings were based on consumer reports on hard landing or crashes, which had our Expert Panel 2 looking for reports like "plastic rotors vs. titanium rotors".
- Satisfaction is a combination of three criteria on our original list: (1) Ease of Use, (2) Compactness, and (3) Maintenance/Warranty. We rated this category between 1 and 100. We also gave scores based on the five-star-rating schema on consumer reports, discussions on product and customer satisfaction, and whether the product carried a limited or an extended warranty.

4.2.2. *Performance*

Five evaluation criteria—Speed, Controller Range, Flight Time, Stability and Resolution—fell under the objective of Performance.

- Speed in which the device can travel. At first this was just a question of how fast the device could travel. Could it follow a motorbike, jet ski, etc.? After we learnt that all our remaining products also had Descend and Ascend Speeds that were all very compatible, we combined this evaluation criteria into one item called Speed.

- Controller Range refers to how far the drone can be away from the person controlling it; this distance is distinctly different within the remaining units and is measured in miles.
- Flight Time is very important as it refers to the duration the unit can maintain flight while filming. The weight of the unit was also factored into this criterion; the lighter the unit the longer the flight time.
- Stability of the drone and camera is important and was ranked on a scale from 1 to 100. The items used to rate the Stability criteria were customer feedback and manufacturers' limitation of flying in windy conditions.
- Camera resolution was based on the megapixels of the camera.

4.2.3. *Features*

Three evaluation criteria—Anti-collision, Water Resistant and 360 Camera—fell under the evaluation objective for Features.

- An anti-collision detection system was ranked on a scale from 1 to 100. Data for this was collected using several different types of technology:
  - LiDAr will allow the drone to track what is beneath the device and adjust the altitude to fly over.
  - Sonar is used to track the front, side or rear, and is able to fly left or right to avoid collisions.

Some devices have only one type of technology, while others have two. Those that have three types are very sophisticated.

- Water Resistant measures how long a drone can fly in the rain, how heavy the rain or storm is, and whether the device is submersible. This criterion was scored between 1 and 100.
- The last criterion, 360 Camera, is the drone's ability to rotate 360 degrees without turning. Some drone cameras were capable of rotating 70 degrees, while others were fixed and could not rotate at all without the drone turning. This was scored directly by the degree in which the camera could rotate.

4.3. Summary of Alternatives with selected criteria

After screening out five products for not meeting the team's minimum screening criteria, we were left with six Follow-Me Drone Camera Systems to evaluate based on the definitions listed earlier. Table 3 shows the criteria ranking of these six units, with scores based on the definitions, explanations and subject matter experts on each of these drone systems.

4.4. HDM

The HDM has four levels—Mission, Objectives, Goals (or Criteria) and Alternatives. Its intent is to link Objectives to Criteria and to get details on the factors that are weighed more than others in order to choose the best Follow-Me Drone Camera System. The details of the HDM and the nodes at each of its level are provided in Figure 3.

Level 1: This represents the Mission of the model—"choosing the best camera drone system"—from a given number of alternative providers.

Level 2: This level represents the major Objectives of our decision-making process. It consists of Performance, Value and Features, which are each further divided into multiple criteria in the 3rd level. This level comprises of a combination of three objectives that decision makers would need to consider before choosing the final product (see Figure 4).

Level 3: Among the large number of criteria, our team came up with 11 major screening criteria for the comparison between different products. This level shows these 11 criteria for selection (see Figure 5).

Level 4: This level comprises six different products/alternatives that provide similar products in terms of the criteria mentioned above (see Figure 6).

Table 3. Products criteria ranking.

Products/13 criteria	Value-Ranking			Performance-Ranking					Features-Ranking		
	Price Scale in USD	Durability Scale (1–100)	Satisfaction Scale (1–100)	Speed Scale in MPH	Control Range Scale in miles	Flight Time Scale in Min	Stability Scale (1–100)	Resolution Scale in MPs	Anti-collision Scale (1–100)	Water Resistant Scale (1–100)	360 Camera Scale in Degrees
DJI Phantom 3	$415	80	60	45	3 miles	23 min	80	1080P& 12MP	0	40	95
DJI Phantom 4	$998	100	100	45	3 miles	28 min	80	1080P& 12MP	60	60	95
DJI Phantom 4 PRO	$1,330	100	80	45	4.3 miles	30 min	80	1080P& 20MP	100	60	360
DJI Mavic	$1,299	60	80	40	4.3 miles	27 min	60	1080P& 12.7MP	60	80	75
Yuneec Typhoon H	$996	80	80	43	1 miles	25 min	60	720P& 12.4 MP	80	80	360
Ehang GhostDrone 2.0	$399	60	80	25	0.6 miles	25 min	60	720P&12MP	0	100	95

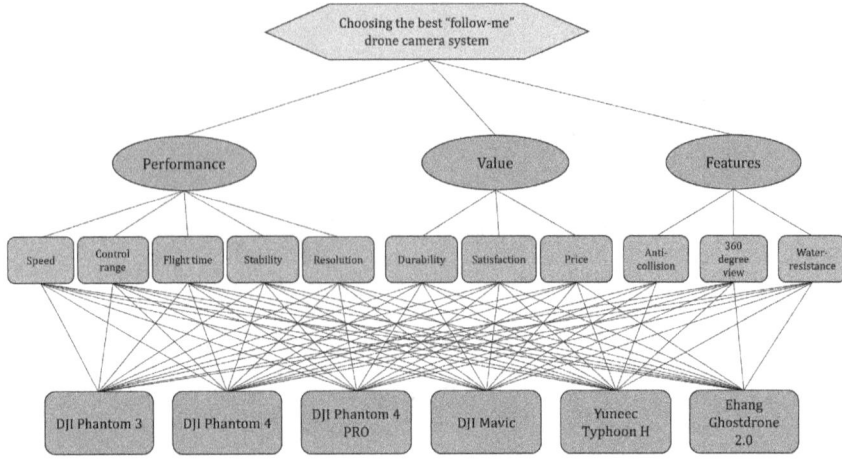

Figure 3. The final HDM.

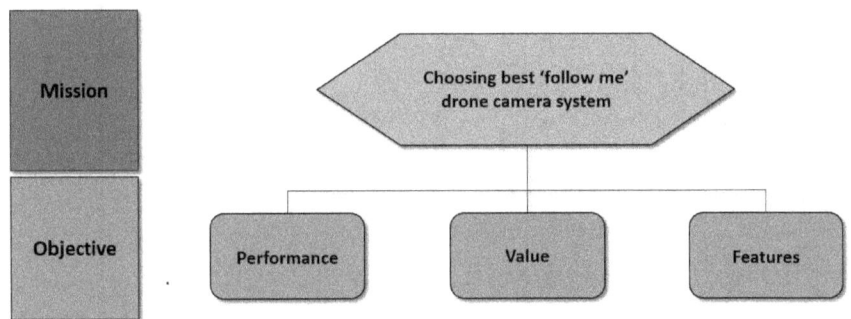

Figure 4. Levels 1 and 2.

4.5. Expert panels

4.5.1. Expert panel 1

As we know from the last few years, drone camera systems have become quite popular in the consumer market. The number of enthusiasts and hobbyists are growing, and so too is the popularity of drones. To choose between multiple drone providers, we enlisted the help of ten experts, of which the majority are active in the field of photography and sports. Some of them are

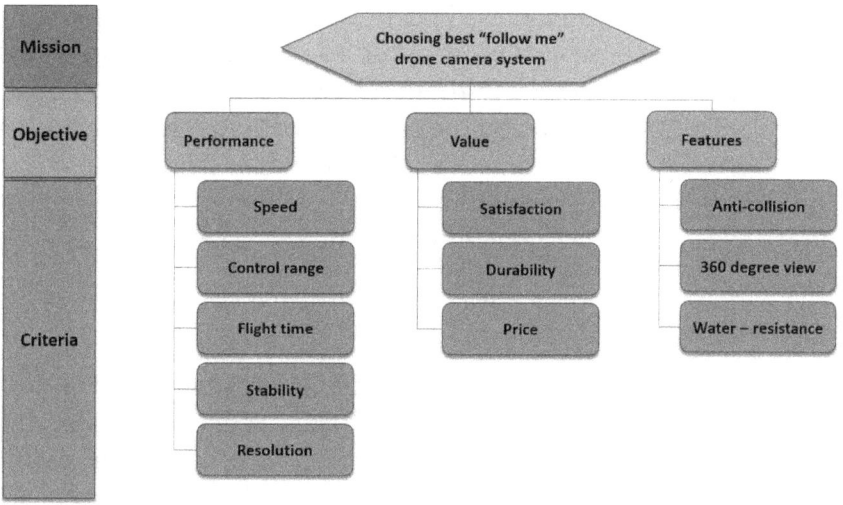

Figure 5. Level 3.

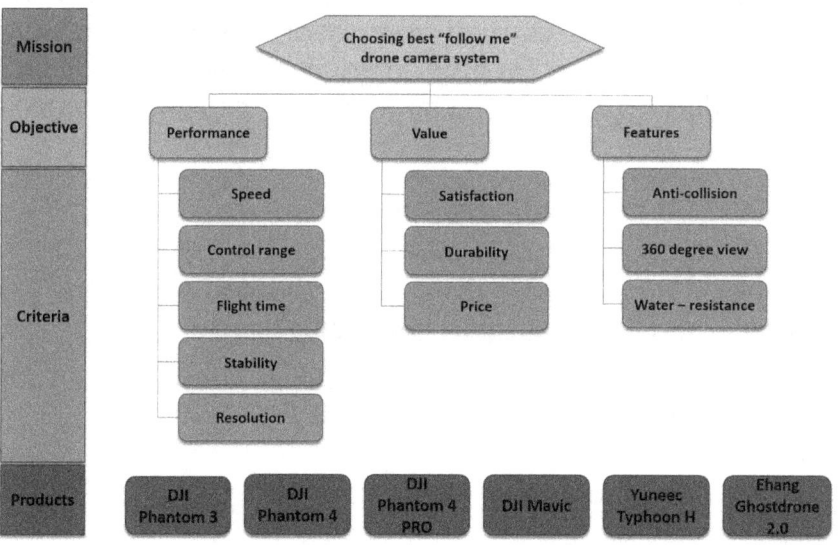

Figure 6. Hierarchical Decision Model.

technically sound and have knowledge about the drone technology. All these experts were asked to give feedback on all the selected criteria and to rate them relative to each other. In other words, all experts answered the question of "what is more important" when it came to comparing between two criteria. However, the experts were unaware of the products they were comparing; they simply rated the value of each criterion over others.

4.5.2. *Expert panel 2*

Since not every expert in the first panel was aware of all the drones available on the market and their related features, the authors of this paper formed Expert Panel 2 with the aim of selecting the best drone camera system with the help of feedback given by Expert Panel 1. They first searched and studied all competitive products. After a close inspection of customer reviews, market numbers and the specifications of each product, they came up with detailed information about each product relative to each other for each criterion. Finally, they compared each product in the HDM and came up with the best amongst all.

5. Data Analysis and Results

5.1. *Level 2 results*

Level 2 of the model includes three objectives:

(1) Performance;
(2) Value;
(3) Features.

5.1.1 *Analysis*

It can be observed from Figure 7 that all the experts rated Performance higher over Value and Features for this product. Almost all the experts wanted a product with high performance and medium value. From the average rating of Feature (17%),

Choosing the best 'follow-me' drone camera system	Performance	Value	Features	Inconsistency
Expert 1	0.51	0.3	0.19	0.01
Expert 2	0.54	0.3	0.16	0
Expert 3	0.54	0.3	0.16	0
Expert 4	0.54	0.31	0.16	0
Expert 5	0.55	0.29	0.16	0.01
Expert 6	0.51	0.3	0.19	0.01
Expert 7	0.51	0.32	0.17	0
Expert 8	0.54	0.3	0.17	0
Expert 9	0.58	0.28	0.14	0
Expert 10	0.5	0.32	0.19	0.01
Mean	0.53	0.3	0.17	
Minimum	0.5	0.28	0.14	
Maximum	0.58	0.32	0.19	
Std. Deviation	0.02	0.01	0.02	
Disagreement				0.015

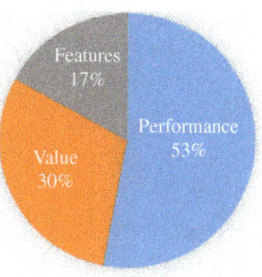

Figure 7. Evaluation of Level 2 (objectives).

we can see that most customers will not accept a product with distinctive features if its performance specifications and value are low.

5.2. *Level 3 results*

Level 3 of the model contains 11 criteria connected to three objectives.

5.2.1. *Performance*

There are a total of five criteria under the Performance objective.

The criteria under Performance are:

1. Speed;
2. Control Range;
3. Flight Time;
4. Stability;
5. Resolution.

5.2.2. *Analysis*

Overall, it can be seen in Figure 8 that the Flight Time criterion has the highest selection (25%) compared to Speed (10%), which has the lowest selection. However, Stability and Control

Experts	Speed	Control range	Flight time	Stability	Resolution	Inconsistency
Expert 1	0.09	0.21	0.27	0.22	0.2	0.07
Expert 2	0.06	0.28	0.36	0.15	0.15	0.03
Expert 3	0.11	0.23	0.37	0.12	0.17	0.01
Expert 4	0.07	0.21	0.28	0.25	0.18	0
Expert 5	0.15	0.2	0.24	0.27	0.14	0
Expert 6	0.11	0.25	0.27	0.18	0.19	0.01
Expert 7	0.04	0.23	0.08	0.26	0.39	0.06
Expert 8	0.12	0.13	0.22	0.2	0.31	0.02
Expert 9	0.06	0.12	0.12	0.3	0.4	0.02
Expert 10	0.23	0.22	0.18	0.18	0.18	0
Mean	0.1	0.21	0.24	0.21	0.23	
Minimum	0.04	0.12	0.08	0.12	0.14	
Maximum	0.23	0.28	0.37	0.3	0.4	
Std. Deviation	0.05	0.05	0.09	0.05	0.09	
Disagreement						0.064

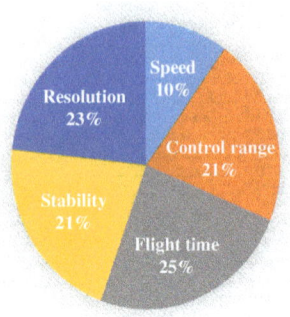

Figure 8. Evaluation of the Performance subcriteria.

Range have the same ranking (21%). Resolution is the second-highest criterion (23%).

Furthermore, it can be seen in the Performance subcriteria evaluation in Figure 8 that five experts ranked the Speed criterion as less than 10% (between 0.04 and 0.09) compared to the four experts who ranked it to be more than 10% (between 0.11 and 0.15). However, the last expert ranked this criterion at 23%, which is the highest weight. Therefore, it can be observed that speed is not a popular feature according to the experts' ranking, with the average preferences being 10%.

In the Control Range criterion, six experts gave it a weight of less than 26% (between 0.21 and 0.25), compared to two experts who ranked this criterion as 13% and 12%. However, one expert gave it a very low weight of 12% and the last expert gave it its highest weight of 28%. Therefore, although this criterion shows a gap in the experts' preferences (2% vs. 28%), most of the preferences are above 20%, thus giving it an average of 21%.

The Flight Time criterion shows that six experts ranked it between 18% and 28%. The lowest weight is 8% (made by just one expert), while the highest weights are 36% and 37% (made by two experts). Therefore, the average among these experts' weights is 24%, after taking the last expert's weight of 12% into consideration.

The Stability criterion shows that four experts gave it a weight of 22% to 27% compared to four other experts (12% to 18%). However, the lowest weight made by two experts was 2% and 3%. Thus, it can be observed that most preferences are above 20%, thus resulting in an average of 21%.

The Resolution criterion shows that six experts gave it a weight between 12% and 19%. However, the highest two weights are 31% and 39%, and the lowest two weights were 2% and 4%. Therefore, it can observed that even though there is a gap in ranking this criterion between the experts and that most of the weights are between 12% and 19%, the high weights made by two experts impacted the overall average and increased it to become 23%.

As a result, the averages in the Flight Time and Resolution criteria are the highest weights among the other criteria in terms of Performance with 24% and 23%, respectively, as compared to Speed, which has the lowest weight of 10%. Also, the two former criteria have the highest standard deviation of 0.09 compared to the other three criteria (0.05). However, the percentage of the disagreement between the experts is 0.064.

5.2.3. *Value*

There are a total of three criteria under the Value objective.
The criteria under Value are:

1. Satisfaction;
2. Durability;
3. Price.

5.2.4. *Analysis*

On average, all subcriteria under the Value Objective (see Figure 9) are almost equally rated: 33% for Satisfaction, 29% for Price and 38% for Durability. However, if we were to observe the table of evaluations, it can be seen that there is considerable

Experts	Satisfaction	Durability	Price	Inconsistency
Expert 1	0.35	0.47	0.18	0
Expert 2	0.23	0.38	0.4	0.03
Expert 3	0.25	0.4	0.35	0
Expert 4	0.35	0.4	0.25	0
Expert 5	0.22	0.33	0.45	0.01
Expert 6	0.33	0.33	0.33	0
Expert 7	0.63	0.24	0.13	0.03
Expert 8	0.25	0.49	0.26	0
Expert 9	0.32	0.43	0.25	0
Expert 10	0.33	0.36	0.31	0
Mean	0.33	0.38	0.29	
Minimum	0.22	0.24	0.13	
Maximum	0.63	0.49	0.45	
Std. Deviation	0.11	0.07	0.09	
Disagreement				0.075

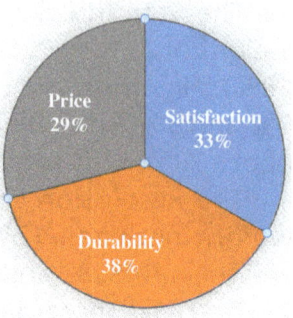

Figure 9. Evaluation of the Value subcriteria.

difference between the average maximum and minimum weight given to each subcriteria, e.g., the minimum weight given for Price is 13% where the maximum is 45%.

In the case of Satisfaction, its average rating was improved because of Expert 7, who gave it a weight of 63%. Otherwise, this criterion would have had an average rating. Overall, Durability was chosen to be the most important criterion above Price and Satisfaction.

5.2.5. *Features*

There are a total of three criteria under the Features objective.
   Criteria under Features

1. Anti-collision;
2. 360 Degree View;
3. Water Resistant.

5.2.6. *Analysis*

It can be observed from Figure 10 that the criterion of Anti-collision was rated highly by the majority of the experts. On average, Anti-collision was rated as 53%, whereas the criteria of Water Resistant and 360 Degree View had similar ratings of 24% and

Experts	Anti-collision	360 Degree view	Water-resistant	Inconsistency
Expert 1	0.34	0.32	0.33	0.04
Expert 2	0.7	0.17	0.13	0
Expert 3	0.7	0.12	0.18	0
Expert 4	0.54	0.18	0.28	0.02
Expert 5	0.37	0.21	0.42	0
Expert 6	0.48	0.31	0.22	0.06
Expert 7	0.84	0.12	0.04	0.04
Expert 8	0.53	0.28	0.19	0.01
Expert 9	0.46	0.29	0.24	0
Expert 10	0.41	0.27	0.32	0
Mean	0.54	0.23	0.24	
Minimum	0.34	0.12	0.04	
Maximum	0.84	0.32	0.42	
Std. Deviation	0.15	0.07	0.1	
Disagreement				0.102

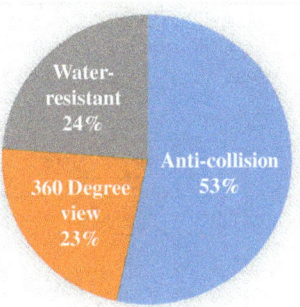

Figure 10. Evaluation of the Features subcriteria.

23%, respectively. However, it is seen from the table in Figure 13 that there is considerable difference between the average minimum and maximum value for each criteria; the maximum difference belongs to the Anti-collision criterion (34% vs. 84%), though it was still given higher weights by many of the other experts.

Hence, it is shown that Anti-collision is the most important feature for follow-me drone technology. For Expert 7, Anti-collision was his/her most important factor with 84% of the weight given to it. In addition, all the other experts rated it highly as well, as compared to the other two criteria. It can be seen from the table that the average minimum value of Anti-collision is higher than the average maximum values of the other two criteria, or the collective sum of them too.

There was a disagreement rate of 0.102 for the ten evaluations. By convention, the tolerance threshold is 10%. Since 0.102 is within the same range, we did not need to ask all experts to evaluate again.

5.3. *Final results*

According to all the evaluations made by the experts in Expert Panel 1, it was clear which objective was more important and which criteria should be rated highly to choose the best product. Expert Panel 2, after studying each product, summarized

Product/Criteria	DJI Phantom 3	DJI Phantom 4	DJI Phantom 4 PRO	DJI Mavic	Yuneec Typhoon H	Ehang GhostDrone 2.0	Inconsistency
Speed	0.19	0.19	0.19	0.16	0.18	0.08	0
Control Range	0.19	0.17	0.27	0.27	0.06	0.04	0
Flight time	0.15	0.18	0.19	0.17	0.16	0.16	0
Stability	0.2	0.2	0.2	0.13	0.13	0.13	0
Resolution	0.15	0.15	0.15	0.15	0.23	0.15	0
Satisfaction	0.11	0.26	0.17	0.17	0.17	0.11	0
Durability	0.16	0.24	0.24	0.1	0.16	0.1	0
Price	0.28	0.12	0.09	0.09	0.12	0.3	0
Anti-Collision	0	0.21	0.32	0.21	0.26	0	0
360 view	0.09	0.09	0.33	0.07	0.33	0.09	0
Water-resistant	0.08	0.12	0.12	0.19	0.19	0.29	0

Figure 11. Expert Panel 2's results.

all the details for each product, which are shown in Figure 11. These products were then compared and rated relative to each other with the help of Expert Panel 1's evaluations.

5.3.1. *Analysis*

As it can be observed from Table 3, the DJI Phantom 4 PRO and DJI Phantom 4 were high on the criteria of Anti-collision, Flight Time, Durability, 360 Degree View and Speed. These criteria were also rated highly by Expert Panel 1. The Anti-collision criterion was the highest rated among them. Also, the Performance factor of the DJI Phantom 4 PRO was the highest and this objective was also rated highly amongst all the experts (see Figure 7). The Yuneec Typhoon H scored highly in the Resolution, Anti-collision and Speed criteria. Thus, these findings made the Phantom 4, Phantom 4 PRO and Yuneec Typhoon H the top three highest ranked products (see Figure 12).

The final results by Expert Panel 2 are shown in Figure 12. It can be analyzed that although there was no large difference between all these ratings, the DJI Phantom 4, Yuneec Typhoon H and Phantom 4 PRO were high on average ratings. The DJI Phantom 4 PRO had the highest average rating of 21%. This was mainly because of its Anti-collision feature, which was the most important criterion for many experts (as can be seen in Figure 13). Also, Performance, with a rating of 53%, was the highest ranked Objective of all the experts. The DJI Phantom 4 PRO proved to be the highest in the Performance

Product	Average Rating
DJI Phantom 3	0.16
DJI Phantom 4	0.18
DJI Phantom 4 PRO	0.21
DJI Mavic	0.16
Yuneec Typhoon H	0.17
Ehang GhostDrone 2.0	0.12
Inconsistency	0

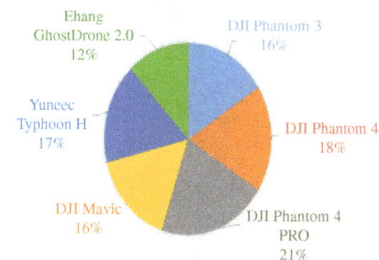

Figure 12. Choosing the best drone.

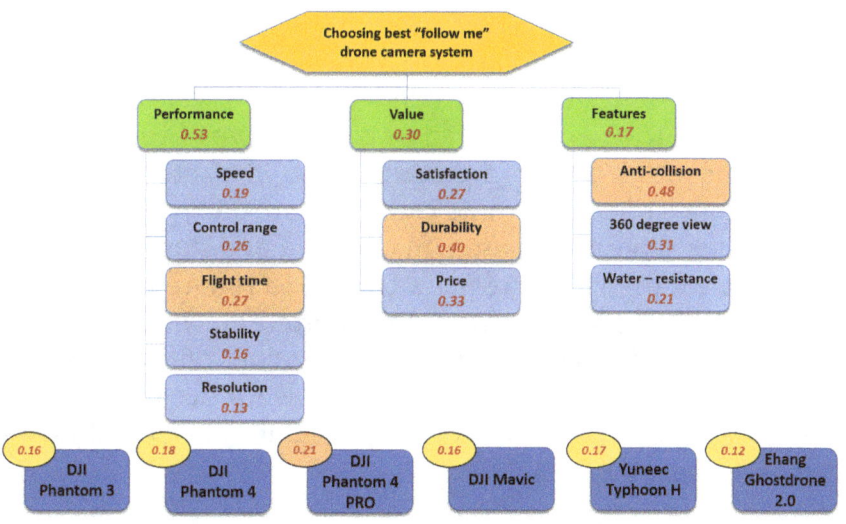

Figure 13. Summary of final results in the HDM.

category, which pushed it ahead of all the other alternative products.

When we analyzed the average rating of each subcriteria, we found out that Flight Time (27%), Control Range (26%) under Performance, Durability (40%) under Value, and Anti-collision (48%) and 360 Degree View (31%) under Features were the most important criteria for the experts. Their average values were the highest among all the subcriteria under their respective Objectives (see Figure 13). Overall, the DJI Phantom 4 proved to be the best for all these subcriteria. Hence, the average value of all rankings by Expert Panel 2 made the Phantom 4 the "Best Drone" in the comparison with a 21% average rating.

6. Limitations and Future Research

6.1. *Limitations*

The products were all very similar and we struggled with some of the criteria—Durability, Satisfaction and Stability. When testing the model we found our results to all be the same in these three categories, Going back to Expert Panel 2's definitions and rankings, we determined we needed to use a scale of 1 to 100 instead of 1 to 5. This allowed us a point spread to differentiate between the different products. Also, the ratings given by Expert Panel 1 in these three categories made it more difficult to get clear separation based on raw data and statistics.

Since we used the screening criteria to reduce our alternatives from 11 to six products, this limited our model to only certain types of drones. It removes the options of less expensive but still good products that do not have a built-in camera. Some customers might be looking for this type of Follow-Me Drone Camera Systems.

6.2. *Future research*

As new products enter into the market, new features are added and new technology becomes available. Thus, new research must be conducted in order to update this model. This model is good only for a short time before it becomes outdated.

Research is also needed is to understand the different types of customers that would use this model. For example, real estate firms film home exteriors to advance the marketability of both their firm and the listed house. How would this model help this type of customer? Other examples of how these types of drone camera systems could be used: industrial inspections, agriculture, remote sensing, and search and rescue operations.

Research in the advancement of artificial intelligence, anti-collection and high tech camera systems like infrared, for example, are things we need to continue to follow and research. There are many different customer groups, different needs and different panel groups, such as remote sensing, search and rescue operations, and industrial surveys.

7. Key Learning Points

The evaluation criteria cannot be ranked between 1 and 5 because the range is too narrow. Testing our model showed us our mistake. It was quickly fixed and we moved forward.

We brought over one criterion that had a "Yes/No" answer. The HDM was able to deal with this type of criterion as long as more than one product had the feature. However, in our case only one item had this one feature, so as soon as Expert Panel 1 placed any weight on that feature it automatically came to the top of the list. This single alternative, being chosen each time, negated the need for a decision-making process.

One should follow the process of building an HDM model and make sure the problem is clearly defined. One should also conduct extensive research on all the products, determine the criteria that needs to be brought into the model and test it before the expert panel stage.

Finally we learned of the importance of writing an instruction sheet for one's expert panel. While software may allow one to choose between Objectives, the expert will not know how to go about doing so without any explicit instructions.

8. Conclusion

The goal of this study was to help select the best Follow-Me Drone Camera System for filming action sports. This goal was achieved by using the HDM. The model was built using criteria gathered through a survey and also through literature review. The alternatives considered represent the diversity of drones that are currently on the market.

After finalizing and running the HDB, the DJI Phantom 4 Pro was considered the best Follow-Me Drone Camera System for filming action sports for hobbyists and enthusiasts. Taking into consideration the judgment of the experts and the criteria listed in the model, the DJI Phantom 4 Pro was the top ranked product, though the other alternatives did not fall behind by a big margin. In addition, the results were remarkably tight. Tight differences between competitors show that there is not much differentiation within the available products currently on the market today. This could be because the drone camera market is very new, unexplored and unknown. Hence, in the near future, when the market is more mature, it should be clearer which products will be the best performers and which will be the worst performers. It will also be clearer who the leader is. The brand that can faster and better understand the needs and characteristics of this new market will definitely have more success and lead the way.

References

1. S. French, "Drone sales in the U.S. more than doubled in the past year", *MarketWatch*, 2017. http://www.marketwatch.com/story/drone-sales-in-the-us-more-than-doubled-in-the-past-year-2016-05-27.
2. T. Innovation, "5 awesome uses for drone technology—iQ by Intel", *iQ by Intel*, 2017. https://iq.intel.com/5-awesome-uses-for-drone-technology/.
3. "The history of drone technology—Redorbit", *Redorbit*, 2017. http://www.redorbit.com/reference/the-history-of-drone-technology/.
4. "The 7 best follow you drones—[2017] follow me drone review", *Dronethusiast*, 2017. http://www.dronethusiast.com/drones-that-follow-you/. Accessed: 17 March 2017.

5. L. Sun, "Drones in 2016: 4 numbers everyone should know", *The Motley Fool*, 2017. https://www.fool.com/investing/general/2016/02/29/drones-in-2016-4-numbers-everyone-should-know.aspx. Accessed: 17 March 2017.
6. N. Majumdar, "The consumer drone market: Trend analysis", *Emberify Blog*, 2017. http://emberify.com/blog/drone-market-analysis/. Accessed: 17 March 2017.
7. "Consumer drone sales to increase tenfold to 67.7 million units annually by 2021", *Tractica.com*, 2017. https://www.tractica.com/newsroom/press-releases/consumer-drone-sales-to-increase-tenfold-to-67-7-million-units-annually-by-2021/.
8. V. Dronelli, "15 best drone that follows you—2017 follow me drones from $261", *Drones Globe*, 2017. http://www.drones-globe.com/guide/follow-me/.
9. E. Aerial, "Ehang Ghostdrone 2.0 Aerial: Features, review, specs, price, competitors", *Mydronelab.com*, 2017. http://mydronelab.com/reviews/ehang-ghostdrone-2-0-aerial.html.
10. "3D robotics Iris Plus drone review", *Sciautonics.com*, 2017. http://www.sciautonics.com/3d-robotics-iris-plus-drone-review/.
11. "3DR Iris—autonomous multicopter ID: 1546—$649.99: Adafruit Industries, unique and fun DIY electronics and kits", *Adafruit.com*, 2017. https://www.adafruit.com/product/1546.
12. D. Kocaoglu, "A participative approach to program evaluation", *IEEE Transactions on Engineering Management*, **30**, 3 (1983) 112–118.
13. S. Munkongsujarit, W. Schweinfort, I. Iskin, R. Colon, N. Tanatammatorn, N. Phopoonsak and A. Almobarak, A., "Decision model for a place to live at PSU: The case of international graduate students", in *Portland International Conference on Management of Engineering and Technology (PICMET)*, 2009, pp. 513–534.
14. R. A. Taha, B. C. Choi, P. Chuengparsitporn, A. Cutar, Q. Gu and K. Phan, "Application of Hierarchical Decision Modeling for selection of laptop", in *Portland International Conference on Management of Engineering and Technology (PICMET)*, 2007, pp. 1160–1175.
15. K. Phan and D. Kocaoglu, (2014). "Innovation measurement framework to determine innovativeness of a company: Case of semiconductor industry", in *Portland International Conference on Management of Engineering and Technology (PICMET)*, 2014, pp. 747–757.

16. M. Alexander, "Decision making using the Analytic Hierarchy Process (AHP) and SAS/IML", 2017. http://analytics.ncsu.edu/sesug/2012/SD-04.pdf.
17. H. Chen and J. Li (2011). "A sensitivity analysis algorithm for the constant sum pair-wise comparison judgments in Hierarchical Decision Models", in *Portland International Conference on Management of Engineering and Technology (PICMET)*, 2011, pp. 1–17.
18. R. Bernhard (1991). "A review of: *Managerial Decisions Under Uncertainty; An Introduction To The Analysis Of Decision Making* by Bruce F. Baird", *The Engineering Economist* **36**, 3 (1991) 264.

Chapter 12

Personal Transformation: Electric Scooter

Esraa Bukhari*, Dana Bakry*, Farshad*, Mert Tonkal* and Tugrul Daim*,†,‡

*Portland State University, Portland, Oregon, USA
†Higher School of Economics, Moscow, Russia
‡Chaoyang University of Technology, Taiwan

Abstract

In Summer 2018, the Portland Bureau of Transportation (PBOT) is conducted a four-month pilot for Shared Electric Scooters (E-Scooters) from 23 July to 20 November 2018. Currently permitted E-Scooter companies in Portland are: Bird, Lime and Skip. The key driver for the adoption of E-Scooters is solving the challenges—increasing traffic congestion woes and environmental concerns—regarding urban commute of an end user [1]. On the other hand, Portlanders are interested in acquiring E-Scooters for their daily use, either by leasing them from a firm or owning them. Thus, an assessment becomes a need to identify the best E-Scooter brand in Portland for an effective, environmentally-friendly, and quick solution for short distance transportation. This assessment was conducted using the Hierarchical Decision Model (HDM). Hence, our study is a methodology to assess the most efficient and affordable candidate technology relevant to all the valid perspectives and criteria. The technology candidates selected for the assessment process were the Xiaomi M365 (Bird), Dockless (Skip) and Ninebot Segway (Lime). According to perspectives and criteria the finding showed that the Xiaomi M365 (leasing by Bird) has the highest mean relative to the other two E-Scooter options.

Keywords: Technology assessment, transportation, scooters.

1. Introduction

There are many ways to get around Portland. People can use the MAX Light Rail, Portland streetcar or bus service. Also, they can get around Portland by car, bike or on foot. Each of these types of transportation has its pros and cons. The most important transport challenges are often related to urban areas and take place when transport systems, for a variety of reasons, cannot satisfy the numerous requirements of urban mobility. Congestion and parking are the most ancient and prevalent transport problems in large urban agglomerations. Also, there are new transport problems such as urban freight distribution or environmental impacts [2].

Recently, there is fast growing interest in electric vehicles because of growing concerns on price fluctuation, depletion of petroleum resources, global warming and environmental and health issues. In Summer 2018, shared E-Scooters started to appear in downtown Portland for leasing. Bird and Skip scooters were the first E-Scooter companies to be issued with permits and allowed to operate in Portland by the Portland Bureau of Transportation (PBOT) [1]. E-Scooters are targeted to solve the challenge regarding urban commute of an end user. Increasing traffic congestion woes, coupled with environmental concerns, is the key driver for the adoption of E-Scooters. The present paper uses Multi-Criteria Decision Making (MCDM) for the optimum E-Scooter brand selection for Portland.

2. Scope

The City's PBOT is a community partner in shaping a livable city. The Bureau plans, builds, manages and maintains an effective and safe transportation system that provides people and businesses with access and mobility. This project will assess the three

E-Scooter brands that are available both for leasing and purchasing in Portland City, and will contribute to the City's mobility, equity, safety and climate action goals [1]. The objective of this assessment is to develop a comprehensive Hierarchical Decision Model (HDM) to evaluate the three E-Scooter brands, Dockless (Skip), Xiaomi M365 (Bird), and Ninebot Segway (Lime), and their benefits and limitations in offering the best usage option to Portlanders. The evaluation is based on the Technical, Economical, Environmental and Social perspectives. This study can be a guide for decision makers to make decisions that suit their goal. There are many criteria and perspectives affecting the decision-making process. The perspectives and criteria used in this model are ranked and weighted by literature review and discussions with experts from the Portland community.

3. Methodology

3.1. *The Hierarchical Decision Model*

In response to the increasing complexity of decision-making problems in a wide variety of environments, multi-attribute hierarchical decision-making tools have been developed. One such method is the HDM, which is a variant of the Analytic Hierarchy Process (AHP) [3].

The HDM/AHP works as a structure of a hierarchy consisting of goals, perspectives, criteria and alternatives. Conducting pairwise comparisons among all variables at every hierarchy of the decision model should be done with respect to each criterion on the prior/higher level. After obtaining the relative judgment weights and checking the consistency, results are evaluated in total. The first step in the application is structuring the decision problem into levels of objectives and their associated criteria. The second step is directing the decision maker (or expert) through pairwise comparisons. The third step is to process the input and then calculating the priorities of the objectives. The final step before analyzing the decision is to check the consistency of input judgment. This is to prevent the occurrence of random and illogical pairwise comparisons [4]. HDM is one of the most common methods for helping the decision making process to quantify and combine quantitative and qualitative judgments into a complex problem [5].

We used the Portland State University Engineering and Technology Management Program's HDM software for creating our model, taking expert judgments in quantities, and then combining them as our results. All calculations were made by the software itself.

4. Technology Gap

According to our research, the requirements, capabilities and the analysis of the technology gap are shown in Table 1.

5. Perspectives

According to the technological gap mentioned above, four high-level perspectives are considered as important determining factors in making decisions regarding to these technologies. The Perspectives are as follows:

- **Technical:** Demonstrates the technical features of each E-Scooter technology [6].
- **Economical:** Takes into account financial and economical factors in these technologies and can include cost structure, affordability and parking costs [7, 8].
- **Environmental:** One of the key aspects of every transportation technology relates to how technology can save energy, reduce pollution, and so on. This perspective is considered because E-Scooters as a means of transportation can potentially help to improve the environment [6].
- **Social:** Social factors show how transportation systems can bring enjoyment, safety, sustainability, etc., to cities [9, 10].

These four perspectives are impacting factors that provide direction for the lower level criterias, which are defined accordingly.

Table 1. Technology gap analysis.

Perspectives	Requirements	Capability	Gap Analysis
Technical [6, 13–15]	Capacity	Weight = 260 pounds and age = 16+ are high satisfactory.	Lack of evidence of usage of capacity required.
	Maximum speed	Maximum speed (14–40 mph) exceeds limit.	Limit the speed within 15 mph as PCO required [8].
	Mileage	Mileage (15–30 miles), moderate high satisfactory.	E-Scooters have a top speed of 40 mph, so their traveling on footpaths poses a safety concern.
	User-friendliness		Capability met the requirement.
Economical [7, 8]	Cost	The cost of vehicles is primarily driven by battery costs, which alone account for up to 30% to 40% (depending on the battery technology) of the vehicle costs. Currently, the charging time averages 1 hour and 40 minutes for 80% of the battery charge.	Forerunners of the lithium-ion battery will anticipate the price of batteries to reduce by half by 2020. By then, even portable battery packs might be available. However, the hassle of carrying the battery before charging remains a clear pain point for end users [6]. Regulation may change it.
	Parking	There is no parking fee.	
	Affordability	Reasonable for all, even low-income people [6].	Capability met the requirement [7].

(Continued)

Table 1. (Continued)

Perspectives	Requirements	Capability	Gap Analysis
Environmental [6]	Air-pollution free	Reduction of CO_2	Capability met the requirement.
	Recycle	Recyclable batteries	A lead acid battery set for an E-Scooter set costs about Rs. 13,000–19,000, depending on its rating, brand, quality and warranties. Lithium-ion batteries do not seem to be available for E-Scooters. Even though these may be better than lead acid batteries, they are far more expensive.
Social [9, 10]	Enjoyment	Public presence	Capability met the requirement.
	Active outdoor lifestyle	It is very difficult to tell the power of a motor from a glance alone.	Capability met the requirement.
	Sustainablity	Low safety	Capability met the requirement.
	Safety		With the influx of E-Scooters that have descended upon streets, walking has become more like navigating an obstacle course.

6. Criteria

The next level after the perspective layer, the criteria level should be designed to be aligned to the upper layer. Different criteria is defined for each of these perspectives: Technical, Economical, Environmental and Social. The criteria and their definitions are categorized as follows:

Technical Perspective's Criteria: Capacity, Maximum Speed, Mileage and User-friendliness.

Capacity: Potential capacity of E-Scooters as a transportation option in the City of Portland [11].

Maximum Speed: The highest speed that each scooter can reach and is measured as mph.

Mileage: The maximum distance that an E-Scooter can travel with a single battery charge. This might be considered as a social impact as well, because lower battery consumption and higher mileage can result in environmental benefits.

User-friendliness: How the product's physical shape, brand and appearance could help shape and better a user's experience [10].

Economical Perspective's Criteria: Purchase Cost, Charging Cost, Parking Cost and Affordability [7, 8].

Purchase Cost: Initial cost to purchase a E-Scooter in US dollars.

Charging Cost: Cost to use the E-Scooter. This pay is based on how long it has been since the scooter was charged. Each E-Scooter has its pricing package in charging users [7].

Parking Cost: Cost of moving a vehicle into a place in a garage or leaving it by the side of the road [12].

Affordability: Demonstrates the E-Scooter as an affordable commute option in the city.

Environmental Perspective's Criteria: Air-pollution Free and Recyclable Battery [11].

Air-pollution Free: E-Scooters produce only about 2% of the CO_2 a car produces while driving per mile [6].

Recyclable Battery: Important from an environmental aspect. Battery replacement and disposal are factors that need to be considered in a more detailed comparison of each E-Scooter.

Social Perspective's Criteria: Enjoyment, Active Outdoor Lifestyle, Sustainability and Safety.

Enjoyment: Understanding the importance of the enjoyment level while using an E-Scooter. Generally, they are fast and easy to use but have a fun factor too.

Active Outdoor Lifestyle: Impact of E-Scooters in facilitating outdoor activities.

Sustainability: Effect of the E-Scooters on repeated social activities such as regular participation in a course, sport or discussion group meetings.

Safety: Scale of importance of safety for E-Scooters riders in everyday use [11].

7. Gathering Data

The Portland State University campus is located downtown. Since the main users of E-Scooters are university students, our experts comprised friends of the Portland community who could both lease and/or buy the product.

8. The HDM Model

After discussions with the key experts, the latest version of the HDM is shown in Figure 1 and diagrammed as: Aim, perspectives and Criteria.

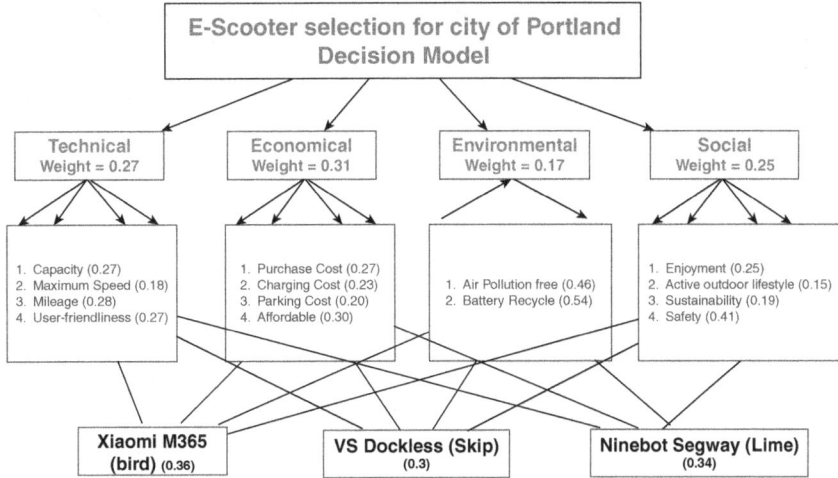

Figure 1. The HDM model.

9. Candidate Technology

There are various types of E-Scooters, each having its unique features. Different E-Scooters have their particular characteristics. However, there are some minor factors that all have in common—they are enjoyable, promote an active outdoor lifestyle, and are environmentally friendly. E-Scooters became popular recently in the City of Portland's market. Portland has three major E-Scooter rental companies: Bird, Lime and Skip. The Xiaomi M365 makes the majority of Bird's scooters, Ninebot Segway makes the majority of Lime's scooters, and Dockless makes the majority of Skip's scooters. Here is a brief description of each company profile (Table 2):

- **Xiaomi M365 (Bird):** Founded in 2017 by Travis Vander Zanden; HQ: Santa Monica; Type: Private; Size: 367 employees; Valuation $2 billion [13].
- **Ninebot Segway (Lime):** Founded in 2017 by Toby Sun; HQ: San Francisco; Type: Private; Size: 333 employees; Valuation: $1.1 billion [13].

Table 2. Candidate technology for E-Scooters in Portland.

Criteria [13–15]	Xiaomi M365 (Bird)	Dockless (Skip)	Ninebot Segway (Lime)
Max. speed	15 mph	18 mph	14.8 mph
Max. distance	18.6 miles	30 miles range	20+ miles
Motor wattage	250 watts	350 watts	250 watts
Weight	26.9 lbs	25 lbs	25 lbs
Max. rider capacity	220 lbs	350 lbs	260 lbs
Price	$1/unlock + $0.15/min to ride, $500 for stolen or lost E-Scooter	$1/unlock + $0.15/min to ride	$1/unlock + $0.15/min to ride, $1,500 for stolen or lost E-Scooter
Safety	Double safety braking system	High intensity LED headlights, tail lights and brake lights to let drivers know where you are	Double brake system
User-friendliness	Application	Application	Application and Bluetooth

- **Dockless (Skip):** Founded in 2017 by Sanjay Dastoor; HQ: San Francisco; Type: Private; Size: 8 employees, Valuation: $100 million [14].

10. Results

- Inconsistency value was 0.05, which is in the acceptable level of inconsistency (10%).
- Disagreement among experts was 0.04, which is in the acceptable level of disagreement (10%).
- The Xiaomi M365 (Bird) has the highest mean value (0.36) relative to the other two technologies.

- The Xiaomi M365 (Bird) also has the highest standard deviation value (0.05).
- The results indicate that the Economical perspective is the most important one for experts. The Technical perspective was in second place.
- From the Economical perspective, Affordability and Purchasing Cost are the highest ranked criteria according to the experts.
- Values of the Economical and Technical criteria show that the weighting of these criteria are approximately evenly distributed. It means that although the Economical and Technical perspectives were chosen as the most important ones, the lower level criteria in both these perspectives are evenly distributed.
- One interesting finding is that among the Social perspective's criteria, Safety has the highest rank, which demonstrates that risk protection should be considered a determining criteria in applying these technologies.
- Another surprising result is that Environmental factors were not considered as important as the team thought that they would be.

11. Discussion and Limitations

After collating the experts' views the results are as follows: the Xiaomi M365 (Bird) has the highest mean relative to the other two E-Scooter options. Moreover, the average Inconsistency rate is less than 10% (see Table 3), which indicates that it is an acceptable level of inconsistency (10%). Additionally, the Disagreement rate among experts was 0.04 (see Table 4), which indicates that it too is an acceptable level of disagreement. However, the Xiaomi M365 not only has the highest mean but also the highest standard deviation.

In conclusion, this project provides valuable insights into shared E-Scooters for the city of Portland. Obviously, the brand

Table 3. The HDM results.

E-Scooter Selection for City of Portland Decision Model	Xiaomi M365(bird)	Dockless (Skip)	Ninebot Segway (Lime)	Inconsistency
Deemah Alassaf	0.41	0.27	0.32	0
Farshad Saadatmand	0.27	0.31	0.42	0.05
Marthed Mohamed	0.33	0.35	0.32	0
Pei Zhang	0.42	0.26	0.32	0.03
Sarah A.badie	0.35	0.32	0.33	0.01
Mean	0.36	0.3	0.34	
Minimum	0.27	0.26	0.32	
Maximum	0.42	0.35	0.42	
Std. Deviation	0.05	0.03	0.04	
Disagreement				0.04

Table 4. Source of variation results.

Source of Variation	Sum of Square	Deg. of Freedom	Mean Square	F-test Value
Between Subjects	0.01	2	0.004	1.11
Between Conditions	0.00	4	0.000	
Residual	0.03	3	0.004	
Total	0.04	14		
Critical F-value with degrees of freedom 2 & 8 at 0.01 level				8.65
Critical F-value with degrees of freedom 2 & 8 at 0.025 level				6.06
Critical F-value with degrees of freedom 2 & 8 at 0.05 level				4.46
Critical F-value with degrees of freedom 2 & 8 at 0.1 level				3.11

selection decision for E-Scooters in this city is multi-dimensional. Only a few criteria are taken into selection for the E-Scooter project with regard to the project's size and limitation of data collection. Also, E-Scooter technology was recently introduced in Portland.

12. Future Studies

Since leasing is more common than ownership in general, our assumption was that people were not fanatical about E-Scooter brands. Thus, we thought that the cost of leasing would be more critical than brand differentiation for our experts. On the other hand, brand turned out to be extremely important for the leasing companies, since to them the cost/effectivity correlation was much more critical in brand selection. Our study's experts are not from leasing firms; they are people who use the product as customers. For further studies, loyalty should be involved in the model. Hence, for a better decision model, loyalty will affect ownership more in the future since people are still in the learning process in Portland.

References

1. "Shared electric scooter pilot", Portland Bureau of Transportation. https://www.portlandoregon.gov/transportation/77294. Accessed: 20 November 2018.
2. G. Duranton and M. A. Turner, "The fundamental law of road congestion: evidence from U.S. Cities", *American Economic Review* **101**, 6 (2011): 2616–2652. doi:10.1257/aer.101.6.2616.
3. M. S. Abbas, "Consistency analysis for judgment quantification in Hierarchical Decision Model", Paper 2699, PhD dissertation, 2016.
4. D. F. Kocaoglu, "A participative approach to program evaluation", *IEEE Transactions on Engineering Management,* **EM-30** (1983) 112–118.
5. T. U. Daim and D. F. Kocaoglu, *Hierarchical Decision Modeling: Essays in Honor of Dundar F. Kocaoglu* (Cham: Springer, 2016).
6. D. Bishop, R. Doucette, D. Robinson, B. Mills and M. McCulloch, "Investigating the technical, economic and environmental performance of electric vehicles in the real-world: a case study using electric scooters", *Journal of Power Sources,* **196**, 23 (2011) 10094–10104.
7. Frost and Sullivan, "2016 Best Practices Award: Gogoro", *2016 European Electric Scooter Technology Innovation Award Report,* 2016. https://ww2.frost.com/files/1314/6428/0919/Gogoro_Award_Write_Up.pdf. Accessed: 24 October 2018.

8. Department For Transport, "Mobility scooters and powered wheelchairs on the road—some guidance for users", March 2015.
9. J. Xu, S. Shang, G. Yu, H. Qi, Y. Wang and S. Xu, "Are electric self-balancing scooters safe in vehicle crash accidents?" *Accident Analysis and Prevention,* **87** (2016) 102–116.
10. S. Seebauer, "Why early adopters engage in interpersonal diffusion of technological innovations: an empirical study on electric bicycles and electric scooters", *Transportation Research Part A,* **78** (2015) 146–160.
11. "Laws applicable to electric scooters in Portland", Portland Bureau of Transportation. https://www.portlandoregon.gov/transportation/article/689878. Accessed: 20 November 2018.
12. "Online parking ticket payments", *Portland me.* https://www.portlandmaine.gov/355/Online-Parking-Ticket-Payments. Accessed: 20 November 2018.
13. "Bird Scooters". https://www.bird.co/. Accessed: 20 November 2018.
14. "Skip Scooters". https://skipscooters.com/. Accessed: 20 November 2018.
15. "Lime Scooters". https://www.li.me/. Accessed: 20 November 2018.

Chapter 13

Personal Transformation: Wireless Services

Asma Razavi*, Prajakta Patil*, Ritu Chaturvedi*, Pallavi Sandanshiv*, Kenny Phan* and Tugrul Daim*,†,‡

*Portland State University, Portland, Oregon, USA
†Higher School of Economics, Moscow, Russia
‡Chaoyang University of Technology, Taiwan

Abstract

This research study explains the decision-making process for the selection of the best mobile service provider. The different types of mobile server providers that were selected are AT&T, Verizon, Sprint, T-Mobile and Cricket. A Hierarchy Decision Model (HDM) was used to construct the decision model. The paper explains the four different criteria as the base criteria for structuring the decision model. We first selected the criteria that are important for the selection of the best mobile service provider. Subcriteria were selected under each of the four major criteria, such that they contribute some part of each major criterion weight or be of relative importance. Once the selection of all four criteria was done, we validated the model using Portland's top five mobile service providers. Data gathered was entered into the model and used to calculate the comparisons between different mobile service providers. Based on this analysis, a recommendation was made for the most popular mobile service provider in Portland.

Keywords: Technology assessment, wireless services, mobile services.

1. Introduction

"Moore's Law" is the observation that over the history of computing hardware the number of transistors in a dense integrated circuit approximately doubles every two years. The observation is named after Gordon E. Moore, co-founder of the Intel Corporation, who first described the trend in a 1965 paper and formulated its current statement in 1975 [1].

In the past two decades Moore's Law has driven innovation in consumer electronics devices to new heights. This has been primarily possible due to the advent of immense computational power on handheld devices. Mobile phones have been used as means of wireless communication long before the introduction of phones.

The advent of smartphones has converted the mobile phone device beyond a portal for just voice and text communication. The increased use of multimedia on such devices has increased the data bandwidth used by telecommunication mobile users. The major mobile service providers have responded to this new market need by providing various kinds of data/voice packages, and customer choose the packages best suited to their needs and affordability.

This report tries to capture facets like speed, connectivity, cost, etc., and how much importance customers give to each of these features.

2. Problem Statement

The objective of this project is to determine the best cellphone service provider in Portland. Our team has collected data on various aspects that customers considered while purchasing or signing up for a mobile service provider. Based on the analysis, the best services provider is recommended.

3. Methodology

Figure 1 shows the research approach that leads to the development of the decision model for selecting the best cellphone service provider.

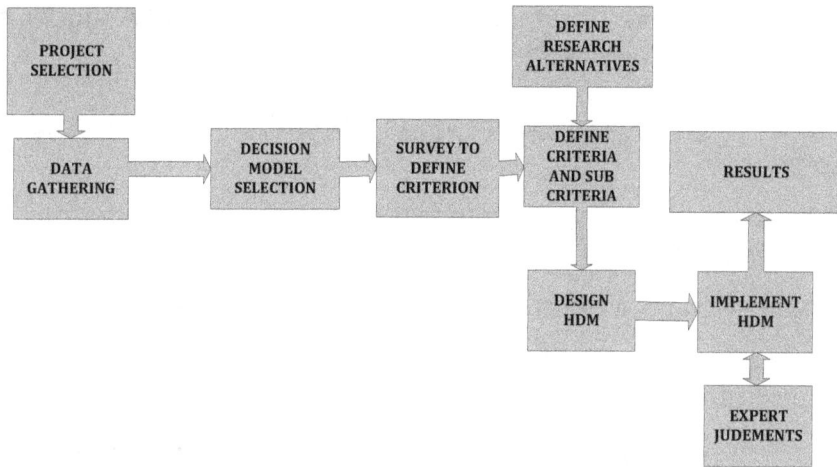

Figure 1. Research methodology.

4. Decision Model

Deciding the best service plan depends on the number of parameters that are subjective in nature based on the preferences of individual users. We did an initial customer survey to identify those preferences (see Appendix C). Hierarchy Decision Model (HDM) is used for the selection process. Pairwise comparisons are used to further evaluate the relative importance of each alternative in terms of each criterion [2].

4.1. *Hierarchy Decision Model*

In the HDM, elements at each level are considered to be preferentially independent. Based on the gathered information about each element at all levels, the impact on decision making between elements at each level is determined. The HDM, developed by David Cleland and Dundar Kocaoglu, is also known as the MOGSA decision hierarchy that consists of Mission,

Objectives, Goals, Strategies and Actions. The MOGSA model is defined as follows [3, 4]:

Mission—What business are we in? What business do we want to be in?

Objectives—What achievements do we have in order to satisfy the mission?

Goals—What are our targets in order to reach our goal?

Strategies—What is the path we need to choose in order to meet our goals?

Actions—What projects do/should we have in order to develop or accomplish our strategies?

In the MOGSA hierarchy (see Figure 2), the Mission is a single node at the top and all the other levels of hierarchy have at least two or more elements. The number of levels in the model ranges from two to more than five. The lines in the model indicate the relationship between the lower hierarchy and the upper hierarchy. All the criteria will contribute to the Mission at

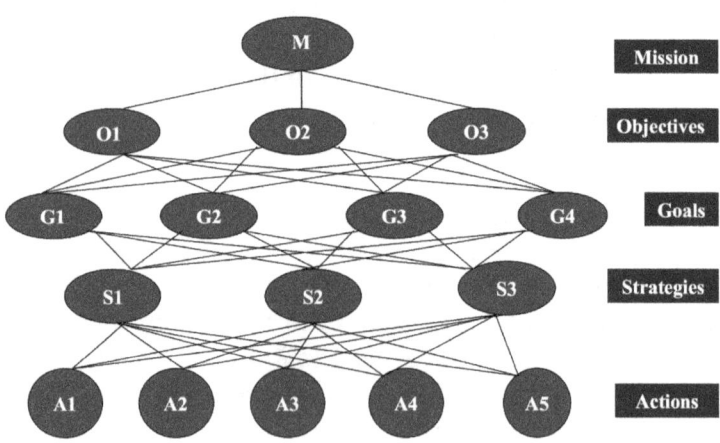

Figure 2. The MOGSA decision hierarchy [3, 4].

different weights. The HDM model can be used for a complex decision process [3].

4.2. *Pairwise comparison*

Pairwise comparison, an approach proposed by Thomas Saaty (1980), is used to determine the relative importance of each alternative in terms of each criterion. The decision maker has to express his/her opinion about the value of one single pairwise comparison at a time. Pairwise comparisons are quantified using a scale (of 1 to 100) and each candidate or alternative is matched one-on-one with each of the other alternatives. This is done for all levels of the HDM model, i.e., for all nodes, and nodes that are compared are always in the same level of hierarchy and are compared with respect to their parent node [5]. Experts are the decision makers who determine the relative importance of each criterion. They are required to distribute points on a scale of 1 to 100 between each pair of comparison to signify their judgment on the importance of those criteria to another element in the preceding level.

4.3. *Criteria and Subcriteria*

To determine the various criteria used, initial web research was done to list down the majority of parameters that can influence any user's decision to select the best suitable service plan and provider. Amongst them, the most important criteria and respective subcriteria were identified that were considered important by any decision maker to evaluate and decide the best suitable alternative. The decision model consists of four levels:

Mission—The mission of the decision model is to highlight the best cellphone service provider by selecting the most suitable plan that meets all criteria depending on the buyer's personal choice.

Criteria and Subcriteria—The four criteria are Cost, Technical Specifications, Customer Service and Support, and Availability. Their respective subcriteria are specified as follows:

4.3.1. *Cost*

The price of different plans of the five mobile service providers. These service providers are AT&T, Verizon, Sprint, T-Mobile and Cricket.

4.3.2. *Technical specification*

For a potential customer the factors that differentiate between "good" and "bad" service are dropped calls, call quality, limited or no coverage, crosstalk, data speeds, etc. There are many more key factors that one can consider when planning to narrow down on a service provider, but we decided on the two most important technical factors that really distinguish between the alternatives. Based on user preferences, they are Network Coverage and Speed.

- *Network Coverage*: Most service providers compete to provide excellent service around the coastline and major cities whereas their coverage is spotty in some areas. If a service provider has great coverage across other places and poor coverage at a customer's residence or work, the customer is most likely not going to continue with the same provider. Therefore, it makes sense for a customer to check how well his/her most frequently visited areas are covered.
- *Speed*: As of today, 4G LTE is the fastest and most reliable form of mobile broadband coverage. For a feature phone user, it may not relate much but for smartphone users, it is the second-most important factor after call quality. If multimedia streaming is a preference, then the user should ensure that there are enough 4G LTE towers around his/her most frequently visited areas.

4.3.3. *Customer service and support*

Customer service is important to an organization because it is often the only contact a customer has with a company. Customers are vital to an organization. Some customers spend hundreds or even thousands of dollars per year with a company. Consequently, when they have a question or product issue, they expect a company's customer service department to resolve their issues. The various ways of providing customer service by cellphone providers are listed below.

- *Online-Support*: Customers may have enquiries about an organization's products or services. It is ideal to provide a self-support service, such as "How To" or FAQ pages, on its official website. Online-Support also helps in determining the Return of Investment (ROI) of an organization to measure the success of its marketing campaigns. Most e-businesses worry about bounce rate percentages because they indicate how many people visit their site but leave immediately. There are a few reasons for this, people do not find what they are looking for or they find the site's navigation too complicated. By offering online support such as a "Live Chat" and e-mail replies, businesses can capture this valuable market research data and hopefully "save the sale" plus make changes to their site to make it more user friendly.
- *Call-Support*: Call-Supports provide quick and rapid solutions to any technical or non-technical problems. Individuals simply call the available hotline number to start the process of getting their issues resolved.
- *Walk-ins*: Walk-ins represent a potential boost to business, provided they are converted to customers. Walk-ins may be a way to check on new devices, plans, discounts, or to seek a quick solution to any issues. Even if satisfactory solutions are not provided, pointers in the correct directions are quick to get.

4.3.4. *Availability of features*

- *Plans*: A plan is a package of services offered by wireless service providers that includes mobile activation, monthly charges, per-minute airtime charges, roaming terms, local service area as well as additional services such as voicemail, data or international roaming [6]. Selecting a right cellphone plan is a difficult and often very baffling task. With so many options available in terms of pricing, data and talk minutes, confusion is normal. It is important to address what elements of a phone plan are important to decide which cellphone plan is right for the individual. Various factors that could be considered for cellphone plan selection are:
    i. *Individual or family plan*: This choice is relatively simple. It depends on the person. If a single line is required then an individual plan can be chosen, or if multiple phone lines are to be combined then family plans give the best bang for the buck.

    An individual plan offers minutes and data for monthly use for only one person.

    A family plan offers a "bucket" of minutes and data for use between members of a family, with each line added to the bucket costing around $10 per month. Family plans can often offer impressive savings, rather than signing up for individual plans for multiple lines in a family [7].
    ii. *Monthly minutes*: Minutes are calls that a person makes and receives. Hence it is better to determine the number of minutes required per month. A look at a couple of past but recent bills helps in the decision.
    iii. *Data*: Both email access and web surfing (including an individual's use of apps on his cellphone like Facebook, maps, etc.) will count towards its monthly usage.
    iv. *Messages*: The number of messages a person sends each month determines the plan for texting. Nowadays, most cellular providers offer bundles for unlimited texting and web surfing.

v. *Device availability*: Sometimes people get confused choosing between service providers or good devices. More and more people chase the phone they want before they start looking at carriers and contracts. If the phone device is to be the priority, one should first research the devices available with each service provider.

- *Discounts*: Cellphone service providers thrive by peddling a plethora of extra features and services on top of basic service plans. Thousands of companies, organizations and educational institutions are in partnerships with wireless carriers to offer discounts to employees, members and students. AT&T, Sprint, T-Mobile and Verizon all have discount pages. Discounts that range from 10% to as much as 25% provide great opportunity to people who can buy them.
- *International roaming*: International mobile roaming is a service that allows mobile users to continue to use their mobile phones or other mobile devices to make and receive voice calls and text messages, browse the internet, and send and receive emails while visiting another country.

Roaming extends the coverage of the home operator's retail voice and messaging services, allowing the mobile user to continue using their home operator phone number and data services within another country. The seamless extension of coverage is enabled by a wholesale roaming agreement between a mobile user's home operator and the visited mobile operator network [8].

4.4. *Alternatives*

We have chosen five alternatives for comparison. All the characteristics with respect to our criteria and subcriteria are listed in Appendix A.

Where, A stands for alternative

A1: Verizon
A2: AT&T

A3: T-Mobile
A4: Sprint
A5: Cricket

4.5. *Assumptions*

- All survey participants are considered experts who have sufficient knowledge about all the alternatives under study.
- Our decision model is not focused towards a particular group of people.
- Our research is limited to only five leading service providers in Portland.

4.6. *Decision model for cellphone service providers*

Our HDM decision model consists of four levels of hierarchy as demonstrated in Figure 3.

5. Implementing the Model

After deciding on the methodology and criteria/subcriteria we created a model in the HDM and sent the survey to different experts who have used or possess knowledge about our alternatives.

The results in Table 1 show the average of each criterion specifying the feature(s) considered important by the experts.

We collected the data of all the users and calculated the average of their scores in all the primary criteria. According to our expert panel the importance for each criterion in terms of the average is shown in Table 1.

Figure 4 graphically summarizes Table 1, where Features is preferred more when choosing the service provider, followed by Cost, Technical Specification and then Customer Service.

Similarly we arranged the data for the subcriteria level and calculated an average for each subcriterion. Table 2 shows the data.

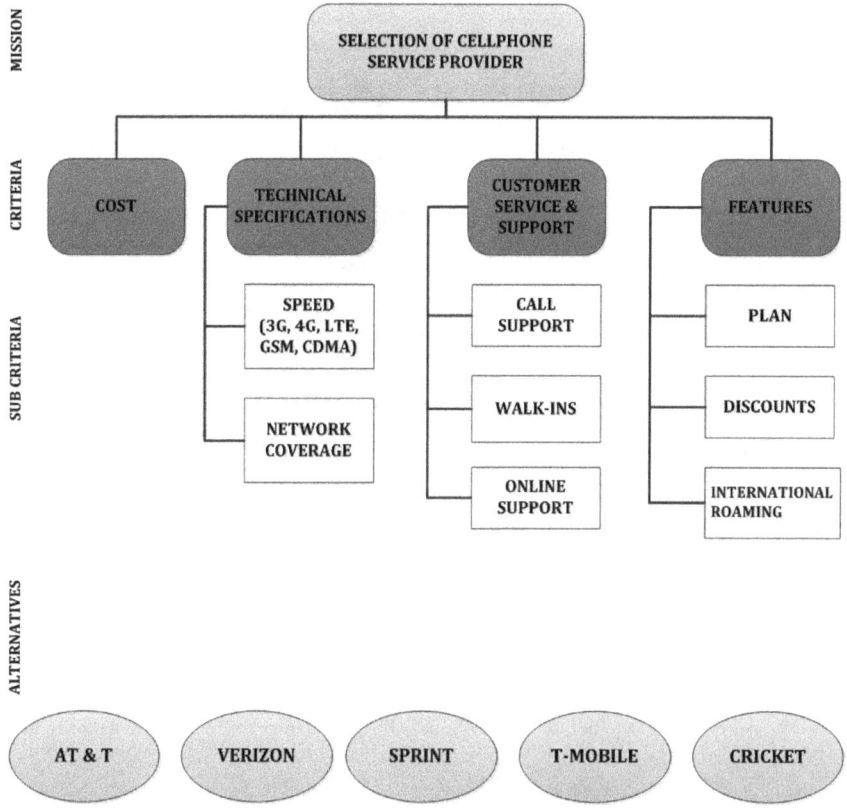

Figure 3. Cellphone service provider decision model.

Table 1. Level 1 (average).

Criteria	Cost	Technical Specification	Customer Service	Features
Average	0.286	0.227	0.1815	0.306

After analyzing the data categorically, we came up with the following charts under the subcriteria.

The chart in Figure 5 exhibits the preferences of experts under Technical Specification. Our analysis shows that experts prefer Network Coverage to Speed.

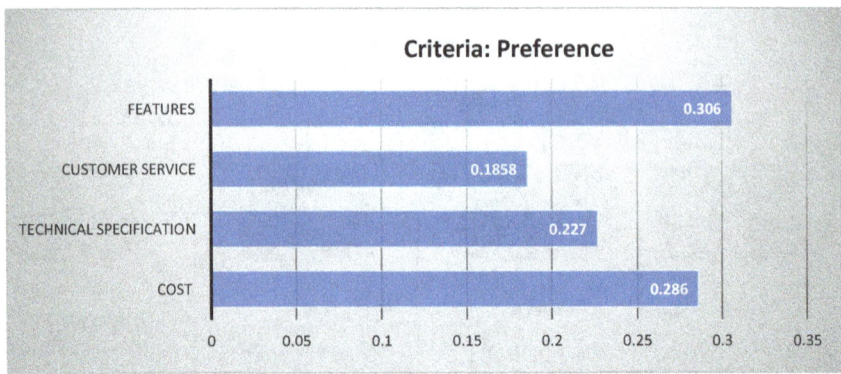

Figure 4. User preferences.

Table 2. Level 2 (average).

Sub criteria	Cost	Speed	Network Coverage	Online Support	Call Support	Walk-ins Support	Plan	Discounts	International Roaming
Average	1	0.43	0.57	0.3015	0.373	0.349	0.415	0.389	0.198

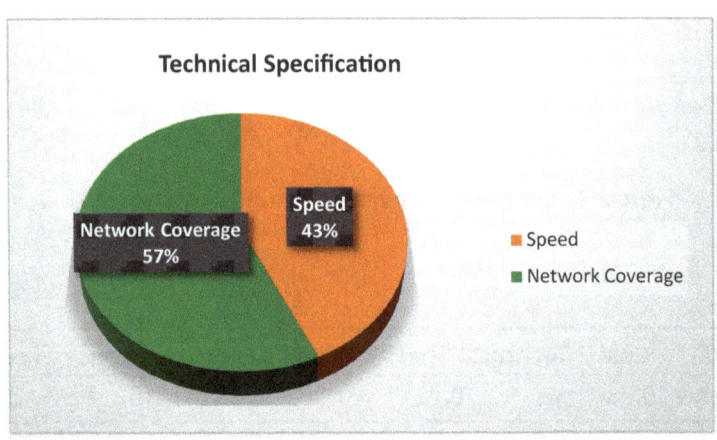

Figure 5. User preference—technical specification.

The chart in Figure 6 shows the preferences of experts under Customer Service and Support. We found out that experts prefer Call Support to Walk-ins and Online Support.

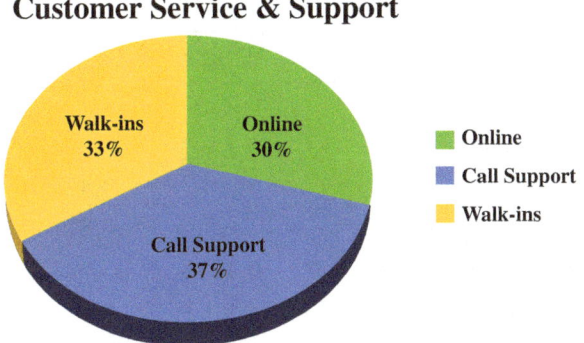

Figure 6. User preference—customer service and support.

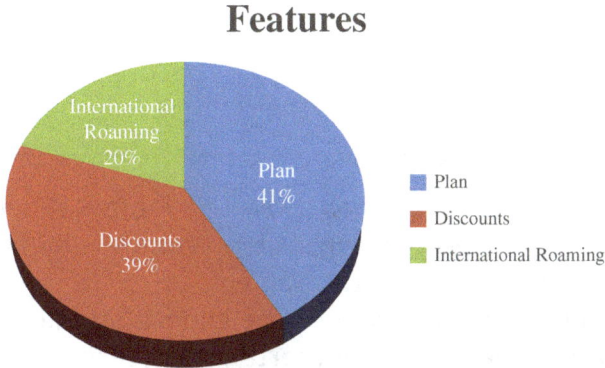

Figure 7. User preference—availability.

The chart in Figure 7 exhibits the preferences of experts under the subcriteria Features. Our analysis shows that experts prefer suitable plans to Discounts, then International Roaming.

The above analysis gave the importance of each criteria and subcriteria as per the expert panel. The comparison determined the relative importance for each criterion with respect to the main objective, and then we determined the relative weight that each subcriterion has from the main criterion.

Table 3. Final output data.

Criteria	Weight	Relative	Final
Cost	0.286		0.28
Technical Specification	0.23		
Speed		0.43	0.0989
Network Coverage		0.57	0.1311
Customer Service and Support	0.1815		
Online Support		0.3	0.05445
Call Support		0.37	0.067155
Walk-ins		0.35	0.063525
Features	0.306		
Plan		0.41	0.12546
Discounts		0.39	0.11934
International Roaming		0.19	0.05814
			0.99807

Multiplying these weights as shown in Table 3 shows us how much each of the defined subcriterion compensate in the decision making for the ultimate goal.

The pairwise comparison of the experts' opinion was used to determine the relative importance of each criterion. The outputs of the comparison are as shown in Table 4. All the pairs have been listed in Appendix B.

Table 4 shows that the 20 expert users had a disagreement value of 0.11. We analyzed the results and found out that three expert users were more inconsistent with respect to others. Those experts are—Expert 8, Expert 12 and Expert 17.

Inconsistency can be explained as:

Imagine that the software asks you to enter three comparisons. You answer:

$$A = 3 * B$$
$$B = 2 * C$$

Table 4. Pairwise comparison of the experts' opinion.

Cellphone Service Provider	Verizon	AT&T	T-Mobile	Sprint	Cricket	Inconsistency
Expert 1	0.22	0.21	0.2	0.18	0.19	0.02
Expert 2	0.85	0.05	0.05	0.03	0.02	0.04
Expert 3	0.29	0.26	0.17	0.15	0.12	0.05
Expert 4	0.24	0.28	0.18	0.17	0.13	0.01
Expert 5	0.21	0.18	0.14	0.11	0.35	0.03
Expert 6	0.25	0.19	0.19	0.18	0.18	0
Expert 7	0.21	0.21	0.21	0.21	0.16	0
Expert 8	0.03	0.07	0.61	0.24	0.05	0.17
Expert 9	0.19	0.19	0.18	0.14	0.3	0.02
Expert 10	0.11	0.13	0.13	0.51	0.12	0.02
Expert 11	0.16	0.45	0.18	0.1	0.1	0.06
Expert 12	0.14	0.12	0.14	0.1	0.1	0.16
Expert 13	0.18	0.21	0.17	0.23	0.2	0.01
Expert 14	0.34	0.27	0.18	0.12	0.1	0.08
Expert 15	0.29	0.4	0.1	0.11	0.11	0.05
Expert 16	0.16	0.18	0.22	0.22	0.22	0.03
Expert 17	0.11	0.23	0.21	0.27	0.19	0.19
Expert 18	0.12	0.21	0.55	0.08	0.04	0.1
Expert 19	0.18	0.23	0.23	0.18	0.18	0.01
Expert 20	0.22	0.2	0.2	0.19	0.18	0
Mean	0.23	0.21	0.21	0.18	0.15	
Minimum	0.03	0.05	0.05	0.03	0.02	
Maximum	0.85	0.45	0.61	0.51	0.35	
Std. Deviation	0.16	0.09	0.13	0.1	0.08	
Disagreement						0.11

The results will be totally consistent only if that answer to that was A = 6 * C. Any other value will be inconsistent (a little, if A = 5 * C but more if A = C) [2].

Hence, in an effort to reduce the disagreement we tried to analyze the results by eliminating the above-mentioned three users. We found that the disagreement value changed to 0.109

Table 5. Effect on disagreement after removing the inconsistent experts.

Cellphone Service Provider	Verizon	AT&T	T- Mobile	Sprint	Cricket	Inconsistency
Expert 1	0.22	0.21	0.2	0.18	0.19	0.02
Expert 2	0.85	0.05	0.05	0.03	0.02	0.04
Expert 3	0.29	0.26	0.17	0.15	0.12	0.05
Expert 4	0.24	0.28	0.18	0.17	0.13	0.01
Expert 5	0.21	0.18	0.14	0.11	0.35	0.03
Expert 6	0.25	0.19	0.19	0.18	0.18	0
Expert 7	0.21	0.21	0.21	0.21	0.16	0
Expert 9	0.19	0.19	0.18	0.14	0.3	0.02
Expert 10	0.11	0.13	0.13	0.51	0.12	0.02
Expert 11	0.16	0.45	0.18	0.1	0.1	0.06
Expert 13	0.18	0.21	0.17	0.23	0.2	0.01
Expert 14	0.34	0.27	0.18	0.12	0.1	0.08
Expert 15	0.29	0.4	0.1	0.11	0.11	0.05
Expert 16	0.16	0.18	0.22	0.22	0.22	0.03
Expert 18	0.12	0.21	0.55	0.08	0.04	0.1
Expert 19	0.18	0.23	0.23	0.18	0.18	0.01
Expert 20	0.22	0.2	0.2	0.19	0.18	0
Mean	0.24823529	0.2264706	0.19294118	0.171176471	0.15882353	
Minimum	0.11	0.05	0.05	0.03	0.02	
Maximum	0.85	0.45	0.55	0.51	0.35	
Std. Deviation	0.16629174	0.0919199	0.10233308	0.102340263	0.08320775	
Disagreement						0.109218541

from 0.11 (see Table 5). Even though the change is minor, we observed that the higher the difference in inconsistency, the higher the disagreement.

We removed the three inconsistent users to develop our understanding about inconsistencies and their effect on the disagreement figure. However, during the actual analysis of the data we did not eliminate the users and used their data as it was.

6. Result

The overall score value is the product of the user preference value and the values of the criteria. Then, they are summed together to give the weighted score and a decision is reached.

Table 6 shows the results and values of each alternative.

The results can be graphically shown in Figure 8.

The final column, which indicates the total value, is the value of the alternative with respect to the perfect situation we have determined. We ranked five alternatives to make an informed decision, as shown in Table 7. The table summarizes that *Verizon* is the best alternative with a weight of 0.22885757.

7. Conclusion

In this paper, we have established a HDM for selecting the best cellphone service provider in Portland. Hierarchy Decision Model (HDM) is used for the selection process. Our hope for this research paper is to find a model that will match the desire of the service provider chooser. Anyone can use this model as a tool to select the desired service provider.

Because of limited time and sample size we could not verify our results on a larger scale. But our model can be modified/altered to suit the preferences of the person(s) interested. The alterations could be: selecting a cellphone service provider on a per-device availability, discounts availability, number of lines with data plan, income specific, and many more. We believe that the model can be finetuned or adjusted in the future once more cases are tested.

8. Further Research

Although we have not included the utility curve in our decision model, using the utility curve methodology can further

Table 6. Result table.

Alternatives\Weight	Cost	Speed	Network Coverage	Online Support	Call Support	Walk-in Support	Plan	Discounts	International Roaming	Total Value
	0.28	0.1311	0.0989	0.05445	0.0671	0.0635	0.12546	0.11934	0.05814	0.99799
Verizon	0.0647	0.0295631	0.0271975	0.012578	0.0150975	0.013081	0.0281	0.0253598	0.0131978	0.22885757
AT&T	0.051	0.0385434	0.0247745	0.012714	0.0161711	0.0148908	0.02559	0.02351	0.013721	0.22087864
T-Mobile	0.0581	0.0226148	0.0181976	0.014293	0.013722	0.014478	0.03111	0.0260161	0.0133722	0.21190783
Sprint	0.0545	0.0234014	0.0176042	0.00795	0.0114406	0.0112713	0.02196	0.0215409	0.0090408	0.17866419
Cricket	0.0561	0.0171086	0.0114724	0.007079	0.0101321	0.0099695	0.01863	0.023152	0.0089826	0.16266645

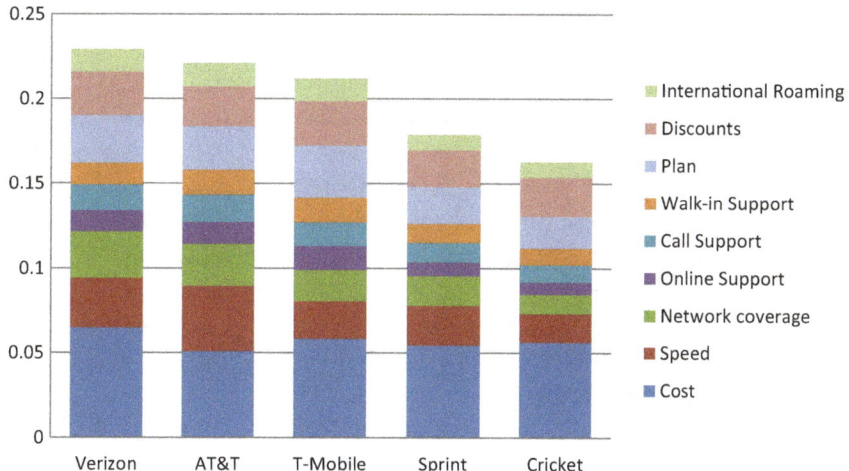

Figure 8. Graphical analysis.

Table 7. Alternatives.

Alternatives	Weight
Verizon	0.22885757
AT&T	0.220878635
T-Mobile	0.211907825
Sprint	0.17866419
Cricket	0.16266645

enhance the usability of our model by adding any additional alternatives, if necessary.

References

1. V. Beal, "Moore's Law". http://www.webopedia.com/TERM/M/Moores_Law.html. Accessed: 1 March 2015.
2. S. H. Mann and E. Triantaphyllou, "Using the Analytic Hierarchy Process for decision making in engineering applications: Some

challenges", *International Journal of Industrial Engineering Applications and Practice,* **2**, 1 (1995) 35–44.
3. D. F. Kocaoglu, "MOGSA decision hierarchy", Portland, 2010.
4. K. Phan, "ETM 530 decision making—MOGSA decision hierarchy", 12 February 2015.
5. "The method of pairwise comparisons". http://www.ctl.ua.edu/math103/Voting/methodpc.htm. Accessed: 5 March 2015.
6. "Service plan", *Phonescoop.* http://www.phonescoop.com/glossary/term.php?gid=35. Accessed: 3 March 2015.
7. "Cell phone plans and getting the right deal", *whistleOut.* http://www.whistleout.com/CellPhones/Cell-Phone-Plans-Buying-Guide. Accessed: 2 March 2015.
8. "International roaming explained", *GSMA,* August 2012. http://www.gsma.com/publicpolicy/wp-content/uploads/2012/09/Africa-International-roaming-explained-English.pdf. Accessed: 1 March 2015.

Appendix A: Service Providers

Service Provider	Pros	Cons
verizon	❖ Unlimited talk time ❖ Unlimited text ❖ Shareable data ❖ Personal hotspot ❖ International messaging ❖ 25 GB cloud storage ❖ Largest coverage	❖ CDMA network: Most of the devices will not work in many places around the world ❖ Costliest plans

Service Provider	Pros	Cons
at&t	❖ Unlimited talk time ❖ Unlimited text ❖ Shareable data ❖ Rollover data ❖ Unlimited international texting ❖ 50 GB cloud storage ❖ GSM network: More success with international usage of devices	❖ Second-most expensive ❖ Comparatively less coverage than Verizon
Sprint	❖ Unlimited talk time ❖ Unlimited text ❖ Shareable data ❖ Switches from Verizon / AT&T networks provide further incentives, such as reducing the bill by half	❖ Limited network, as opposed to Verizon and AT&T
T-Mobile	❖ GSM network ❖ Prepaid plans ❖ Cheaper plans ❖ Unlimited data service	❖ Network suffers as compared to Verizon and AT&T ❖ Its family plan is the most expensive among the service providers

Service Provider	Pros	Cons
cricket wireless	❖ Unlimited talk time ❖ Good coverage in Portland, but limited national coverage ❖ Pay-you-go plans	❖ Roaming/long-distance plans not available ❖ Plans not available for smartphones

Appendix B

Criteria comparison

Experts: _____

Allocate

Use pairwise comparison to quantify your judgment. Allocate a total of 100 points to express your judgment about the ratio of one criterion to the other one in pair.

Criterion	Weight	Weight	Criterion
Cost			Technical Specification
Cost			Customer Service and Support
Cost			Availability
Technical Specification			Customer Service and Support
Technical Specification			Availability
Customer Service and Support			Availability

- Technical specification—it includes network and speed.
- Customer service and support—it includes walk-ins, online support and call support.
- Availability—it includes, plan, discount and international roaming.

Subcriteria comparison
(with respect to Technical Specification)

Criterion	Weight	Weight	Criterion
Speed			Network Coverage

- Speed—3G, 4G, LTE, GSM and CDMA
- Network coverage

Subcriteria comparison
(with respect to Customer Service and Support)

Criterion	Weight	Weight	Criterion
Online Support			Walk-in
Call Support			Online Support
Walk-in			Online Support
Walk-in			Call Support

- Online Support
- Call Support
- Walk-in

Subcriteria comparison
(with respect to Availability)

Criterion	Weight	Weight	Criterion
Plan			Discount
Plan			International Roaming
Discount			International Roaming

- Plan
- Discount
- International Roaming

Appendix C

Cellphone Service Provider Survey

Please take a moment to help us gather information on your cell provider. Please note that this information will be kept strictly confidential.

Name

Age

Gender (Optional)

1. How important are/were each of the following attributes in your decision for plan selection? In the order of your preference rate the below on **scale of 1 to 10 (10 being highest and 1 being the lowest)**

Attributes	Scale of Importance
Network Coverage	
Plan Cost	
High Speed Data	
Talk Time	
Text	
Quality of Service	
Discounts (Corporate, Student, Senior Citizen)	
Customer Service	

Part 3
Organizational Transformation

Chapter 14

Organizational Transformation: Semiconductors

Tejas Deshpande* and Tugrul Daim*,†,‡

*Portland State University, Portland, Oregon, USA
†Higher School of Economics, Moscow, Russia
‡Chaoyang University of Technology, Taiwan

Abstract

This study aims to calculate the innovativeness index of Lam Research Corporation, a global supplier and one of the largest manufacturers of semiconductor processing equipment since 1980. This study proposes to apply Phan's innovation measurement framework [1] to Lam Research to estimate its innovativeness index.

The methodology used to construct this framework is a Hierarchical Decision Model (HDM). The HDM used in this research basically divides the problem into three hierarchies—the first level is its mission, i.e., to find the innovativeness index of a company; the second level uses different output criteria contributing to the innovativeness of a company; the third level subcategorizes these output criteria into sub-factors. The HDM uses three expert panels to identify different output criteria contributing to the innovativeness of a company, to evaluate their importance relative to each other, and provide desirability values for each of them. All the experts selected for this study are experienced professionals in the semiconductor industry with different areas of specializations.

Application of this model to Lam Research includes collecting data with respect to all the output factors described in the framework that contribute to the innovativeness of the company, and calculating the score using desirability values and weights estimated by the experts in an innovation measurement framework. The innovativeness index of Lam Research was calculated as 68.96, which is a good score compared to the other companies mentioned in dissertations such as Intel and AMD. Further analysis of these results shows that Lam Research shows strength in Revenue and Market share by new products as well as number of new products introduced every year. However, the highest possible score for Lam Research could have been 73.18 but for data regarding some indicators, such as the number of paper presentations, awards and honors won by Lam Research employees, not being available for evaluation. In addition to this analysis, the innovativeness index of this company can be improved through more publications and patents, and by encouraging a greater number of scientists to conduct researches in different fields of study.

Keywords: Technology assessment, semiconductor manufacturing, innovation.

1. Introduction

Measuring the innovativeness of any company is a subjective approach and has been attracting many researchers to come up with a structured and standardized framework. As innovation is one of the most important factors for surviving in the business environment these days, companies need to know where they stand in terms of innovation capabilities. Companies cannot improve their innovativeness until they are able to measure it [2].

Considering the fast-paced growth of the semiconductor industry for the last 50 years, the most important factor for this study is product innovation. Companies in the semiconductor industry spend significant amount of their profits on Research and Development (R&D) to gain competitive advantage and to survive in the race [3, 4]. Hence, it becomes important to measure the innovativeness of a company in order to provide insights regarding a company's innovation capabilities and where it stands in the competition in terms of innovation and development practices [2, 5].

This study operates within Phan's innovation measurement framework. This framework hopes to calculate the innovativeness index of Lam Research Corporation, a global supplier of water fabrication equipment and services and one of the world's leading companies in the semiconductor industry. The paper computes the score of the company's innovation capabilities by collecting its data with respect to described factors contributing to innovation. It also provides some insights as to how the company can improve its score.

The innovation measurement framework uses the HDM methodology to quantify the company's innovation index. The methodology comprises of three main stages: the first stage involves the formation of expert panels to decide the output criteria contributing to the innovativeness of a company and then ranking the importance of these factors by weighting them using pairwise comparisons in the HDM; the second stage includes collecting the company's data in terms of the factors described in the framework of innovation measurement; the last stage includes defining the desirability function for each output criteria and computing the company's innovativeness using a weighted sum. The experts recording their responses to design the framework are experienced professionals with different areas of specializations in the semiconductor industry around the world.

1.2. *Project objective*

The objective of this project is to apply Phan's research model/framework to determine the innovativeness of Lam Research Corporation.

The purpose of this study is to collect data with respect to all the output criteria described in the framework from Lam Research and to calculate its innovativeness index. This study further analyzes the results built by the framework under certain scenarios and provides insights to Lam Research on how it can improve its innovativeness index. This study aims to analyze Lam Research's innovation competencies and offers suggestions on how the company can improve these competencies.

2. Background and Framework Overview

The study focuses on product innovation [31]. The results of this research not only provide some insights to the company about their innovation activities, areas for improvement but also help them compare their score with competitors.

The framework described in the research paper calculates the innovativeness of a company in a numeric form. The main purpose of this research was to focus on technological product innovation in a technology-driven industry, such as that of semiconductors. The model described in the study is standard and can be applied to many other companies within the industry. It can also provide comparisons between different competitors. The demonstration in Phan's research paper will help us to understand how this model works by applying it to a few hypothetical semiconductor companies and comparing their scores to see where they stand in terms of innovation.

2.1. *Framework of innovation measurement*

Innovation is vital in the business environment as it gives a company an edge over its competitors. Because of globalization and technological and knowledge revolutions, innovation has become the most important factor in strategic planning for any business [6]. Companies, especially in industries such as semiconductors where product innovation is crucial to survival, are spending tremendous amount of resources to innovate [3, 5]. However, it is important to have a consistent and general framework for innovation measurement, in order to allow companies to measure their innovativeness in comparison with other companies and thus improve accordingly. Hence, quantifying and measuring the innovativeness of a company have attracted much interest from researchers, who have tried to develop many mathematical models and performance metrics to measure innovation in a structured way [2].

Being a complex phenomenon, innovation cannot be measured though a single indicator. Additionally, there are different types of innovation including products, processes, marketing innovation, etc. [7]. This study only focuses on product innovation. Phan's research paper compiles a large number of output indicators outlining innovativeness, amongst which the most important ones are selected using the experts' judgments, so as to avoid the complexity of the model. The framework developed for the semiconductor industry where product innovation is so important uses output indicators such as the number of new products invented, actual revenue, market share and patents granted, which can show the results of a company's innovation efforts compared to using input indicators such as a company's investments in R&D. Many researchers until now have been using input indicators to know the innovation capabilities of a company. However, input indicators can give biased results because a company can control these, (i.e., although a company's spending on R&D can be an input indicator to gauge the innovativeness of a company, it does not guarantee actual innovative output/products no matter how much resources are spent). In addition, finding all the input factors is not straightforward and can lead to biased results. Hence, outputs that transform a company's innovation inputs into economical values are more appropriate measures of innovativeness [2].

2.2. *Methodology used*

The HDM is used to quantify the experts' judgments into an innovation index.

This methodology comprises of three stages:

1. HDM development;
2. Indicator evaluations;
3. Innovativeness measurement.

2.1.1. *How does the HDM work?*

The HDM is a subjective approach that uses the principle of decomposing a problem into hierarchies. It is a comprehensive, logical and organized framework used to quantify an expert's judgment by using pairwise comparison. The pairwise comparison method weighs each criterion in each hierarchy relative to each other [2]. The constant sum method is used for judgment quantification where each expert assigns their subjective judgments by distributing a total of 100 points between one pair of output factors at a time. The result of this comparison is tested for inconsistencies in the experts' judgment. The recommended value of inconsistency for this model is between 0.0 and 0.1.

Agreement/disagreements between the experts are tested using an intraclass correlation coefficient and an *F*-test as well. The higher the intraclass correlation coefficient is, the higher the level of agreement between experts. It can fall between +1 and 0; all negative values are considered "0" to define disagreements between experts. The *F*-test is performed by defining a null hypothesis where the critical value of 0.01 is used to ensure the high level of agreement between experts [2].

2.1.2. *Expert panels*

This study used three expert panels comprising experienced professionals in different areas of specializations in the semiconductor industry around the world. These experts, who have expertise in the field of innovation in the semiconductor industry, consist of researchers, scientists, managers, engineers and government personnel.

1. Expert Panel 1 (30 experts) will identify all the output indicators to be used in this framework.
2. Expert Panel 2 (36 experts) will provide quantified judgments (or weights) to decide the relative importance/ranking of all the factors in each hierarchy level.
3. Expert Panel 3 will develop desirability functions by providing their responses on the scale of 0 to 100 for each criterion.

To test the validity of this research and model, the experts were asked to verify if this model could be generalized and applicable to measure the innovativeness index of any company. The Delphi method was used in order to structure a systematic communication between the three expert panels. The purpose of this method is to prevent individual biases on the results/model. It facilitates several iterations of the judgments until it becomes stable [2].

The distribution and background of all the experts used in this paper are shown in Appendices A–C.

2.1.3. *HDM development*

The HDM was designed by collecting the responses recorded by Expert Panel 1, whose members identified all the important output indicators and sub-factors contributing to the innovativeness of a company.

The breakdown of the HDM is shown in Figure 1. The first level of an HDM is the mission of the study, i.e., finding the innovativeness index of a company. The second level shows all the output indicators contributing to a company's innovativeness and the last level shows the breakdown of all output indicators into sub-factors, which will give a more comprehensive measure of innovativeness.

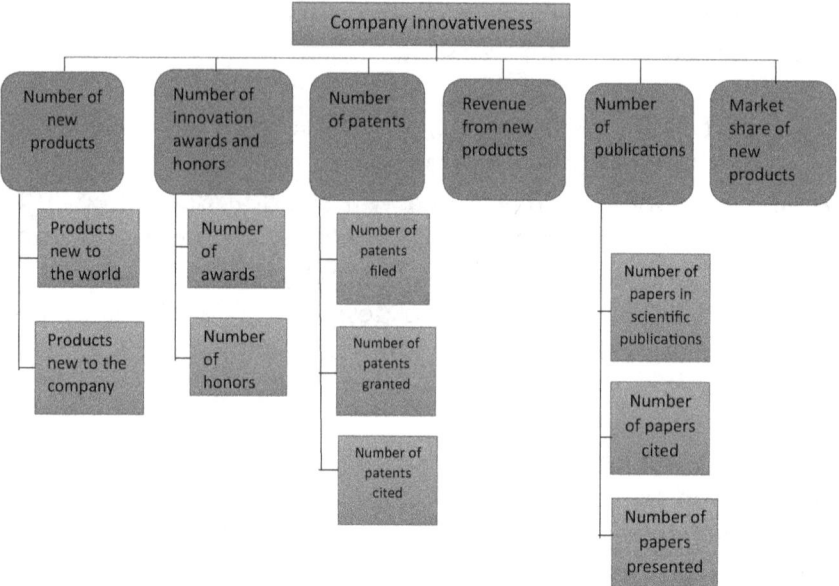

Figure 1. The HDM to find the innovativeness index of a company.

2.1.4. *Indicator evaluation and weights*

Expert Panel 2 was asked to rank the importance of each output indicator and sub-factor using pairwise comparison in the HDM. After collecting data from Expert Panel 2 and then testing it for inconsistency in the judgments, weights were determined for each indicator and sub-factor. Table 1 shows the relative importance or ranking of all the estimated output indicators.

2.1.5. *Desirability values*

Expert Panel 3 was asked to provide the desirability values for each sub-factor in the HDM. The value given was between 0 and 100. To estimate the final desirability value for each factor in the model, the arithmetic mean of all judgments was used and tested for statistical significance using the *F*-test [2].

Appendix D shows all the desirability curves derived from Expert Panel 3's results. These desirability functions are used to

Table 1. Relative importance/ranking of all the output factors and sub-factors.

Output Indicators	Value	Sub-Factors		Sub-Factors Value to the Innovativeness Index
Number of New Products	0.2	New to the World	0.66	0.132
		New to the Company	0.34	0.068
Number of Innovation Awards	0.09	Number of Awards	0.5	0.045
		Number of Honors	0.5	0.045
Number of Patents	0.14	Patents Granted	0.6	0.084
		Patents Filed	0.26	0.036
		Patents Cited	0.14	0.02
Revenue From New Products	0.28	Revenue from New Products	1	0.28
Number of Publications	0.07	Papers Published	0.55	0.039
		Papers Presented	0.27	0.019
		Papers Cited	0.17	0.012
Market Share of New Products	0.21	Market Share of New Products	1	0.21
Total	1			1

calculate Lam Research's innovativeness index according to respective values on the curve.

2.1.6. *Sensitivity analysis*

A sensitivity analysis was performed to find the allowable range of tolerance and perturbations, and the sensitivity coefficient derived represents the acceptable range for each output indicator to maintain the priority of each sub-factor [2]. Table 2 shows the sensitivity for each output factor in the

Table 2. Sensitivity coefficients for all factors.

	Number of New Products	Number of Innovation Awards	Number of Patents	Revenue from New Products	Number of Publications	Market Share of New Products
Relative Importance	0.2	0.09	0.14	0.28	0.08	0.21
Allowable Ranges of Perturbations	[–0.051, 0.0079]	[–0.0092, 0.0433]	[–0.0453, 0.0487]	[–0.0542, 0.72]	[–0.07, 0.0091]	[–0.00798, 0.0517]
Tolerance	[0.149, 0.2079]	[0.0808, 0.1333]	[0.0947, 0.1887]	[0.2258, 1]	[0, 0.0791]	[0.202, 0.2617]
Sensitivity Coefficient	16.78	19.05	10.64	1.292	12.64	16.75

framework. It also allows the framework to analyze whether changes in that indicator affect the ranks of other output factors.

2.2. *Demonstration of the model*

A demonstration of this research is nothing but the application of a model to different technology-driven companies where performance metrics are developed and the innovativeness index of any given company is calculated using desirability values on the curve for each indicator.

A simulated application of the model illustrates five hypothetical semiconductor companies with different profiles. Using desirability values and weights of the respective performance metric (weighted sum), the innovativeness index of each company is calculated. To check whether the changes of priorities/importance of output indicators affect the innovativeness of a company, extreme scenarios are applied on the model by changing the weights of different factors. For some factors and sub-factors, a minor change in the weights has a high impact on the results [2].

3. Lam Research Overview

Lam Research Corporation is a global supplier in the semiconductor industry engaged in the design and manufacture of semiconductor processing equipment used in the fabrication of integrated circuits [8]. Being a fundamental enabler of silicon technology, almost all the advanced chips used to make electronic devices are built using Lam technology. The company was founded in 1980 and is headquartered in Fremont, California. As of now, it is one of the largest manufacturers in the industry [8–10].

The company's innovative water fabrication equipment and services have allowed chipmakers to develop smaller, faster and better functioning electronic devices. Its mission is to be dedicated to the success of its customers by being a world-class provider of innovative technology and productivity solutions to the semiconductor industry [8].

Lam Research Corporation always strives to provide their customers with the most efficient and updated technology and solutions through its constant innovation and improvement processes. Today, its semiconductor practices and solutions used for chipmaking are among some of the most advanced and challenging processes in the industry that push the limits of physics and chemistry with extremely progressed features such as complex 3D structures, nano scalability, etc. [8]. To ensure new processes and technologies, Lam Research invests a significant amount of their profits in R&D, making sure they deliver the most updated innovative products and solutions to their customers and partners [11]. As a result, there is constant growth in their revenue, market shares and other output factors, thus justifying their innovativeness [12–14]. But, it is also important to know Lam Research's innovativeness index so that the company can improve its innovation practices. Hence, this framework was proposed to the company's management as being able to calculate the company's innovativeness score and provide insights and recommendations, accordingly. To compute the innovativeness index of Lam Research, we focused on the output factors described in the framework for the company, with respect to innovation activities. The following tangibles were considered as the results/output criteria to define the company's innovation capabilities.

3.1. Growth in revenue and market share

According to Lam Research's records, their revenue grew more than three times the industry growth in 2016 and continued to grow substantially in 2017 and 2018 [14–17]. The company's effective analysis of industry growth, challenges and opportunities made them invest significantly in R&D, which delivered advanced, innovative products and solutions for their customers [12, 18]. In 2016, Lam Research achieved the number one position in the market by achieving more than 50% of the market share. The company's expertise in multiple patterning and continuous focus on reducing complexity has made it a world leader—currently, 95% of wafers in the world are being processed with Lam Research's products and technology [13, 19, 20]. The company recorded $5.9 billion, $8 billion and $11 billion in revenue in 2016, 2017 and 2018, respectively. Lam Research's average stock price has grown substantially from 2016 to 2018, from $85 to $154, then $176 [12, 13, 20, 21].

3.2. Research and development

The semiconductor equipment capital market is considered highly competitive because of rapid technological changes and product innovation. Lam Research's ability to gain competitive advantage solely depends on its continuous and timely development of new products and solutions, for which the company devotes a significant amount of their resources to R&D programs. It also seeks innovation constantly. The company's R&D investments in 2016, 2017 and 2018 were $914 million, $1 billion and $1.2 billion, respectively, and are projected to increase each year to ensure advanced and innovative products to their customers. The majority of the company's R&D spending has been focused on deposition, etching, cleaning and other semiconductor manufacturing products as well [11].

3.3. *Skilled talent pool, patents and publications*

In the semiconductor and semiconductor capital equipment industry, competition for highly skilled employees is strong. Additionally, the success of a company primarily depends on its ongoing research processes, the development in its engineering and manufacturing practices, and improved services. Hence, its ability to attract and retain a large pool of talented and highly skilled employees is crucial. As of 2018, Lam has 10,900 employees globally, amongst which 25% of them possess higher degrees in their respective fields of study such as science, mathematics, engineering and management. These employees are engaged in continuous research in different areas of specialization in the semiconductor field [23–26]. The company believes in encouraging its employees to publish and innovate, and its policy is to seek patents on new or enhanced products/solutions as a part of its ongoing research activities. As a result, Lam Research has filed for 857 patents in the last three years, i.e., from 2016 to 2018, whereas the number of patents granted to Lam Research by the US government was 660 in the same period of time [26–29]. Additionally, Lam Research's employees have authored a total of 824 publications (research papers and articles) in the last three years, amongst which many of them have been cited more than 50 times [30–32].

3.4. *New products and solutions*

As a result of its highly skilled, efficient workforce and its global, high-quality, multi-product manufacturing capability, Lam Research has introduced more than 20 new products on an annual basis and has maintained an approximately 98% on-time delivery performance [33]. The company's innovation and development competencies have allowed them to increase factory efficiency and automation tremendously. Its ongoing commitment to protecting customer IP, ISO 27001, has gained them customer trust and many awards and honors, such as Intel's "Preferred quality supplier award" and Samsung's "Best in Value appreciation" [34–38]. As of now, the company offers a wide range of products and solutions, as can be seen in Table 3 [39].

Table 3. Lam Research's product offerings.

Product Family	Number of Product Offerings
ALTUS product family	6
CORONUS Product family	2
DSIE product family	3
DV-PRIME and DA VINCI	2
EOS	1
FLEX	6
GAMMA	3
KIYO	5
METRYX	2
RELIANT Clean	2
RELIANT Deposition products	5
RELIANT ETCH	6
SABRE 3D	2
SABRE	3
SOLA	2
SP Series	5
Speed product	2
Striker	1
SYNDION	2
VECTOR	5
VERSYS METAL	4

4. Data Collection and Calculations

Lam Research Corporation is a global supplier of semiconductor equipment. The company's consistent R&D practices and its effective analysis of industry opportunities and applications have allowed them to be innovative in its products, solutions and processes. However, knowing the innovativeness index of the company can provide it with a better understanding of its innovation competencies and areas for improvement. The innovativeness index of the company can also be compared with other companies in the industry to gain insights on the output indicators/criteria company that it should excel in to improve its score.

We approached Dr Easwar Srinivasan, Managing Director of Lam Research's Engineering and Global Product Support, with this framework and explained its usefulness to a company. He liked our proposal of applying the innovation measurement framework to Lam Research and finding out the company's innovativeness index. He guided us to some of the data required for this study on his company's behalf. All the information and tangibles collected for this study are taken from the company's website, blogs and newsletters, and are available for public consumption. In addition, information regarding patents and publications was taken from the US government's website of patents and trademarks, as well as from Google Scholar.

This study measures the innovativeness index of Lam Research based on data from the last three years.

Assumptions based on the information and financial report analysis:

1. The total number of researchers in Lam Research makes up 25% of its total employees. According to Dr Srinivasan, almost 25% of Lam Research's total employees possess higher degrees in their respective fields of study. Hence, it is assumed that the total number of researchers and scientists involved in paper publications, patents and research make up 25% of the total employees in Lam Research.

2. According to the *2016 Investor and Analyst Meeting* [12] (slide 51) report, Lam Research introduces more than 20 products per year. Hence, it is assumed that almost 25 new products are introduced every year, where as much as 20% of these products can be also new to the world.
3. The total number of products in three years is assumed to be 100 by considering the current range of product offerings and the introduction of new solutions and processes every year.
4. Since Lam Research products are innovative and highly efficient, and the technology they introduce is fairly new to the world, it can be assumed that the market share and revenue from its new products is almost 75% of the total figure.

Table 4 describes some initial findings of the study.

4.1. *Distribution values and innovativeness index*

The distribution values for each output indicator and sub-factor are estimated from the distribution curve/function derived by the experts in Expert Panel 3. The distribution curves for all indicators are attached in the appendices for reference.

With respect to all the factors in the HDM, the numbers in Table 5 can be estimated by using the data in Table 4. Then, the distribution values for their respective indicators on the curve are identified.

As the weights or rankings for all output indicators and sub-factors have been estimated by the experts in Expert Panel 2 (shown in Table 1) who used distribution values for each indicator, the innovativeness index of Lam Research can be calculated using a weighted sum.

As we can see from Table 6, the information about some output factors is not available. Hence, we ignore these factors in the model and recalculate the score by normalizing the weights or rankings for all the remaining output factors.

Hence, the final model for estimating the innovativeness index of Lam Research is shown in Tables 7 and 8. The innovativeness index of Lam Research is 68.96.

Table 4. Lam Research's company profile.

Output Criteria	2016	2017	2018	Current Total/ Average Number
Total revenue (in billions)	5.9	8	11	24.9
Average stock price	84.1306	153.8091	175.5543	137.8313333
Total number of employees	7,500	9,400	10,900	10,900
Total number of employees with advanced degrees/ researchers (25%)	1,875	2,350	2,725	2,725
Patents issued	189	215	256	660
Patents filed	412	315	130	857
Patents cited	No data available			
Number of papers published				824
Number of papers presented	No data available			
Number of papers cited				824
Total number of products				100
New products (to the company)	25	25	25	75
New products (to the world)	20	20	20	60
The ratio of number of awards to total researchers	No data available			
Number of honors to total researchers	No data available			
Market share of new products				75%
Revenue from new products as percentage of total revenue				75%

Table 5. Lam Research's performance metrics.

No.	Output Criteria	Total/Average Number	Desirability Values
1	New products new to the world as the percentage of total products	60%	75
2	New products new to the company as the percentage of total products	75%	62
3	The ratio of number of awards to total researchers	No data	NA
4	Number of honors to total researchers	No data	NA
5	Number of patents granted per researcher	(1/5)	70
6	Number of patents filed per researcher	(1/4)	72
7	Number of patents cited per researcher	No data	NA
8	Revenue from new products as percentage of total revenue	75%	68
9	Number of papers published per researcher	(1/4)	32
10	Number of papers presented per researcher	No data	NA
11	Number of papers cited per researcher	(1/4)	20
12	Market share of new products	75%	82

5. Results and Analysis

The results in Table 8 show that Lam Research has a good score of 68.96 with respect to its innovativeness. However, data regarding some output factors is not available to evaluate the whole model. Further analysis of this model shows that the

Table 6. Initial score.

No.	Output Criteria	Total/Average Number	Weights	Desirability Values	Initial Score
1	New products new to the world as the percentage of total products	60%	0.13	75	9.75
2	New products new to the company as the percentage of total products	75%	0.07	62	4.34
3	The ratio of number of awards to total researchers	No data	0.05	NA	
4	Number of honors to total researchers	No data	0.05	NA	
5	Number of patents granted per researcher	(1/5)	0.08	70	5.6
6	Number of patents filed per researcher	(1/4)	0.04	72	2.88
7	Number of patents cited per researcher	No data	0.02	NA	
8	Revenue from new products as percentage of total revenue	75%	0.28	68	19.04
9	Number of papers published per researcher	(1/4)	0.04	32	1.28
10	Number of papers presented per researcher	No data	0.02	NA	
11	Number of papers cited per researcher	(1/4)	0.01	20	0.2
12	Market share of new products	75%	0.21	82	17.22
				Innovativeness index	60.31

innovativeness index of Lam Research can be improved up to 73.18 (shown in Table 9).

1. Tables 8 and 9 show that revenue from new products has the highest weightage (0.28) given by all experts. Considering Lam Research's profile and the number of new products

Table 7. Lam Research's innovativeness index.

No.	Output Criteria	Total/Average Number	Desirability Values	Weights	New Weights	New Score
1	New products new to the world as the percentage of total products	60%	75	0.13	0.18	13.5
2	New products new to the company as the percentage of total products	75%	62	0.07	0.12	7.44
3	The ratio of number of awards to total researchers	No data	NA	0.05	0	0
4	Number of honors to total researchers	No data	NA	0.05	0	0
5	Number of patents granted per researcher	(1/5)	70	0.08	0.1	7
6	Number of patents filed per researcher	(1/4)	72	0.04	0.04	2.88
7	Number of patents cited per researcher	No data	NA	0.02	0	0
8	Revenue from new products as percentage of total revenue	75%	68	0.28	0.28	19.04
9	Number of papers published per researcher	(1/4)	32	0.04	0.04	1.28
10	Number of papers presented per researcher	No data	NA	0.02	0	0
11	Number of papers cited per researcher	(1/4)	20	0.01	0.03	0.6
12	Market share of new products	75%	82	0.21	0.21	17.22
					Innovativeness index	68.96

Table 8. Innovativeness index and final model for Lam Research.

No.	Output Criteria	Total/Average Number	Desirability Values	New Weights	New Score
1	New products new to the world as the percentage of total products	60%	75	0.18	13.5
2	New products new to the company as the percentage of total products	75%	62	0.12	7.44
3	Number of patents granted per researcher	(1/5)	70	0.1	7
4	Number of patents filed per researcher	(1/4)	72	0.04	2.88
5	Revenue from new products as percentage of total revenue	75%	68	0.28	19.04
6	Number of papers published per researcher	(1/4)	32	0.04	1.28
7	Number of papers cited per researcher	(1/4)	20	0.03	0.6
8	Market share of new products	75%	82	0.21	17.22
			Innovativeness index		68.96

launched by the company, Lam Research has scored well in this criterion—75% of the company's revenue comes from its new products, which makes the innovativeness index increase significantly as this criterion is chosen to the most important one contributing to the innovativeness of a company.

2. The market share from new products is the second-most important criterion with the weightage of 0.21, and Lam Research is assumed to be having 75% of its market share coming from new products, which is a pretty good number to have contributing to the innovativeness of a company, although the highest possible score for this factor would be 80% of its market share achieved by new products.

3. The third important criterion chosen is the number of new products that are new to the world. Though 60% is an

Table 9. Maximum possible score for Lam Research.

No.	Output Criteria	Current Performance Metrics	Current Desirability Values	Possible Weights	Possible Desirability Values	Possible Performance Metrics	Max. Possible Score
1	New products new to the world as the percentage of total products	60%	75	0.18	75	60%	13.5
2	New products new to the company as the percentage of total products	75%	62	0.12	63	80%	7.56
3	Number of patents granted per researcher	(1/5)	70	0.1	77	More than 1 researcher	7.7
4	Number of patents filed per researcher	(1/4)	72	0.04	76	More than 1 researcher	3.04
5	Revenue from new products as percentage of total revenue	75%	68	0.28	68	75%	19.04
6	Number of papers published per researcher	(1/4)	32	0.04	50	(1/20)	2
7	Number of papers presented per researcher	No data	No data	0.02	52	(1/20)	1.04
8	Number of papers cited per researcher	(1/4)	20	0.01	40	(1/20)	0.4
9	Market share of new products	75%	82	0.21	90	80%	18.9
							73.18

assumption based on the company's financial reports and newsletters, Lam Research is seen to have the highest desirability value for this criterion.

4. The number of products that are new to the company represents the next important factor contributing to the

innovativeness of a company, and Lam Research shows strength in this criterion with a good desirability value for 75% of its new products. However, this score can be improved if the number of products introduced by the company is increased by 5%.
5. The ratio of the number of patents granted/filed to the number of researchers in the company represents a good figure, i.e., on average a patent is filed per five researchers in a company. This number can also be improved by encouraging more researchers or scientists in a company to conduct research in different fields. These individuals are then also encouraged to publish their work. The maximum possible score in this criterion can be achieved if there is more than one patent per researcher in a company.
6. The number of papers published by researchers and the citations made show a relatively lower value in contribution to the innovativeness index of a company. Value can be added to its innovation capabilities by having employees with higher degrees and company scientists that conduct different researches and then publishing their work. Hence, it is important for a company to promote work published by its researchers. As of 2018, Lam Research has 10,900 employees globally, amongst which 25% are assumed to be researchers and scientists possessing higher degrees. This percentage can be improved by encouraging and providing resources to a greater number of employees to specialize in different fields of study, innovate, and then publishing their work.
7. In the case of "number of papers presented by Lam Research employees in scientific conferences/seminars", Lam Research has no score. This can be improved by more presentations and publications.
8. Overall, Lam Research has a good innovativeness index compared to the other companies described in Phan's research paper. However, data regarding awards and honors won by researchers in a company could have added more value to the innovativeness index.

6. Conclusion and Recommendations

Overall, Lam Research shows a good score in terms of innovativeness of the company. The company's innovativeness index is estimated to be 68.96 (Table 8), which is above average and better than the other companies described in Phan's research paper. However, the company's maximum possible innovativeness index can be 73.18 (Table 9) if more publications, patents and new products are launched every year.

1. Twenty-five per cent of Lam Research's total number of employees possess higher degrees and are involved in R&D. Though this percentage is above average when compared to other companies in the industry, encouraging a greater number of employees to specialize in their respective fields of study can certainly improve this percentage. Ultimately, the number of publications, paper presentations, patents and inventions through different researches and studies can also be improved. It can be seen from Table 9 (points 6, 7 and 8) that the company needs to have a greater number of researchers to improve its score.
2. Encouraging a greater number of employees to specialize in different areas of study and applying their knowledge to innovate can result in better R&D practices. Seeking patents for different inventions and enhanced studies by researchers can inspire others to do likewise. These new inventions and researches can ultimately boost the innovation process and the company can introduce a greater number of products each year in order to gain the highest possible share through new products.
3. Additionally, the number of paper publications by Lam Research scientists represents a relatively less score. The value for the "number of papers presented" criterion is also seen to be "0", as data for this factor is not available for evaluation. Hence, it is recommended that the company encourage its researchers to present more papers and

publications at seminars and conferences. By doing so, it can motivate other employees and improve the innovativeness score of the company as well.

It can be concluded that Lam Research shows strength in most of the output factors that contribute to its innovativeness index, as compared to the other companies mentioned in Phan's research paper [31]. However, the highest possible score for the company could have been 73.18 had there been a greater number of researchers, publications, patents and awards won by the employees in the company. The company can consider working on different policies to motivate their scientists and researchers to publish and present their works. This will not only improve its innovativeness index but also help in motivating other employees to specialize in different areas and add value to innovation through their researches.

7. Limitations of the Study

The framework of innovation measurement can structure the way the innovativeness of a company should be measured. Applying this framework to Lam Research certainly provided some insights on how the company could improve its innovation practices and development activities. However, there are some limitations to this study:

1. The model considers a three-year time frame to calculate the innovativeness of a company. However, it can be possible for the company to have breakthrough innovations outside the years considered in the study. As a result, the company will miss out on some important data and output indicators that could have been measured otherwise.
2. Acquiring complete information on all the output indicators mentioned in the model is not straightforward. For example, data regarding some of the output criteria

contributing to the innovativeness index of the company may not be available. Hence, the model/ framework needs to be modified or normalized against the available data.
3. The model only considers output indicators to measure the innovativeness index of the company. However, combining it with input indicators such as the company's investments on innovation activities can give better results.

References

1. K. Phan, "Innovation measurement: A decision framework to determine innovativeness of a company", PhD dissertation, Portland State University, 2013. https://core.ac.uk/download/pdf/37767873.pdf.
2. "Economic impact of measurement in the semiconductor industry", Planning Report 07-2, National Institute of Standards & Technology, December 2007. https://www.nist.gov/sites/default/files/documents/director/planning/report07-2.pdf.
3. H. Gruber, "The semiconductor industry: Review of the sector and financing opportunities", European Investment Bank. https://www.eib.org/attachments/pj/semiconductor_industry_en.pdf.
4. "Lam Research", *Wikipedia*. https://en.wikipedia.org/wiki/Lam_Research.
5. "15 types of innovation", *The Gentle Art of Smart Stealing*. https://thegentleartofsmartstealing.wordpress.com/types-of-innovation/.
6. "Innovation", International Association of Innovation Professionals (IAOIP). https://www.iaoip.org/page/What_Why.
7. W. Ballhaus, A. Pagella and C. Vogel, "A change of pace for the semiconductor industry?" *PricewaterhouseCoopers*, November 2009. http://maltiel-consulting.com/Semiconductor_Industry_After_Economic_Upheaval_maltiel_semiconductor.pdf.
8. "Lam Research overview", *Glassdoor*. https://www.glassdoor.com/Overview/Working-at-Lam-Research-EI_IE1582.11,23.htm.
9. "Company overview", Lam Research Corporation. https://www.lamresearch.com/company/company-overview/.
10. "Products overview", Lam Research Corporation. https://www.lamresearch.com/products/products-overview/.
11. https://investor.lamresearch.com/static-files/82321fa5-0664-4bac-9eb7-e81875118c0f.

12. "Lam Research Corporation common stock (LRCX) revenue EPS", *Nasdaq.* https://www.nasdaq.com/symbol/lrcx/revenue-eps.
13. "Lam Research—stock price history", *Macrotrends.* https://www.macrotrends.net/stocks/charts/LRCX/lam-research/stock-price-history.
14. K. Ferrell, "Lam recognized by Infineon as best front-end equipment supplier", *Lam Blog*, 27 March 2017. https://blog.lamresearch.com/lam-recognized-by-infineon-as-best-front-end-equipment-supplier/.
15. "Stock price", https://investor.lamresearch.com/system/files-encrypted/nasdaq_kms/assets/2018/09/27/18-15-49/Lam%20Research%20618470_002_Web_BMK.pdf.
16. https://investor.lamresearch.com/system/files-encrypted/nasdaq_kms/assets/2018/09/27/18-15-49/Lam%20Research%20618470_002_Web_BMK.pdf.
17. "287. Lam Research", *Fortune.* http://fortune.com/fortune500/lam-research/.
18. "Lam Research—revenue", *Macrotrends.* https://www.macrotrends.net/stocks/charts/LRCX/lam-research/revenue.
19. https://investor.lamresearch.com/static-files/f7e97ebf-6862-4465-8dd8-0de8aa598ed7.
20. "2018 investor meeting", Lam Research Corporation, 6 March 2018. https://investor.lamresearch.com/static-files/82321fa5-0664-4bac-9eb7-e81875118c0f.
21. "Financial report", https://investor.lamresearch.com/static-files/82321fa5-0664-4bac-9eb7-e81875118c0f.
22. "Annual report 2016", Lam Research Corporation. https://investor.lamresearch.com/static-files/b5f81325-f7ea-41e0-8a0e-cd300d925756.
23. "Annual report 2017", Lam Research Corporation. https://investor.lamresearch.com/static-files/18d95fe4-f4ea-4275-a17d-2c4003a1ae91.
24. "Annual report 2018", Lam Research Corporation. https://investor.lamresearch.com/system/files-encrypted/nasdaq_kms/assets/2018/09/27/18-15-49/Lam%20Research%20618470_002_Web_BMK.pdf.
25. "Continued scaling with multiple patterning", *Lam Blog.* https://blog.lamresearch.com/continued-scaling-with-multiple-patterning/.
26. "Patents issued", United States Patent and Trademark Office. http://patft.uspto.gov/netacgi/nph-Parser?Sect1=PTO2&Sect2=HITOFF&p=1&u=%2Fnetahtml%2FPTO%2Fsearch-bool.html&r=0&f=

S&l=50&TERM1=lam+research&FIELD1=AANM&co1=AND&TERM2=2016&FIELD2=ISD&d=PTX.
27. "Top 300 organizations granted US patents in 2016", Intellectual Property Owners Association. https://www.ipo.org//wp-content/uploads/2017/05/2016_Top-300-Patent-Owners.pdf.
28. "Top 300 organizations granted US patents in 2017", Intellectual Property Owners Association. https://www.ipo.org//wp-content/uploads/2018/06/2017_Top-300-Patent-Owners.pdf.
29. "Patents assigned to Lam Research". https://patents.google.com/?assignee=lam+research&oq=lam+research.
30. "Lam Research profiles", *Google Scholar*. https://scholar.google.com/citations?view_op=view_org&hl=en&org=945303617654309516&after_author=ge5tABz___8J&astart=30.
31. "Lam Research profiles", *Google Scholar*. https://scholar.google.com/citations?view_op=view_org&hl=en&org=945303617654309516&before_author=WpLc_xcDAAAJ&astart=0.
32. "Paper publications by Lam research scientists", https://scholar.google.com/citations?user=pmjPM1oAAAAJ&hl=en.
33. "2016 investor and analyst meeting", Lam Research Corporation, 18 November 2016. https://investor.lamresearch.com/static-files/f7e97ebf-6862-4465-8dd8-0de8aa598ed7.
34. "Lam receives 2017 Intel's preferred quality supplier award", *Lam Blog*, 7 March 2018. https://blog.lamresearch.com/lam-receives-2017-intels-preferred-quality-supplier-award/.
35. "Lam receives Intel's 2016 preferred quality supplier award", *Lam Blog*, 9 March 2017. https://blog.lamresearch.com/lam-receives-intels-2016-preferred-quality-supplier-award/.
36. K. Ferrell, "Lam joins prestigious Forbes and Fortune lists", *Lam Blog*, 28 March 2016. https://blog.lamresearch.com/lam-joins-prestigious-forbes-and-fortune-lists/.
37. "Lam receives supplier award from TowerJazz", *Lam Blog*, 14 December 2015. https://blog.lamresearch.com/lam-receives-supplier-award-from-towerjazz/.
38. https://investor.lamresearch.com/static-files/f7e97ebf-6862-4465-8dd8-0de8aa598ed7.
39. "Investor and analyst report_Lam research", https://investor.lamresearch.com/static-files/f7e97ebf-6862-4465-8dd8-0de8aa598ed7.

Appendix A: Distribution and Background of Expert Panel 1 [1]

Expert	Industry	Government	Academia	Affiliation	Country
1			x	University of Bamberg	Germany
2			x	Delft University of Technology	Netherlands
3			x	University of Bamberg	Germany
4			x	Erasmus University	Netherlands
5			x	University of Exeter	UK
6			x	German Graduate School of Management & Law	Germany
7			x	University of Manchester	UK
8			x	INRS	Canada
9			x	University of Bologna	Italy
10			x	Melbourne Institute of Applied Economic and Social Research	Australia
11			x	German Graduate School of Management & Law	Germany
12			x	Fuzhou University	China
13			x	Innovation IMS Instruction	USA
14			x	Korea University	South Korea
15	x			PwC	USA
16	x			IPR & Innovation at Crompton Greaves Ltd	India
17	x			TriQuint Semiconductor	USA
18	x			Intel Corporation	USA
19	x			Cascade Mictotech	USA
20	x			FEI Company	USA
21	x			Lattice Semiconductor	USA
22	x			Intel Corporation	USA
23	x			TOK America	USA
24	x			Intel Corporation	USA
25	x			Intel Corporation	USA

(*Continued*)

Appendix A: (*Continued*)

Expert	Industry	Government	Academia	Affiliation	Country
26	×			Tektronix, Inc.	USA
27	×			Intel Corporation	USA
28	×			Tektronix, Inc.	USA
29	×			Novellus System	USA
30			×	Italian National Research Council	Italy

Appendix B: Distribution and Background of Expert Panel 2 [1]

Expert	Industry	Government	Academia	Affiliation	Country
1			×	Delft University of Technology	Netherlands
2			×	INRS	Canada
3			×	German Graduate School of Management & Law	Germany
4			×	German Graduate School of Management & Law	Germany
5			×	University of Bamberg	Germany
6			×	University of Bamberg	Germany
7			×	Korea University	South Korea
8			×	University of Bologna	Italy
9			×	Fuzhou University	China
10			×	Erasmus University	Netherlands
11			×	Indian Institute Technology	India
12			×	University of Exeter	UK
13			×	University of Manchester	UK
14			×	Innovation IMS Instruction	USA
15	×			Samsung Electronic Research Institute	South Korea
16	×			Lattice Semiconductor	USA
17	×			FEI Company	USA
18	×			TOK America	USA

Appendix B: (Continued)

Expert	Industry	Government	Academia	Affiliation	Country
19	x			Tektronix, Inc.	USA
20	x			Tektronix, Inc.	USA
21	x			Tektronix, Inc.	USA
22	x			Tektronix, Inc.	USA
23	x			Intel Corporation	USA
24	x			Intel Corporation	USA
25	x			Intel Corporation	USA
26	x			Intel Corporation	USA
27	x			TriQuint Semiconductor	USA
28	x			TriQuint Semiconductor	USA
29	x			TriQuint Semiconductor	USA
30	x			PwC	USA
31	x			Cascade Mictotech	USA
32	x			Novellus System	USA
33	x			IPR & Innovation at Crompton Greaves Ltd	India
34	x			Texas Instruments	USA
35		x		Italian National Research Council	Italy
36		x		Oregon Business Innovation Council	USA

Appendix C: Distribution and Background of Expert Panel 3 [1]

Expert	Industry	Government	Academia	Affiliation	Country
1			x	Delft University of Technology	Netherlands
2			x	INRS	Canada
3			x	German Graduate School of Management & Law	Germany
4			x	Gentian Graduate School of Management & Law	Germany

(Continued)

Appendix C: (*Continued*)

Expert	Industry	Government	Academia	Affiliation	Country
5			×	University of Bamberg	Germany
6			×	University of Bamberg	Germany
7			×	Korea University	South Korea
8			×	University of Bologna	Italy
9			×	Fuzhou University	China
10			×	Erasmus University	Netherlands
11			×	University of Exeter	UK
12			×	University of Manchester	UK
13			×	Innovation EMS Instruction	USA
14	×			Samsung Electronic Research Institute	South Korea
15	×			TOK America	USA
16	×			Tektronix, Inc.	USA
17	×			Tektronix, Inc.	USA
18	×			Tektronix, Inc.	USA
19	×			Tektronix, Inc.	USA
20	×			Intel Corporation	USA
21	×			Intel Corporation	USA
22	×			TriQuint Semiconductor	USA
23	×			TriQuint Semiconductor	USA
24	×			PwC	USA
25	×			Cascade Mictotech	USA
26	×			Novellus System	USA
27	×			IPR & Innovation at Crompton Greaves Ltd	India
28	×			Texas Instruments	USA
29		×		Italian National Research Council	Italy
30		×		Oregon Business Innovation Council	USA

Appendix D: Desirability Curves for All Indicators [1]

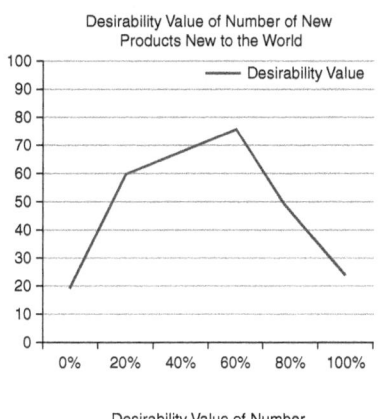

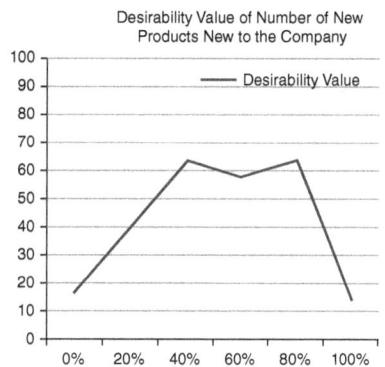

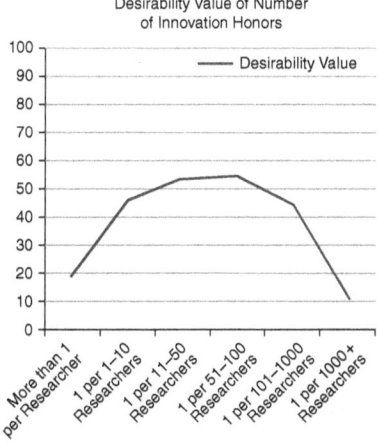

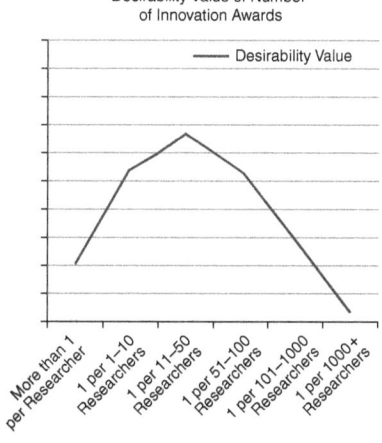

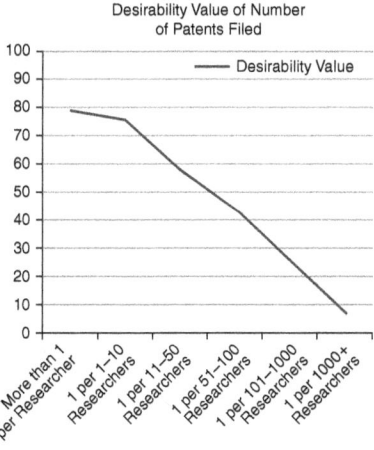

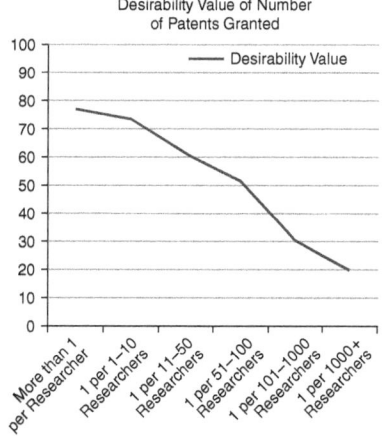

478 Digital Transformation

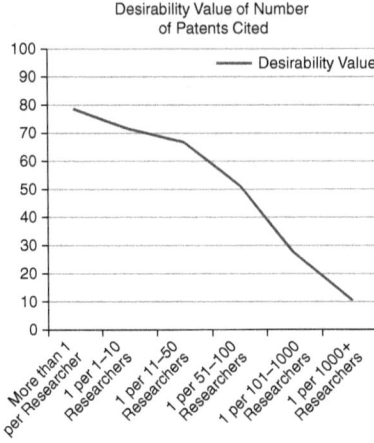
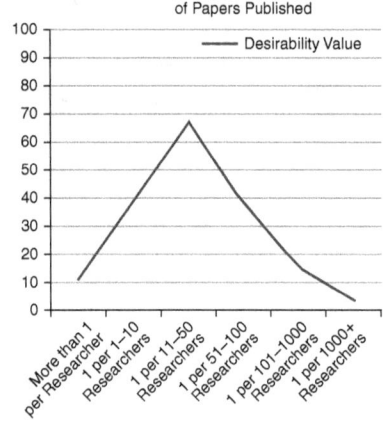
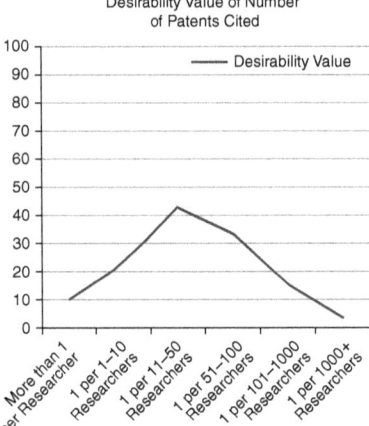
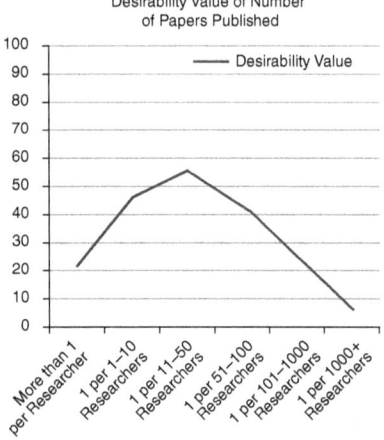
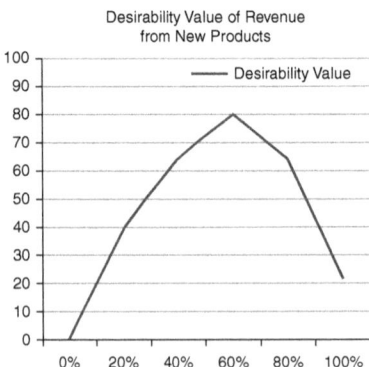
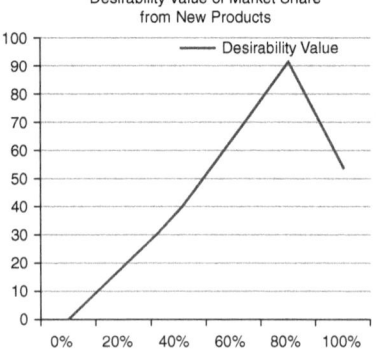

Chapter 15

Organizational Transformation: Universities

Ahmed Bohliqa*, Corey White*, Srujana Penmetsa*, Sara Bahreini*, Zeina Boulos* and Tugrul Daim*,†,‡

*Portland State University, Portland, Oregon, USA
†Higher School of Economics, Moscow, Russia
‡Chaoyang University of Technology, Taiwan

Abstract

Most of the time, institutions are faced with the dilemma of choosing from a vast selection of Learning Management System (LMS) providers that will suit their needs and requirement. This report identifies the criterions and factors that an educational institute looks for in an LMS and develops a hierarchical decision model using them. This model will help decision makers to make the correct decision in picking an LMS based on the institution's priorities and expert judgments on LMS alternatives. We have applied our proposed model to two case studies and have done a sensitivity analysis to see how changes in priorities would affect the choice of an LMS.

Keywords: Technology assessment, university, learning management systems.

1. Introduction

Recent technological advances have increased the amount of information available and improved accessibility to said information, while at the same time the costs of publishing information have decreased. These general shifts throughout society are mimicked in education and have caused students to be more demanding and more knowledgeable about alternatives for their education [1]. Increasingly, many institutions and schools around the globe are offering online education. However, in order to manage, track and deliver courses and training programs, they are required to use a Learning Management System (LMS). With many LMS providers available, institutions are faced with the dilemma of deciding which system to obtain. This paper proposes a decision-making model for selecting an LMS. A case study of Portland State University (PSU) is used to test the model.

2. Literature Review

2.1. *Online education*

We derive our definition about online learning from Ref. [2], in which the 2nd Annual Study of the State of Online Higher Education in the United States was supported by a grant from the Sloan Foundation, in collaboration with the Sloan Consortium and the Sloan Center for On-Line Education (SCOLE), collocated at Babson College and Franklin W. Olin College of Engineering.

This survey for establishing a baseline about metrics involving online learning in higher education started with an e-mail containing a link to a web-based survey form that was sent to Chief Academic Officers at degree-granting institutions of higher education in the United States. In cases where the Chief Academic Officer was responsible for more than one campus, the survey was sent to the primary campus only. If there was no designated Academic Officer, the survey was sent to the President of the institution. The recipient's e-mail address was linked by a unique identifier in order to track metrics. Of the 3,033 surveys sent, 994 responses were received, representing a 32.8% response rate in 2002. This process has carried on since then [2–9, 11].

The metrics established from the said survey serves as a basis for our understanding of online education. To begin the discussion of "online education", we must first have an understanding of the methods of educational delivery that have been utilized by higher learning institutions [2]. We identify the standard definitions of the four types of educational delivery methods—traditional, web facilitated, blended/hybrid and online. Table 1 gives definitions to the aforementioned delivery method.

Table 1. Educational Delivery Methods.

Proportion of Content Delivered Online	Type of Course	Typical Description
0%	Traditional	Course where no online technology used. Content is delivered in writing or by mouth.
1 to 29%	Web facilitated	Course that uses web-based technology to facilitate what is essentially a face-to-face course. May use a course management system (CMS) or web pages to post the syllabus and assignments.
30 to 79%	Blended/ hybrid	Course that blends online and face-to-face delivery. A substantial proportion of the content is delivered online and typically uses online discussions and has a reduced number of face-to-face meetings.
80+%	Online	A course where most or all of the content is delivered online. It typically has no face-to-face meetings.

Taking into consideration the information provided in the focus of this paper is Online Education, or education in which 80% of the content is delivered online.

With this understanding, one might then ask: what is the significance of online learning? According to Ref. [10], less than half of all higher education institutions reported that online education was critical to their long-term strategy since the report series began in 2002. That number reached 69.1% in 2012.

- The proportion of chief academic leaders that say online learning is critical to their long-term strategy is now at 69.1%—the highest it has been for this ten-year period.
- Likewise, the proportion of institutions reporting that online education is not critical to their long-term strategy has dropped to a new low of 11.2%.

With an ever-increasing presence of students learning through online means, the ability to effectively administer education via the internet requires institutions to make investments into LMSs that deliver the content and facilitate the necessary teacher-student interactions.

2.2. Learning management systems

An LMS provides the platform for the enterprise's online learning environment by enabling the management, delivery and tracking of blended learning (i.e., online and traditional classroom) for employees, stakeholders and customers. A robust LMS should integrate with other departments, such as human resources, accounting, e-commerce and information technology, so administrative and supervisory tasks can be streamlined and automated, and the overall cost and impact of education can be tracked, quantified and managed [11].

2.3. Learning content management systems

While LMSs help manage learners and keep records, Learning Content Management Systems (LCMSs) provide course management content, which is stored as learning objects in a learning object repository database. The difference between an LMS and LCMS is becoming almost negligible since many LCMSs offer course authoring and LMS capabilities. Systems known as Virtual Learning Environments (VLE) and Course Management Systems are education-oriented LCMSs [12]. The list below provides many of the known LCMSs [12]:

1. Ambone (Amvonet (AVE INTERVISION LLC)), http://www.amvonet.com/.
2. Any-3 LCMS (iPerformance) (Any-3 Ltd), http://www.any-3.com (based in UK).
3. Assima Training Suite (Assima Plc), http://www.assima.net.
4. Aunwesha LearnITy Enterprise Suite LMS/LCMS, http://www.aunwesha.com.
5. Blackboard Learn (Blackboard, Inc), http://www.blackboard.com.
6. Chalk Pushcast Software (Chalk Media Corporation), http://www.chalk.com.
7. Composica (Composica), http://www.composica.com/.
8. ConnectEDU Educate Hub (ConnectEDU Inc), http://connectedu.com/.
9. Conzentrate Learning Arena (Conzentrate), http://www.conzentrate.com/.
10. Desire2Learn Learning Repository (Desire2Learn, Inc), http://www.desire2learn.com.

3. Problem Statement

Increasingly, many institutions and schools around the globe are offering online education. However, in order to manage, track and deliver courses and training programs, they are required to use an LMS. With many LMS providers available, institutions are faced with the dilemma of deciding which system to obtain. This paper proposes a decision-making model for selecting an LMS. A case study of PSU is used to test the model.

4. Methodology

4.1. *Technology evaluation*

The equation below presents a quantitative model used for evaluating the alternatives in an HDM model.

$$TV_n = \sum_{K=1}^{K} \sum_{JK=1}^{JK} W_k \cdot F_{jk} \cdot V(t_{n,jk,k}).$$

The formula to evaluate an alternative is based on obtaining a set of an expert's opinions and subjective judgments through various quantitative and qualitative measures, e.g., relative priority criteria, relative impact of factors, etc. The most common approach is to interview experts and reduce the outcome to numerical values where it enables us to use the mathematical formula to quantitatively determine the impact of the alternatives on the mission.

5. Proposed Model

To prepare our model we reviewed multiple LMS evaluation criteria in the literature [13–18] as well as proposals by universities to LMS providers. Then, we had multiple iterations of validation and we finalized our model as can be seen in Figure 1.

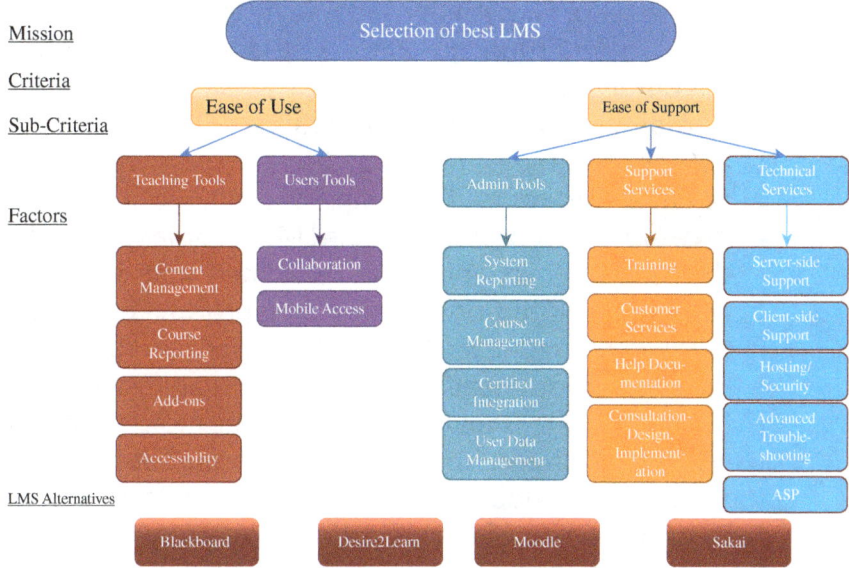

Figure 1. Decision Model.

5.1. *Model definitions*

The following is a brief definition for each criterion on the model along with its subcriteria and factors.

- **Ease of use:** The ease of using the LMS tools, which are used by teachers and other users.
  - *Teaching tools*: This section compacts all the required tools and concepts that are needed to design learning courses. For example, establishing the course interface, students accessibility, assignment, exams creator, drop box, etc.
    - *Content management*: Ability and quality to organize files and repositories as well as multimedia files such as powerpoint slides, videos, audio and flash files, and PDFs. This includes school, department, and course and term levels.

- *Course reporting*: Ability and quality to do assessments, analytics and rubrics, and track students' progress and measurements.
- *Add-ons*: Availability and quality of add-ons like authoring, grade book, functioning tools, drop box, calendar and course building.
- *Accessibility*: Accessibility of the LMS. As of January 2012, the federal government has mandated that all higher education institutions must either provide or have a tangible plan to provide accessibility services for all students of various needs. Because of this, accessibility has become a huge issue in the higher education world.
  - *Users tools*: This section addresses tools that are used by users, which includes collaboration tools such as emails, file exchange, virtual classes and mobile accessibility.
    - *Collaboration*: Ability and quality to collaborate, such as file/data sharing, chats/teleconferencing, possibly social networking functionality, and the ability for a student to initiate activities on the system.
    - *Mobile access*: Ability and quality to access the LMS via different mobile browsers and the availability of apps for smartphones on different platforms.

- **Ease of support:** The ease and comprehensiveness of services and support provided by the LMS provider.
  - *Admin tools*: This section includes administrative tools, such as compact reporting tools system and securing exams tools that allows the administrator to organize aspects of online courses.
    - *System reporting*: Availability and quality of tools for reporting system statistics, student and instructor statistics on course and term levels, and other standard reporting tools used by administrators.
    - *Course management*: Ability and quality to manage courses by term, department and college. Also the ability to create customized roles, assign permissions,

provide guest access, enable self-enrollment and cross list courses.
- *User data management*: Ability and quality to manage user data including customization of user account data, creation of user profiles and identification of inactive users.
- *Certified integration*: Ability and quality to integrate with third party software such as student integration systems, direct database access, exam creation tools, plagiarism detection/prevention tools, content management systems, identity management systems and so on.

○ *Support services*: This section emphasizes vender support and all required documented instructions for help/FAQs. This includes proper implementation of direct assistance in a regular base and disaster recovery, in addition to providing any needed training.
- *Help documentation*: Availability and quality of documentation for system help (online or printed).
- *Training*: Availability and quality of training to administrators, staff or trainers.
- *Consultation—design, implementation*: Availability and quality of service in helping an institution implement the new LMS, which includes configuring institutional and user roles, data conversion and transfer from a previous LMS, and on-going consultation services for determining the best use practices for their system.
- *Customer services*: Availability and quality of general customer services, which includes non-technical to basic technical services.

○ *Technical services*: Includes all the technical aspects of support, such as troubleshooting, server support or client support.
- *Server-side support*: The availability and quality of tools that are specific to the server where the LMS is

hosted. For example, integration abilities between the LMS system and other systems like DBMS and the Web/Application Server.
- *Client-side support*: The availability and quality of tools specific to the user machine (client). For example, integration with different browsers, operating systems and mobile devices.
- *Hosting/security*: Availability and quality of hosting and security issues related to an LMS. For example, is the LMS hosted by the institution, company or a third-party entity?
- *Advanced troubleshooting*: Availability and quality of 24/7 advanced technical support and other services for administrators and users by the LMS provider.
- *ASP*: Ability and quality to have disaster recovery, off-site backup, uptime monitoring and virus scanning support.

5.2. *Normalization values*

We developed the following scoring to be able to compute the technology value for each LMS alternative.

Score	Definition
0	If the criterion is not available and does not have ability
1	If the LMS provides only a few options related to the criterion and they are of poor quality
2	If the LMS provides only a few options related to the criterion and they are of fair quality
3	If the LMS provides only a few options related to the criterion and they are of best quality
4	If the LMS provides all options related to the criterion and they are of fair quality
5	If the LMS provides all options related to the criterion and they are of best quality

Two experts gave each factor in our model a score of between 0 and 5 in accordance to each LMS alternative. We used these scores and the weights of the factors obtained from the pairwise comparison to calculate the value of the LMS tool.

5.3. *Model validation*

We had our model validated by two experts. One was part of a decision-making committee who had chosen an LMS for a university prior to this. The other was an instructional designer, who was also a teacher and an IT professional who was part of a support team for an LMS in another university.

5.4. *Data collection*

To collect the experts' evaluations of the model, we used the HDM software developed by the Engineering and Technology Management Department at PSU. The software creates a special record to collect evaluations from each participant, and displays for participants a pairwise comparison for every level of the model. The participant can choose to either use the slider or enter values into a text box as shown in Figure 2.

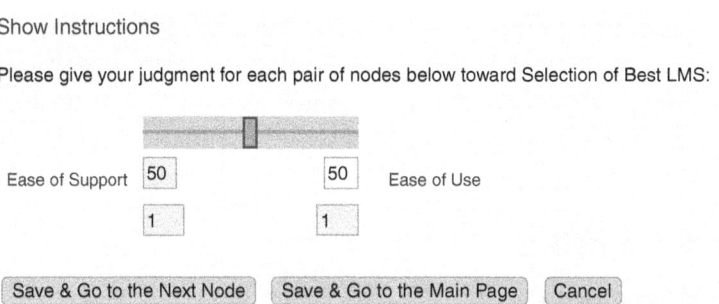

Figure 2. Evaluation Tool.

Table 2. Number of Experts.

	For Ease of Use	For Ease of Support	For the Entire Model
Contacted	9	1	6
Received	4	1	2

We contacted nine professors and instructors from PSU to help us evaluate the subcriteria and factors under the Ease of Use criterion. We also contacted an IT individual at PSU to help us evaluate the same thing. Finally, we contacted six individuals who had both IT and teaching backgrounds and asked them to help us evaluate the entire model. Table 2 shows the number of contacts and received evaluations. More details about our contacted experts can be found in Appendix A.

In order to collect our normalization values, we created an online matrix-shaped data collection tool using Qualtrics software. The tool included instructions on how to assign values from the explained scoring system above, the different LMS alternatives, and all the factors including their descriptions. Each description includes different options for the factor. For example, to assign values for Course Reporting, an expert will look at its description, which is the same as the definitions above. The description for Course Reporting is the "ability and quality to do assessment, analytics, rubrics, track students' progress and measurements". If the LMS had all of these, plus if they were all of high quality, then it would get a value of 5. In another example, if it had only assessment and analytics (both fair quality) functions, then it would get a "2". Appendix B has a snapshot of the normalization values collection tool.

6. Case Studies

6.1. *Portland State University*

PSU is a public institution that was founded in 1946. It gained full university status in 1969. Now, PSU is an urban research university spanning eight colleges, 226 degree programs, 29,000 students, including 1,700 international students from 91 different countries, and 126,000 alumni members.

The mission of PSU is to enhance various qualities of urban life by providing access to a quality education for undergraduates and an array of professional and graduate programs relevant to metropolitan areas.

In order for PSU to achieve its mission, the university needs a better LMS for the online education it offers. Currently, PSU uses Desire2Learn. It used BlackBoard in the past. In this case study we identify which LMS will suit the needs of PSU based on the expert judgments given by PSU experts. After consulting the PSU experts, we were able to identify four LMSs that the university is considering. These are BlackBoard, Desire2Learn, Moodle and Sakai.

6.2. *Results and analysis*

From the expert judgments of the proposed model, we could calculate the weightage of each criteria, subcriteria and factors with respect to the mission, criteria and subcriteria, respectively.

Figure 3 indicates the weightages of Ease of Use and Ease of Support with respect to the mission of selecting the best LMS.

Figure 4 indicates the weightages of Teaching Tools and User Tools with respect to the Ease of Use.

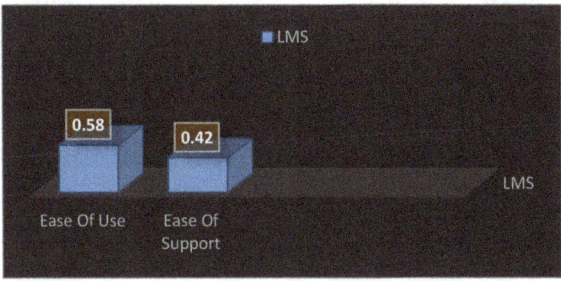

Figure 3. LMS.

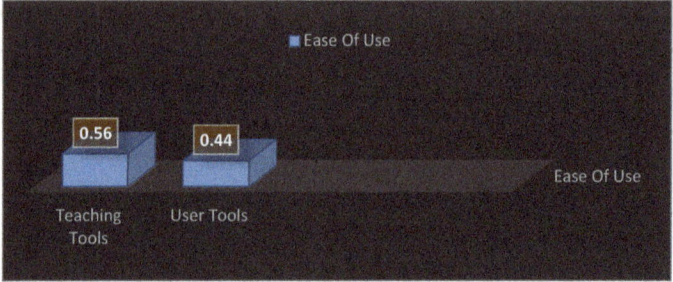

Figure 4. Ease of Use.

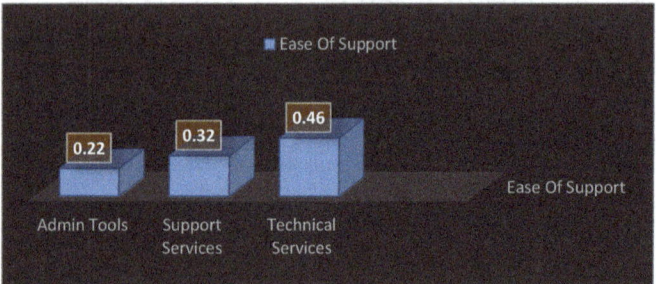

Figure 5. Ease of Support.

Figure 5 shows the importance of Technical Services over the other subcriteria with respect to Ease of Support.

Figure 6 shows that Content management is the most important factor and Accessibility is the next most important factor in choosing the LMS with respect to the subcriteria Teaching Tools.

The importance of being able to use the LMS in collaboration with other functionalities over the mobile accessibility is depicted in Figure 7.

Similarly, the importance of other factors with respect to the subcriteria Admin Tools, Support Services and Technical Services are indicated in Figures 8–10.

Organizational Transformation: Universities 493

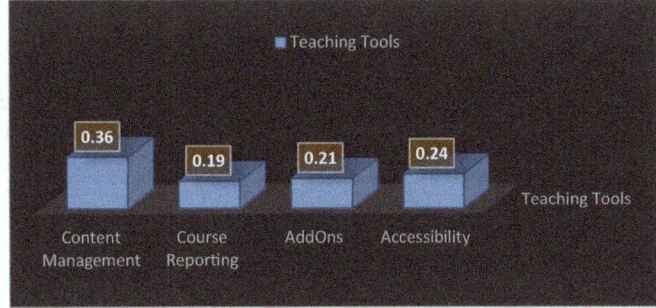

Figure 6. Teaching Tools.

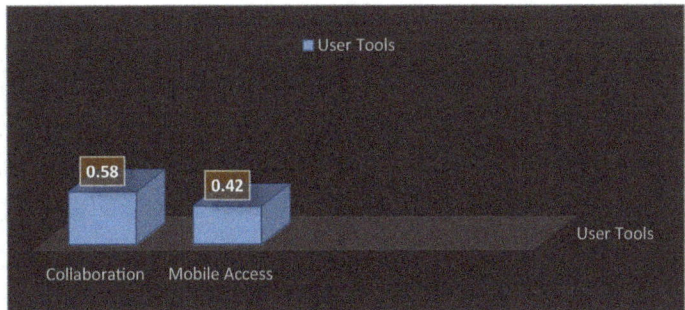

Figure 7. User Tools.

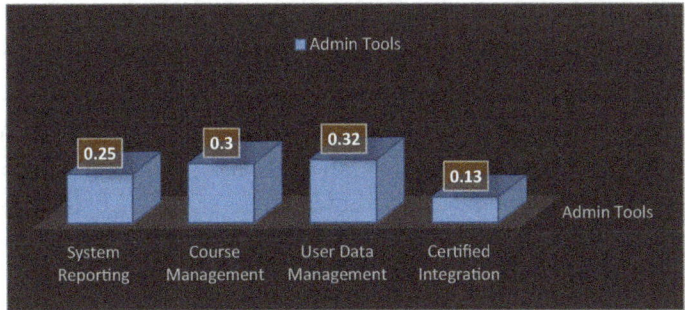

Figure 8. Admin Tools.

494   Digital Transformation

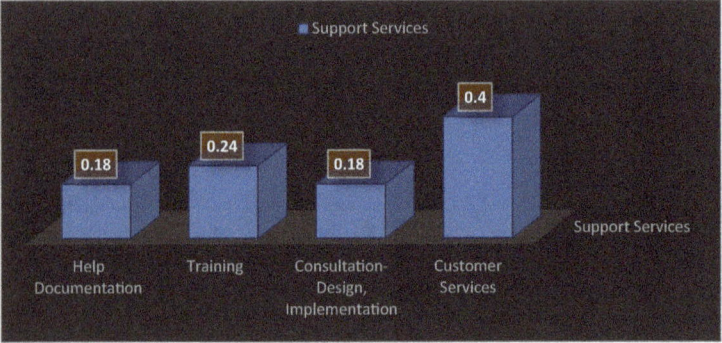

Figure 9. Support Services.

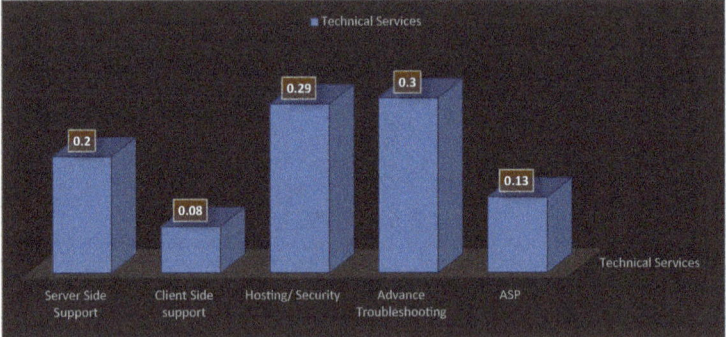

Figure 10. Technical Services.

The priorities of each criteria, subcriteria and factors with respect to the LMS are indicated in the proposed model in Figure 11.

As the priorities of the factors with respect to the LMS are known, we have asked the experts to score the four LMS alternatives with respect to each factor on a scale of 0 to 5 as mentioned earlier. The mean scores of the LMS alternatives given by the experts are shown in Table 3.

We then normalized the scores from the scale of 0 to 5 to a scale of 0 to 100. For example, if the experts gave BlackBoard a score of 4 with respect to content management, we normalized

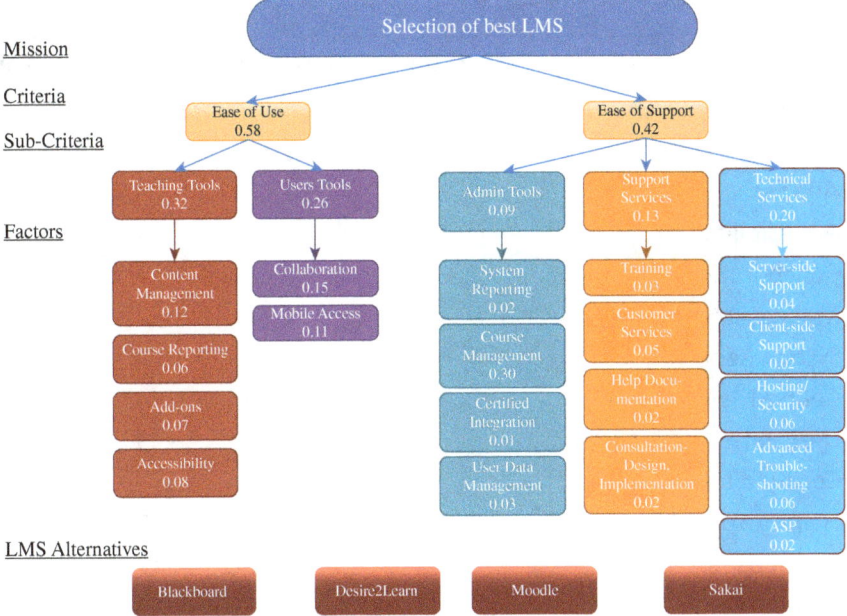

Figure 11. Final Model — PSU.

it and gave BlackBoard a score of 80 instead with respect to the content management in the technology value calculations.

Once we had all the normalized scores for the alternatives, we calculated the technology value for each alternative as shown in Table 4 using values from Figure 12.

From the technology value calculations, it is evident that BlackBoard suits PSU's mission and requirements with a technology value of 66.2. The ranking of the LMS alternatives from the calculations with respect to PSU requirements and priorities is shown in Table 5.

6.3. *Community college*

The college's enrollment figure in 2012–2013 was 2,607. The college is currently using Angel 8.0 as their LMS, which was bought over by Blackboard in 2009. The end of its lifespan is 2014, but the college wants to switch before then.

Table 3. Factor Scores.

Factors	BlackBoard	Desire2learn	Moodle	Sakai
Content Management	4	3	3.5	3.5
Course Reporting	3.5	2.5	2	2.5
Addons	2.5	2.5	4.5	3.5
Accessibility	3.5	4.5	3.5	2.5
Collaboration	2	1.5	2	3
Mobile Access	3	2.5	2.5	2.5
System Reporting	4	2	3.5	4
Course Management	4.5	4.5	4	4
User Data Management	3	2.5	3.5	3
Certified Integration	5	4	5	4.5
Help Documentation	2.5	2.5	3.5	2.5
Training	3.5	3.5	2	2.5
Consultation—Design, Implementation	4.5	4	2	1.5
Customer Service	4.5	3	2	2
Server Side Support	3.5	4	3.5	3.5
Client Side Support	3	4.5	3.5	3
Hosting/Security	3.5	4.5	3.5	3.5
Advanced Troubleshooting	3.5	3.5	2	2
ASP	4.5	4	2	2.5

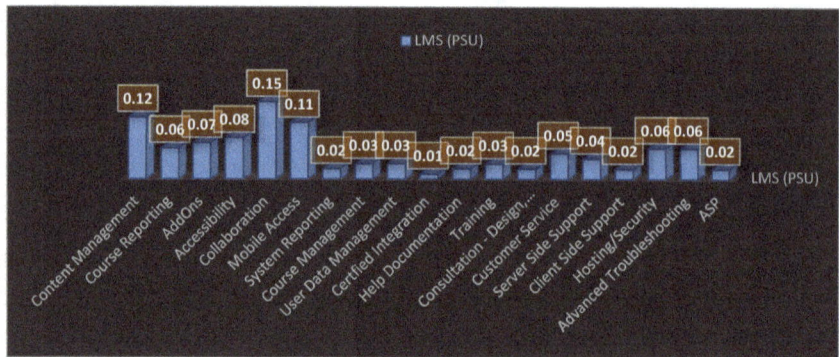

Figure 12. LMS (PSU).

Table 4. Weighted Scores — PSU.

Factors	Weightage	Blackboard			Desire2Learn			Moodle			Sakai		
		Ranking	Desirability Value	Weight* Desirability Value	Ranking	Desirability	Weight* Desirability Value	Ranking	Desirability	Weight* Desirability Value	Ranking	Desirability	Weight* Desirability Value
Content Management	0.12	4	80	9.6	3	60	7.2	3.5	70	8.4	3.5	70	8.4
Course Reporting	0.06	3.5	70	4.2	2.5	50	3	2	40	2.4	2.5	50	3
Addons	0.07	2.5	50	3.5	2.5	50	3.5	4.5	90	6.3	3.5	70	4.9
Accessibility	0.08	3.5	70	5.6	4.5	90	7.2	3.5	70	5.6	2.5	50	4
Collaboration	0.15	2	40	6	1.5	30	4.5	2	40	6	3	60	9
Mobile Access	0.11	3	60	6.6	2.5	50	5.5	2.5	50	5.5	2.5	50	5.5
System Reporting	0.02	4	80	1.6	2	40	0.8	3.5	70	1.4	4	80	1.6
Course Management	0.03	4.5	90	2.7	4.5	90	2.7	4	80	2.4	4	80	2.4
User Data Management	0.03	3	60	1.8	2.5	50	1.5	3.5	70	2.1	3	60	1.8
Certified Integration	0.01	5	100	1	4	80	0.8	5	100	1	4.5	90	0.9
Help Documentation	0.02	2.5	50	1	2.5	50	1	3.5	70	1.4	2.5	50	1
Training	0.03	3.5	70	2.1	3.5	70	2.1	2	40	1.2	2.5	50	1.5
Consultation – Design, Implementation	0.02	4.5	90	1.8	4	80	1.6	2	40	0.8	1.5	30	0.6
Customer Service	0.05	4.5	90	4.5	3	60	3	2	40	2	2	40	2
Server Side Service	0.04	3.5	70	2.8	4	80	3.2	3.5	70	2.8	3.5	70	2.8
Client Side Service	0.02	3	60	1.2	4.5	90	1.8	3.5	70	1.4	3	60	1.2
Hosting/Security	0.06	3.5	70	4.2	4.5	90	5.4	3.5	70	4.2	3.5	70	4.2
Advanced Troubleshooting	0.06	3.5	70	4.2	3.5	70	4.2	2	40	2.4	2	40	2.4
ASP	0.02	4.5	90	1.8	4	80	1.6	2	40	0.8	2.5	50	1
Technology Value				66.2			60.6			58.1			58.2

Table 5. Alternative Ranking — PSU.

LMS Alternatives	Ranking
BlackBoard	1
Desire2learn	2
Moodle	4
Sakai	3

Some of the main challenges the college is facing in making a decision over the selection of an LMS are:

1. Finding an LMS that migrates their current courses as well as it being easy to use for beginners.
2. Deciding whether to self-host or vendor-host.

Currently, its faculty, staff and students are involved in the decision-making process and a recommendation will be made to the Executive Council. The current decision-making process involves:

1. An LMS Alliance consisting of an LMS Subcommittee, Faculty Cohort and the campus's community at large will work together to find the best platform for their institution.
2. They are using a modified rubric of Longsight LLC (Education Consulting Company).

6.4. *Results and analysis*

The experts from the community college provided weightages of 70 for Ease of Use and 30 for Ease of Support, while having the same weightages for the subcriteria and factors. The priorities of all the criteria, subcriteria and factors with respect to the LMS are indicated in Figures 13 and 14.

Using the same scores provided by the experts for the LMS alternatives and normalizing their scores, the technology value for the alternatives is calculated as shown in Table 6.

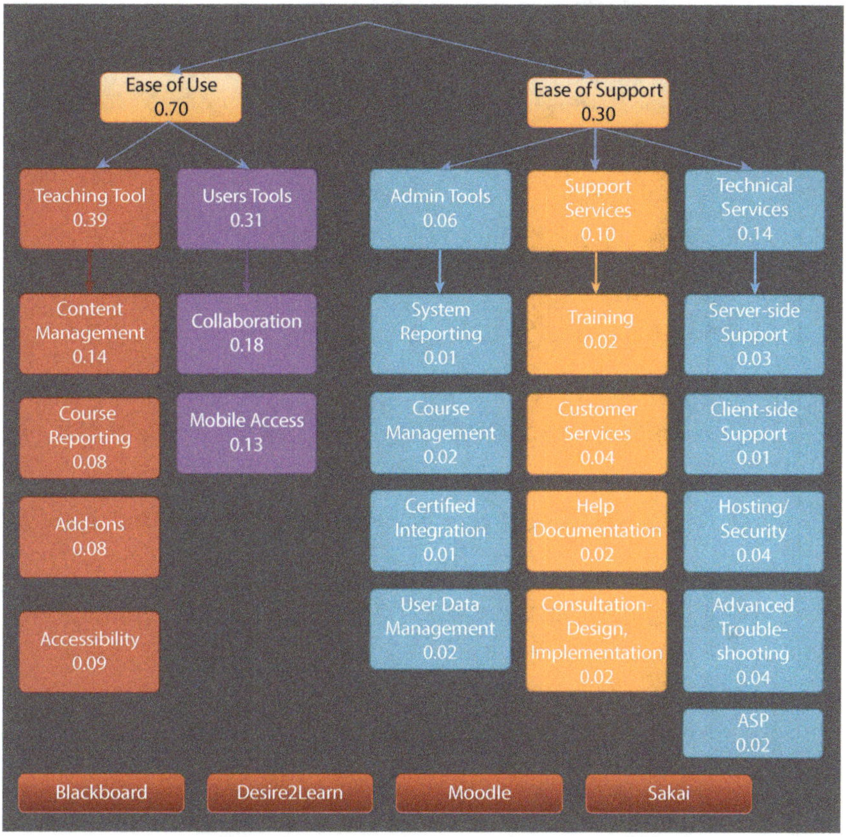

Organizational Transformation: Universities 499

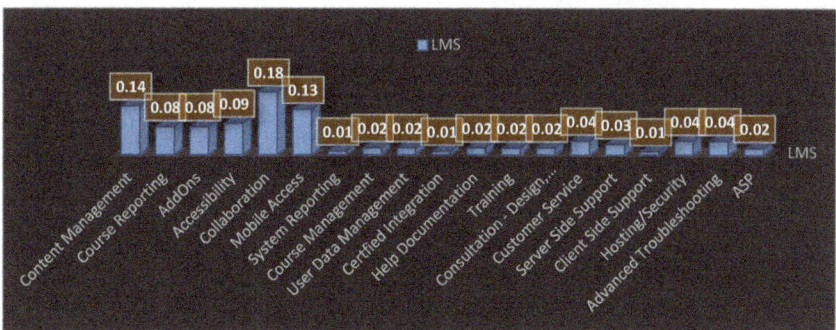

Figure 13. Final Model — Community College.

Figure 14. LMS Community College.

Table 6. Weighted Scores — Community College (CC).

Factors	Weightage	Blackboard			Desire2Learn			Moodle			Sakai		
		Ranking	Desirability Value	Weight* Desirability Value	Ranking	Desirability Value	Weight* Desirability Value	Ranking	Desirability Value	Weight* Desirability Value	Ranking	Desirability Value	Weight* Desirability Value
Content Management	0.14	4	80	11.2	3	60	8.4	3.5	70	9.8	3.5	70	9.8
Course Reporting	0.08	3.5	70	5.6	2.5	50	4	2	40	3.2	2.5	50	4
AddOns	0.08	2.5	50	4	2.5	50	4	4.5	90	7.2	3.5	70	5.6
Accessibility	0.09	3.5	70	6.3	4.5	90	8.1	3.5	70	6.3	2.5	50	4.5
Collaboration	0.18	2	40	7.2	1.5	30	5.4	2	40	7.2	3	60	10.8
Mobile Access	0.13	3	60	7.8	2.5	50	6.5	2.5	50	6.5	2.5	50	6.5
System Reporting	0.01	4	80	0.8	2	40	0.4	3.5	70	0.7	4	80	0.8
Course Management	0.02	4.5	90	1.8	4.5	90	1.8	4	80	1.6	4	80	1.6
User Data Management	0.02	3	60	1.2	2.5	50	1	3.5	70	1.4	3	60	1.2
Certified Integration	0.01	5	100	1	4	80	0.8	5	100	1	4.5	90	0.9
Help Documentation	0.02	2.5	50	1	2.5	50	1	3.5	70	1.4	2.5	50	1
Training	0.02	3.5	70	1.4	3.5	70	1.4	2	40	0.8	2.5	50	1
Consultation — Design, Implementation	0.02	4.5	90	1.8	4	80	1.6	2	40	0.8	1.5	30	0.6
Customer Service	0.04	4.5	90	3.6	3	60	2.4	2	40	1.6	2	40	1.6
Server Side Support	0.03	3.5	70	2.1	4	80	2.4	3.5	70	2.1	3.5	70	2.1
Client Side Support	0.01	3	60	0.6	4.5	90	0.9	3.5	70	0.7	3	60	0.6
Hosting/Security	0.04	3.5	70	2.8	4.5	90	3.6	3.5	70	2.8	3.5	70	2.8
Advanced Troubleshooting	0.04	3.5	70	2.8	3.5	70	2.8	2	40	1.6	2	40	1.6
ASP	0.02	4.5	90	1.8	4	80	1.6	2	40	0.8	2.5	50	1
Technology Value				64.8			58.1			57.5			58

Table 7. Alternative Ranking CC.

LMS Alternatives	Ranking
BlackBoard	1
Desire2learn	2
Moodle	4
Sakai	3

From the technology value calculations, it is evident that BlackBoard has the highest technology value of 64.8 while all the other LMS alternatives have somewhat similar technology values. The ranking of the LMS alternatives from the calculations with respect to community college priorities is shown in Table 7.

7. Sensitivity Analysis

A sensitivity analysis has been carried out in order to understand the effect of changes in the priorities of the criterion and subcriterion on the technology value and the ranking of the LMS alternatives. To analyze the effects, we changed the weights on a few criteria and calculated the new priorities of the factors to the LMS. We used the new priorities to calculate the technology value for the LMS alternatives.

7.1. *Sensitivity analysis 1*

For sensitivity analysis 1, we changed the weights for Ease of Use to 30 and Ease of Support to 70. We then calculated the priorities of the criterion, subcriterion and factors, and the priorities of the factors to the LMS are shown in Figure 15.

It is clear from Table 8 that when Ease of Support is more important than Ease of Use, the rankings for Moodle and Sakai are interchanged. It is also clear that the technology values for

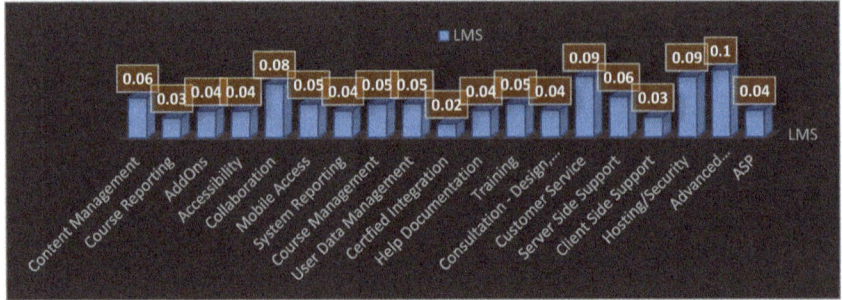

Figure 15. LMS Sensitivity Analysis 1.

Table 8. Sensitivity Analysis 1.

LMS Alternatives	Technology Value	Ranking
BlackBoard	70.7	1
Desire2learn	65.7	2
moodle	58	3
sakai	57.4	4

Blackboard and Desire2Learn have increased, which indicates that the two LMSs have better support than ease of use.

7.2. *Sensitivity analysis 2*

For sensitivity analysis 2, we have changed the weights for Teaching Tools to 80 and User Tools to 20. We then calculated the priorities of the criterion, subcriterion and factors, and the priorities of the factors to the LMS are shown in Figure 16.

Using these priorities, the technology value of the LMS alternatives is calculated and they are ranked based on the technology values shown in Table 9.

From Table 9 it is clear that when Teaching Tools is more important than User Tools, the rankings for Moodle and Sakai

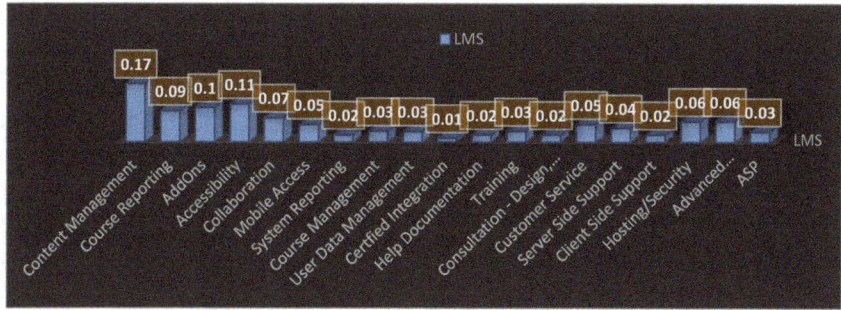

Figure 16. LMS Sensitivity Analysis 2.

Table 9. Sensitivity Analysis 2.

LMS Alternatives	Technology Value	Ranking
BlackBoard	70.4	1
desire2learn	64.7	2
moodle	61.9	3
sakai	59.5	4

are interchanged. It is also clear that the technology values for Blackboard and Desire2Learn have increased, which indicates that the two LMSs have better tools for teaching.

7.3. *Sensitivity analysis 3*

For sensitivity analysis 3, we changed the weights for Teaching Tools to 20 and User Tools to 80. We then calculated the priorities of the criterion, subcriterion and factors, and the priorities of the factors to the LMSs are shown in Figure 17.

Using these priorities, the technology value of the LMS alternatives is calculated and they are ranked based on the technology values shown in Table 10.

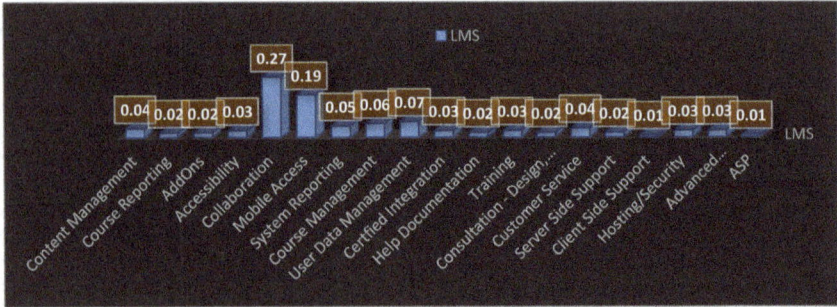

Figure 17. LMS Sensitivity Analysis 3.

Table 10. Sensitivity Analysis 3.

LMS Alternatives	Technology Value	Ranking
BlackBoard	63.4	1
Desire2learn	54.2	4
moodle	55.8	3
sakai	59.7	2

From the ranking in Table 9, it is clear that when User Tools is more important than Teaching Tools, the rankings for Desire2learn, Moodle and Sakai are interchanged. It is also clear that Blackboard and Sakai have high technology values, which indicate that the two LMSs have better tools for users.

8. Comparison of Results

By comparing the technology values of LMS alternatives for various case studies and sensitivity analysis, we can see the trends of changes in the technology value and ranking when the weights on the criterion and subcriterion change. Figure 18 shows the changes in the technology value in each case.

Table 11 shows the changes in ranking of the LMS alternatives with the changes in the priorities of the criterion and subcriterion.

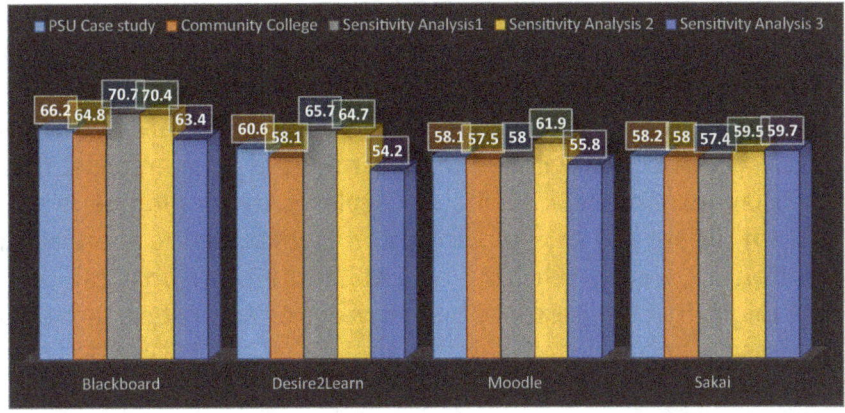

Figure 18. Comparison of the Technology Values.

Table 11. Comparison of Analyses.

LMS Alternatives	Ranking				
	PSU	Community College	Sensitivity Analysis 1	Sensitivity Analysis 2	Sensitivity Analysis 3
BlackBoard	1	1	1	1	1
Desire2learn	2	2	2	2	4
Moodle	4	4	3	3	3
Sakai	3	3	4	4	2

9. Conclusions

With many LMS providers available, institutions are faced with the dilemma of deciding which system to go for. The LMS selection process is often more about selecting a path than a product. In this case the LMS will be selected based on institutional values and leadership. There is no perfect LMS to be selected for all universities, for the variety of needs and function methods might differ from one university to the other. For example, some institutions place a high value on control, stability and the ability to customize their applications and shape the future of the LMS they select. Others are more comfortable with a product that is defined and supported by a commercial firm that innovates on behalf of their clients. Finding the LMS that best meets your organization's needs is not easy. There are hundreds of LMS products available. The investment of time and cost that organizations make in LMSs and related technologies is significant, as are the risks of disruption from selecting the wrong solution. However, selecting an LMS could be a political issue that might involve personal bias towards or against certain LMS systems and providers. From the analysis of the PSU case study, it is evident that Blackboard has the highest technology value while the remaining three alternatives have similar technology values. From the sensitivity analysis, it is clear that the technology values are more or less the same. There is no visible effect on them.

10. Limitations

Time was a limited factor in this study. Conversing with experts and collecting their evaluations took time. We had to forgo some evaluations due to time constraints on the project. More time would have also allowed us to have a better and larger selection of experts, and to collect better information and background information about them. Having better information about our experts could help us target the needed expertise. Finally, the model does not consider the financial aspects of buying and maintaining an LMS tool. Currently, this aspect is blended into the Hosting/Security factor since this is where the main difference in expenses lies.

11. Future Research

Future research could consider inviting more experts from local and international universities. It could also include the financial aspect in developing the model.

References

1. Z. Berge, "Barriers to online teaching in post-secondary institution: can policychanges fix it?" *Online Journal of Distance Learning Administration*, 1, 2 (Summer 1998).
2. E. Allen and J. Seaman, *Entering the Mainstream: The Quality and Extent of Online Education in the United States, 2003 and 2004* (United States: Sloan Consortium, 2004).
3. E. Allen and J. Seaman, *Seizing the Opportunity; The Quality and Extent of Online Education in the United States, 2002 and 2003* (United States: Sloan Consortium, 2003).
4. E. Allen and J. Seaman, *Growing by Degrees: Online Education in the United States, 2005* (United States: Sloan Consortium, 2005).
5. E. Allen and J. Seaman, *Making the Grade: Online Education in the United States, 2006* (United States: Sloan Consortium, 2006).
6. E. Allen and J. Seaman, *Online Nation: Five Years of Growth in Online Learning* (United States: Sloan Consortium, 2007).
7. E. Allen and J. Seaman, *Staying the Course: Online Education in the United States, 2008* (United States: Sloan Consortium, 2008).
8. E. Allen and J. Seaman, *Learning on Demand: Online Education in the United States, 2009* (United States: Sloan Consortium/Babson Survey Research Group, 2009).
9. E. Allen and J. Seaman, *Class Differences: Online Education in the United States, 2010* (United States: Sloan Consortium/Babson Survey Research Group, 2010).
10. E. Allen and J. Seaman, *Changing Course: Ten Years of Tracking Online Education in the United States* (United States: Sloan Consortium/Babson Survey Research Group/Quahog Research Group, 2013).
11. J. Hall, "Assessing learning management systems", *PPT Media*, 2003. http://pttmedia.com/newmedia_knowhow/KnowHow_Deploy/LMS/Docs/Assessing_MS.doc.
12. D. McIntosh, "Vendors of learning management and e-learning products", *Trimeritus eLearning Solutions*, 2013. http://www.trimeritus.com/vendors.pdf.
13. E. Kurilovas, "Methods of multiple criteria evaluation of the quality of learning management systems for personalised learners needs", January 2009.

14. S. Ozkan and R. Koseler, "Multi-dimensional students' evaluation of e-learning systems in the higher education context: an empirical investigation", *Computers & Education 53*, 4 (2009) 1285–1296.
15. E. Kurilovas and V. Dagiene, "Learning objects and virtual learning environments technical evaluation criteria", *Electronic Journal of e-Learning 7*, 2 (2009) 127–136.
16. S. W. Kim and M. G. Lee, "Validation of an evaluation model for learning management systems", *Journal of Computer Assisted Learning 24*, 4 (2008) 284–294.
17. J. Itmazi and M. Megías, "Survey: comparison and evaluation studies of learning content management systems", in *International Conference of Learning and Working in New Media Environments*, Innsbruck, Austria, 2005, pp. 1–8.
18. P. Sturgess and F. Nouwens, "Evaluation of online learning management systems", *Turkish Online Journal of Distance Education*, 5, 3 (July 2004).
19. S. Foreman, "Five steps to evaluate and select an LMS: proven practices", *Learning Solutions*, 5 June 2013. http://www.learningsolutionsmag.com/articles/1181/five-steps-to-evaluate-and-select-an-lms-proven-practices.

Appendix A: Expert List

Experts who were contacted to do the entire model

Expert	Role
1	Instructional Designer
2	Computer Sciences Professor
3	Instructional Designer
4	Educational Technologist
5	Instructional Designer
6	Instructional Designer

Experts who were contacted to do the ease of use criterion

Expert	Role
1	Distance Learning Librarian
2	School of Business Professor
3	Systems Science Professor
4	Conflict Resolution Professor
5	Applied Linguistics Professor
6	School of Business Professor
7	School of Education Professor
8	University Studies instructor
9	Biology Instructor

Experts who were contacted to do the ease of support criterion

Expert	Role
1	Senior Instructional Designer

Appendix B: Normalization Values Collection Tool

Please use the following rating system to rate each of the Learning Management System providers below in terms of each criterion :

0 if the criterion is not available and does not have ability
1 if the LMS provides only a few options related to the criterion and they are of poor quality
2 if the LMS provides only a few options related to the criterion and they are of fair quality
3 if the LMS provides only a few options related to the criterion and they are of best quality
4 if the LMS provides all options related to the criterion and they are of fair quality
5 if the LMS provides all options related to the criterion and they are of best quality

You can refer to the description of each criterion to know what possible options for it

	Blackboard	Desire2Learn	Moodle	Sakai
Content Management - Ability and quality to organize files and repositories as well as multimedia such as ppt, videos, audio, flash PDF. This would at school, dept., course and term level.	☐	☐	☐	☐
Course Reporting - Ability and quality to do assessment, analytics, rubrics, track students progress, and measurements.	☐	☐	☐	☐
Add-ons-Availability and quality of add-ons like authoring, grade book, functioning tools, drop box, calendar, and course building	☐	☐	☐	☐
Accessibility - Accessibility of the LMS. As of January of 2012, the federal government has mandated that all higher education institutions must either provide or have a tangible plan to provide accessibility services for all students of various needs. Because of this, accessibility has become a huge issue in the higher education world.	☐	☐	☐	☐
Collaboration - Ability and quality to collaborate which is more than just a means for dialog, such as file/data sharing, chat/teleconferencing, possibly social networking functionality, and the ability for student's to initiate activities on the system.	☐	☐	☐	☐
Mobile Access - Ability and quality to access the LMS via different mobile browsers and availability of apps for smartphones on different platforms.	☐	☐	☐	☐
System Reporting - Availability and quality of tools for reporting system statistics, student and instructor statistics	☐	☐	☐	☐

Appendix C: Case Study and Sensitivity Analysis Calculations

Calculations for priorities and weights

Weightage Calculations

Criteria	Weights	Sub-Criteria	Weights	Priorities	Factors	Weights	Priorities
Ease of Use	0.58	teaching tools	0.56	0.325	content management	0.36	0.117
Ease of Support	0.42	user tools	0.44	0.256	course reporting	0.19	0.062
		admin tools	0.22	0.093	addons	0.21	0.069
		support services	0.32	0.135	accessibility	0.24	0.078
		technical services	0.46	0.194	collaboration	0.58	0.149
				1.003	mobile access	0.42	0.108
					system reporting	0.25	0.024
					course management	0.3	0.028
					user data management	0.32	0.03
					certified integration	0.13	0.013
					help documentation	0.18	0.025
					training	0.24	0.033
					consultation — design, implementation	0.18	0.025
					customer services	0.4	0.054
					server side support	0.2	0.039
					client side support	0.08	0.016
					hosting/security	0.29	0.057
					Advanced troubleshooting	0.3	0.059
					ASP	0.13	0.026
							1.012

Weightage Calculations

Criteria	Weights	Sub-Criteria	Weights	Priorities	Factors	Weights	Priorities
Ease of Use	0.3	teaching tools	0.56	0.168	content management	0.36	0.061
Ease of Support	0.7	user tools	0.44	0.132	course reporting	0.19	0.032
		admin tools	0.22	0.154	addons	0.21	0.036
		support services	0.32	0.224	accessibility	0.24	0.041
		technical services	0.46	0.322	collaboration	0.58	0.077
				1	mobile access	0.42	0.056
					system reporting	0.25	0.039
					course management	0.3	0.047
					user data management	0.32	0.05
					certified integration	0.13	0.021
					help documentation	0.18	0.041
					training	0.24	0.054
					consultation — design, implementation	0.18	0.041
					customer services	0.4	0.09
					server side support	0.2	0.065
					client side support	0.08	0.026
					hosting/security	0.29	0.094
					Advanced troubleshooting	0.3	0.097
					ASP	0.13	0.042
							1.01

Weightage Calculations

Criteria	Weights	Sub-Criteria	Weights	Priorities	Factors	Weights	Priorities
Ease of Use	0.7	teaching tools	0.56	0.392	content management	0.36	0.142
Ease of Support	0.3	user tools	0.44	0.308	course reporting	0.19	0.075
		admin tools	0.22	0.066	addons	0.21	0.083
		support services	0.32	0.096	accessibility	0.24	0.095
		technical services	0.46	0.138	collaboration	0.58	0.179
				1	mobile access	0.42	0.13
					system reporting	0.25	0.017
					course management	0.3	0.02
					user data management	0.32	0.022
					certified integration	0.13	0.009
					help documentation	0.18	0.018
					training	0.24	0.024
					consultation — design, implementation	0.18	0.018
					customer services	0.4	0.039
					server side support	0.2	0.028
					client side support	0.08	0.012
					hosting/security	0.29	0.041
					Advanced troubleshooting	0.3	0.042
					ASP	0.13	0.018
							1.012

Weightage Calculations

Criteria	Weights	Sub-Criteria	Weights	Priorities	Factors	Weights	Priorities
Ease of Use	0.58	teaching tools	0.8	0.464	content management	0.36	0.168
Ease of Support	0.42	user tools	0.2	0.116	course reporting	0.19	0.089
		admin tools	0.22	0.093	addons	0.21	0.098
		support services	0.32	0.135	accessibility	0.24	0.112
		technical services	0.46	0.194	collaboration	0.58	0.068
				1.002	mobile access	0.42	0.049
					system reporting	0.25	0.024
					course management	0.3	0.028
					user data management	0.32	0.03
					certified integration	0.13	0.013
					help documentation	0.18	0.025
					training	0.24	0.033
					consultation — design, implementation	0.18	0.025
					customer services	0.4	0.054
					server side support	0.2	0.039
					client side support	0.08	0.016
					hosting/security	0.29	0.057
					Advanced troubleshooting	0.3	0.059
					ASP	0.13	0.026
							1.013

Weightage Calculations

Criteria	Weights	Sub-Criteria	Weights	Priorities	Factors	Weights	Priorities
Ease of Use	0.58	teaching tools	0.2	0.116	content management	0.36	0.042
Ease of Support	0.42	user tools	0.8	0.464	course reporting	0.19	0.023
		admin tools	0.5	0.21	addons	0.21	0.025
		support services	0.25	0.105	accessibility	0.24	0.028
		technical services	0.25	0.105	collaboration	0.58	0.27
				1	mobile access	0.42	0.195
					system reporting	0.25	0.053
					course management	0.3	0.063
					user data management	0.32	0.068
					certified integration	0.13	0.028
					help documentation	0.18	0.019
					training	0.24	0.026
					consultation — design, implementation	0.18	0.019
					customer services	0.4	0.042
					server side support	0.2	0.021
					client side support	0.08	0.009
					hosting/security	0.29	0.031
					Advanced troubleshooting	0.3	0.032
					ASP	0.13	0.014
							1.008

Calculations for technology values

Factors	Weightage	Ranking	Blackboard Desirability Value	Weight*Desirability Value	Ranking	Desire2Learn Desirability Value	Weight*Desirability Value	Ranking	Moodle Desirability Value	Weight*Desirability Value	Ranking	Sakai Desirability Value	Weight*Desirability Value
Content Management	0.06	4	80	4.8	3	60	3.6	3.5	70	4.2	3.5	70	4.2
Course Reporting	0.03	3.5	70	2.1	2.5	50	1.5	2	40	1.2	2.5	50	1.5
AddOns	0.04	2.5	50	2	2.5	50	2	4.5	90	3.6	3.5	70	2.8
Accessibility	0.04	3.5	70	2.8	4.5	90	3.6	3.5	70	2.8	2.5	50	2
Collaboration	0.08	2	40	3.2	1.5	30	2.4	2	40	3.2	3	60	4.8
Mobile Access	0.05	3	60	3	2.5	50	2.5	2.5	50	2.5	2.5	50	2.5
System Reporting	0.04	4	80	3.2	2	40	1.6	3.5	70	2.8	4	80	3.2
Course Management	0.05	4.5	90	4.5	4.5	90	4.5	4	80	4	4	80	4
User Data Management	0.05	3	60	3	2.5	50	2.5	3.5	70	3.5	3	60	3
Certified Integration	0.02	5	100	2	4	80	1.6	5	100	2	4.5	90	1.8
Help Documentation	0.04	2.5	50	2	2.5	50	2	3.5	70	2.8	2.5	50	2
Training	0.05	3.5	70	3.5	3.5	70	3.5	2	40	2	2.5	50	2.5
Consultation — Design, Implementation	0.04	4.5	90	3.6	4	80	3.2	2	40	1.6	1.5	30	1.2
Customer Service	0.09	4.5	90	8.1	3	60	5.4	2	40	3.6	2	40	3.6
Server Side Support	0.06	3.5	70	4.2	4	80	4.8	3.5	70	4.2	3.5	70	4.2
Client Side Support	0.03	3	60	1.8	4.5	90	2.7	3.5	70	2.1	3	60	1.8
Hosting/Security	0.09	3.5	70	6.3	4.5	90	8.1	3.5	70	6.3	3.5	70	6.3
Advanced Troubleshooting	0.1	3.5	70	7	3.5	70	7	2	40	4	2	40	4
ASP	0.04	4.5	90	3.6	4	80	3.2	2	40	1.6	2.5	50	2
Technology Value				70.7			65.7			58			57.4

Organizational Transformation: Universities 517

Factors	Weightage	Blackboard Ranking	Blackboard Desirability Value	Blackboard Weight*Desirability Value	Desire2Learn Ranking	Desire2Learn Desirability Value	Desire2Learn Weight*Desirability Value	Moodle Ranking	Moodle Desirability Value	Moodle Weight*Desirability Value	Sakai Ranking	Sakai Desirability Value	Sakai Weight*Desirability Value
Content Management	0.168	4	80	13.44	3	60	10.08	3.5	70	11.76	3.5	70	11.76
Course Reporting	0.089	3.5	70	6.23	2.5	50	4.45	2	40	3.56	2.5	50	4.45
AddOns	0.098	2.5	50	4.9	2.5	50	4.9	4.5	90	8.82	3.5	70	6.86
Accessibility	0.112	3.5	70	7.84	4.5	90	10.08	3.5	70	7.84	2.5	50	5.6
Collaboration	0.068	2	40	2.72	1.5	30	2.04	2	40	2.72	3	60	4.08
Mobile Access	0.049	3	60	2.94	2.5	50	2.45	2.5	50	2.45	2.5	50	2.45
System Reporting	0.024	4	80	1.92	2	40	0.96	3.5	70	1.68	4	80	1.92
Course Management	0.028	4.5	90	2.52	4.5	90	2.52	4	80	2.24	4	80	2.24
User Data Management	0.03	3	60	1.8	2.5	50	1.5	3.5	70	2.1	3	60	1.8
Certified Integration	0.013	5	100	1.3	4	80	1.04	5	100	1.3	4.5	90	1.17
Help Documentation	0.025	2.5	50	1.25	2.5	50	1.25	3.5	70	1.75	2.5	50	1.25
Training	0.033	3.5	70	2.31	3.5	70	2.31	2	40	1.32	2.5	50	1.65
Consultation—Design, and Implementation	0.025	4.5	90	2.25	4	80	2	2	40	1	1.5	30	0.75
Customer Service	0.054	4.5	90	4.86	3	60	3.24	2	40	2.16	2	40	2.16
Server Side Support	0.039	3.5	70	2.73	4	80	3.12	3.5	70	2.73	3.5	70	2.73
Client Side Support	0.016	3	60	0.96	4.5	90	1.44	3.5	70	1.12	3	60	0.96
Hosting/Security	0.057	3.5	70	3.99	4.5	90	5.13	3.5	70	3.99	3.5	70	3.99
Advanced Troubleshooting	0.059	3.5	70	4.13	3.5	70	4.13	2	40	2.36	2	40	2.36
ASP	0.026	4.5	90	2.34	4	80	2.08	2	40	1.04	2.5	50	1.3
Technology Value				70.43			64.72			61.94			59.48

Factors	Weightage	Blackboard			Desire2Learn			Moodle			Sakai		
		Ranking	Desirability Value	Weight* Desirability Value	Ranking	Desirability value	Weight* Desirability Value	Ranking	Desirability value	Weight* Desirability Value	Ranking	Desirability value	Weight* Desirability Value
Content Management	0.042	4	80	3.36	3	60	2.52	3.5	70	2.94	3.5	70	2.94
Course Reporting	0.023	3.5	70	1.61	2.5	50	1.15	2	40	0.92	2.5	50	1.15
AddOns	0.025	2.5	50	1.25	2.5	50	1.25	4.5	90	2.25	3.5	70	1.75
Accessibility	0.028	3.5	70	1.96	4.5	90	2.52	3.5	70	1.96	2.5	50	1.4
Collaboration	0.27	2	40	10.8	1.5	30	8.1	2	40	10.8	3	60	16.2
Mobile Access	0.195	3	60	11.7	2.5	50	9.75	2.5	50	9.75	2.5	50	9.75
System Reporting	0.053	4	80	4.24	2	40	2.12	3.5	70	3.71	4	80	4.24
Course Management	0.063	4.5	90	5.67	4.5	90	5.67	4	80	5.04	4	80	5.04
User Data Management	0.068	3	60	4.08	2.5	50	3.4	3.5	70	4.76	3	60	4.08
Certified Integration	0.028	5	100	2.8	4	80	2.24	5	100	2.8	4.5	90	2.52
Help Documentation	0.019	2.5	50	0.95	2.5	50	0.95	3.5	70	1.33	2.5	50	0.95
Training	0.026	3.5	70	1.82	3.5	70	1.82	2	40	1.04	2.5	50	1.3
Consultation — Design, and Implementation	0.019	4.5	90	1.71	4	80	1.52	2	40	0.76	1.5	30	0.57
Customer Service	0.042	4.5	90	3.78	3	60	2.52	2	40	1.68	2	40	1.68
Server Side Support	0.021	3.5	70	1.47	4	80	1.68	3.5	70	1.47	3.5	70	1.47
Client Side Support	0.009	3	60	0.54	4.5	90	0.81	3.5	70	0.63	3	60	0.54
Hosting/Security	0.031	3.5	70	2.17	4.5	90	2.79	3.5	70	2.17	3.5	70	2.17
Advanced Troubleshooting	0.032	3.5	70	2.24	3.5	70	2.24	2	40	1.28	2	40	1.28
ASP	0.014	4.5	90	1.26	4	80	1.12	2	40	0.56	2.5	50	0.7
Technology Value				63.41			54.17			55.85			59.73

Selection of Best LMS	Ease of Use	Ease of Support	Inconsistency
Expert 1	0.75	0.25	0
Expert 2	0.4	0.6	0
Mean	0.58	0.43	
Minimum	0.4	0.25	
Maximum	0.75	0.6	
Std. Deviation	0.18	0.18	
Disagreement			0.18

Ease of Support	Admin Tools	Support Services	Technical Services	Inconsistency
Expert 1	0.14	0.43	0.43	0
Expert 2	0.31	0.2	0.49	0.02
Mean	0.23	0.32	0.46	
Minimum	0.14	0.2	0.43	
Maximum	0.31	0.43	0.49	
Std. Deviation	0.09	0.12	0.03	
Disagreement				0.08

Users Tools	Collaboration	Mobile Access	Inconsistency
Expert 3	0.22	0.78	0
Expert 4	0.55	0.45	0
Expert 1	0.5	0.5	0
Expert 5	0.73	0.27	0
Expert 6	0.8	0.2	0
Expert 2	0.66	0.34	0
Mean	0.58	0.42	
Minimum	0.22	0.2	
Maximum	0.8	0.78	
Std. Deviation	0.19	0.19	
Disagreement			0.19

Support Services	Help Documentation	Training	Consultation-Design, Implementation	Customer Services	Inconsistency
Expert 1	0.22	0.39	0.1	0.29	0.03
Expert 2	0.13	0.09	0.27	0.51	0.01
Mean	0.18	0.24	0.19	0.4	
Minimum	0.13	0.09	0.1	0.29	
Maximum	0.22	0.39	0.27	0.51	
Std. Deviation	0.05	0.15	0.09	0.11	
Disagreement					0.1

Ease of Use	Teaching Tools	Users Tools	Inconsistency
Expert 3	0.6	0.4	0
Expert 4	0.65	0.35	0
Expert 1	0.5	0.5	0
Expert 5	0.5	0.5	0
Expert 6	0.7	0.3	0
Expert 2	0.41	0.59	0
Mean	0.56	0.44	
Minimum	0.41	0.3	
Maximum	0.7	0.59	
Std. Deviation	0.1	0.1	
Disagreement			0.1

Teaching Tools	Content Management	Course Reporting	Add-ons	Accessibility	Inconsistency
Expert 3	0.43	0.19	0.24	0.14	0.03
Expert 4	0.38	0.19	0.12	0.3	0
Expert 1	0.46	0.18	0.18	0.18	0.15
Expert 5	0.43	0.13	0.27	0.16	0.01
Expert 6	0.2	0.28	0.1	0.42	0.06
Expert 2	0.26	0.18	0.32	0.24	0.07

(*Continued*)

(Continued)

Teaching Tools	Content Management	Course Reporting	Add-ons	Accessibility	Inconsistency
Mean	0.36	0.19	0.21	0.24	
Minimum	0.2	0.13	0.1	0.14	
Maximum	0.46	0.28	0.32	0.42	
Std. Deviation	0.1	0.04	0.08	0.1	
Disagreement					0.08

Admin Tools	System Reporting	Course Management	User Data Management	Certified Integration	Inconsistency
Expert 1	0.12	0.48	0.27	0.12	0.03
Expert2	0.37	0.11	0.37	0.15	0.01
Mean	0.25	0.3	0.32	0.14	
Minimum	0.12	0.11	0.27	0.12	
Maximum	0.37	0.48	0.37	0.15	
Std. Deviation	0.13	0.19	0.05	0.02	
Disagreement					0.09

Technical Services	Server-side Support	Client-side Support	Hosting/ Security	Advanced Troubleshooting	ASP	Inconsistency
Expert 1	0.27	0.05	0.34	0.22	0.11	0.04
Expert 2	0.13	0.13	0.23	0.37	0.14	0.08
Mean	0.2	0.09	0.29	0.3	0.13	
Minimum	0.13	0.05	0.23	0.22	0.11	
Maximum	0.27	0.13	0.34	0.37	0.14	
Std. Deviation	0.07	0.04	0.06	0.08	0.02	
Disagreement						0.05

Chapter 16

Organizational Transformation: Consumer Goods

Yogi Hamdani* and Tugrul Daim*,†,‡

*Portland State University, Portland, Oregon, USA
†Higher School of Economics, Moscow, Russia
‡Chaoyang University of Technology, Taiwan

Abstract

The importance of packaging has proved to be essential for consumers which has brought more focus on packaging, especially in the Fast-moving Consumer Goods environment, where a huge number of product alternatives are available for consumers. Therefore, the understanding of the packaging itself and its role in the business is essential, which then leads back to the main question, *what is the purpose of packaging in consumer goods product?* This thesis tries to briefly explain the importance of packaging and points out the critical part of the packaging system in FMCG industries that enables a sustainable business practice in a highly competitive market environment. The findings show that packaging is not only for the protection of the product but also as an information channel and marketing tool. These needs have led to the existence of the label on the packaging that serves as a communication media. However, the very same labels also have both a positive and a negative impact toward other attributes of the packaging. *Therefore, the objective of the thesis is to assess and find out the best label alternative which is the most suitable in the FMCG environment and applicable throughout the supply chain, from manufacturer to the end consumer.*

Keywords: Technology assessment, consumer goods, supply chain.

1. Introduction

Packaging is essential for consumers to choose among fairly similar products [1]. It is also believed to be the activator that leads consumers' preference, set at the moment to buy the product [2]. Another study also revealed that packaging is able to make the difference in the competition between consumer goods [3]. These claims have brought more attention to packaging, especially in a Fast Moving Consumer Goods (FMCG) environment, where a huge number of product alternatives are available to consumers. Therefore, the understanding of the packaging itself and its role in business is essential, which then leads back to the main question: what is the purpose of packaging in a consumer goods product?

This research paper tries to explain briefly the importance of packaging and points out the critical part of the packaging system in FMCG industries that enables a sustainable business practice in a highly competitive market environment. The findings show that packaging is not only for the protection of the product, but also acts as an information channel and marketing tool [4, 5]. These needs have led to the existence of the label on the packaging that serves as a communication media. However, the very same label also has both positive and negative impacts toward other attributes of the packaging. *Therefore, the objective of the research paper is to assess and find the best label alternative, which will be the most suitable in the FMCG environment and applicable throughout the supply chain, from the manufacturer to the end consumer.*

The research paper's methodology is based on the works of Tugrul Daim, regarding technology assessment and acquisition [6] and cooperation with one of the leading FMCG industry partner. Firstly, the gap analysis was conducted to understand the problem clearly and identify the gap between the needs of the label from the manufacturer to the end consumers, and its current capability in labeling across the supply chain in Europe. The gap analysis was also used as a filter to screen out the inapplicable labels in the given current infrastructure. Secondly, the Hierarchical Decision Model (HDM) was developed with expert opinion input from the various departments of different expertise and position of an industry partner. The results reflect and represent the best label alternatives for primary packaging in the FMCG environment.

1.1. *Purpose of packaging*

Several studies show that packaging for consumer goods has more purposes rather than simply just being a cover. However, it serves mainly to fulfill its role in either logistic or marketing purposes [6]. As a logistic tool, the core purposes of packaging are to protect the product itself from the outside environment to maintain the product quality. At the same time, packaging also serves as a communication medium to carry information from the manufacturer across the supply chain to the consumer [7]. From a marketing perspective, packaging creates the end of the "promotion-chain" [8] and behaves as the "silent salesman" to attract and hold consumers' attention for a few seconds, long enough for him/her to buy the product [5].

Moreover, the complexity of integrating packaging and logistic is also rather high [9] because packaging directly impacts main logistic activities, such as transportation, inventory and warehousing, and communication [10]. In addition, to protect the product, the overall packaging must be easy to handle with the right amount and size throughout the supply chain [7]. The packaging also has to ensure a secure, efficient and effective delivery to maintain the product in its best condition to be offered to the end consumer [11]. Any disorder or misfit may lead to an incorrect handling process or even instability of the whole packaging, which can cause product damage and rejection from customers.

Packaging is also responsible for the information flow regarding the product held inside. More information available on the package will reduce shipment delays, handling process cost and also decrease tracking effort, either as forward delivery or backward flow (i.e., recalls) [12]. This will enable a quick response in the supply chain as it is deemed to be essential, especially in FMCG industries [13]. Moreover, consumers also use packaging as an extrinsic cue to infer the intrinsic quality [14] and communicate the benefit of the product [15].

In the end, the consumer will be the key factor in the future demands of product packaging [16], as it transforms from a physical box (or any other covering) protecting the product into a "silent salesman" [5]. Procter & Gamble came out with this term "First Moment of Truth" (FMOT) to describe the moment that occurs at the store shelf when consumers decide to buy a brand [17]. It is a critical moment that is enabled by the packaging to obtain the consumers' attention [15]—consumers only spend a few seconds to make choices among products at the point of purchase [5, 14]. Therefore, the packaging not only serves the logistical perspective but also supports sales. Particularly within FMCG industries, packaging impacts the overall supply chain's effectiveness as it represents an interface from the manufacturer—the start of the chain—to serve the ultimate end chain, the end consumers [18].

1.2. *Type of packaging*

In general, there are three layers of packaging: primary, secondary and tertiary [7, 10]. Primary packaging is the first package that is directly touching the product. It is often called the selling unit (or consumer packaging), as this is what the consumer usually takes home. If the product's primary packaging is too small, the secondary packaging comes into place to gather several primary packagings into one big pack. Then, they are all combined into a large-sized tertiary packaging for transportation, usually in the form of a pallet. Because the nature of the tertiary packaging is to cater to transportation and distribution needs, the tertiary packaging is sometimes also referred to as a shipping unit.

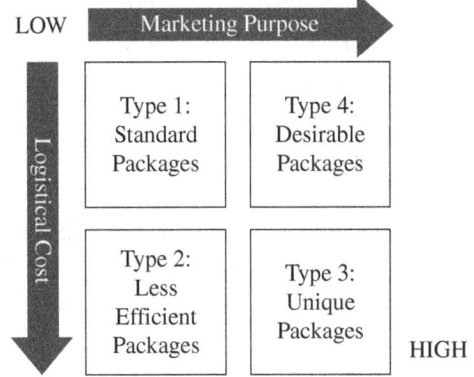

Figure 1. Different types of packaging [18].

According to Saghir [18], there are four different area types of packaging to cater to different logistic needs and marketing purposes, as shown in Figure 1. In an FMCG environment, packages have a short lifespan with high turnover rates [19]. They also serve a marketing purpose [4], which explains why most FMCG products found in a retail store have fancy and nice-looking artwork at the expense of marketing purpose, with a standardized dimension to reduce the overall cost of the product.

This research paper focused on the study of the label on the primary packaging because it has a direct impact on the end consumers. Besides, the product that was studied is large and sustainable enough not to require a secondary packaging. Hence, the product in the primary packaging is palletized directly as a shipping unit instead. The nature of the FMCG industries has also brought attention to standardizing labels on packaging across different product platforms and users, with only minor changes to minimize the cost.

1.3. *Labels on packaging*

Labels have been an integral part of the packaging industry for over a hundred years [20]. Labels exist to enable packaging to carry information about the product throughout the supply chain. Several types of labels have been introduced on the market to fit the packaging, and 86.29% of them are self-adhesive label [21]. This label has become the most popular due to its wide range of applications, fast design changeover and rapid labeling process [20].

In term of digitalization, there have also been numerous technologies developed, such as Point of Sale tracking, Electronic Data Interchange (EDI), and many others [22]. This issue has put labels into a more important role, as producers and retailers have to keep track of each Stock Keeping Unit (SKU) efficiently and reliably to support tracking and EDI. It enables the product to be tracked easily and delivered from the production sites to the end consumers in an autonomous system [23]. For instance, in an automated distribution center and warehouse, all the autonomous conveyor belts and the storage-and-retrieval system read a certain label of the finished product to identify and transport it to the right place at the right time.

The complex logistic flows and varieties of many involved parties point toward a need for easy physical identification of logistics units, which GS1, a non-profit organization, offers a standard to help to solve the issue [24]. GS1 was founded in the United States in 1973 and was the first to be administered with a Universal Product Code (UPC) [25]. Nowadays, GS1 has been transformed into a worldwide organization comprising members and contributors from various multinational enterprises in numerous sectors. GS1 established GS1 Logistics Label, a standardized label for the logistic unit that comprises tertiary packaging and/or secondary and primary packaging. This standard label allows all users throughout the supply chain to trace and identify each unit uniquely [24].

The GS1 Logistics Label is derived into two basic forms of information:

1. Information read by people.
2. Information used by machines and automated systems.

The first basic form of information consists of Human Readable Interpretation, non-HRI text and graphics. This information is necessary in the event of manual handling or in processes that do not use any machines. Also, in the event of machine failure or any other error that causes the logistic unit to be rejected from the automatic process flow, an operator can still identify and process the product manually, as the information is readable to the human eye. The second basic form of information is coded information, which is read specifically by machines. It is designed such that the machine can identify each logistic unit rapidly, for example, scanning at the high-speed conveyor belt. In conclusion, both these forms of information must come together in the GS1 Logistics Label [24].

The UPC is the original format for the commercial product barcode, which is standardized by GS1. It consists of a five-digit manufacturer code, a five-digit product item code, and two check digits [26]. As demand in Europe, Asia and Australia grew two digits of country code were added to the first string of the code, which GS1 later referred to as European Article Numbers (EANs) or International Article Numbers [27]. However, in the United States and Canada, the original UPC is still used, which is translatable with a country code of zero in the EAN code. The EAN is also a GS1 standardized barcode, which exists in most of the commercialized products in stores worldwide [26]. Besides the EAN and UPC variant, there are also other standardized barcodes for other purposes, such as ITF-14, GS1-128 and Omnidirectional. Alongside the standardized 1D barcode, GS1 also has the standard for smaller 2D codes—the QR Code and 2D Data Matrix code (Figure 2) [25].

In addition to the general information used by all users in the supply chain, there is extra information required internally by the consumer goods manufacturer. As was studied in the industry partner, this label on the packaging must also have information about the batch and order numbers. This information is part of a quality assurance requirement where in cases of quality incidence, the product can be traced back to its origin,

Figure 2. GS1 standardized code examples [25].

i.e., the production plant and line it came from, and also the date and time when the product was produced. For the sake of reliability, the information has to be available in HRI and code from the package, although it is not possible to print this information onto the artwork of the package as it requires live timestamp information. Therefore, this information is printed online on the production line through a self-adhesive label, which is then placed on the primary packaging.

1.4. *Labels dilemma*

Being the "silent salesman", the packaging on a product has a huge responsibility to sell out the encased product. Moreover, a study has revealed that brand loyalty in consumer goods is hard to obtain [28]. This increases the onus on packaging to boost the marketability of the product. According to research, packaging is one of the important factors that determine whether consumers accept a certain product [29]. Visual graphics on the packaging are intended to attract consumers [30], and elements such as text, color and shape are important for consumer perception [31]. In the end, packaging has become a part of the communicational dimension factor that affects the buying behavior of consumers [32].

As in the FMCG industry, where firms are able to build brands based on a specific product [33], many companies have put in much effort to design the best artwork and deliver a nice-looking packaging to attract consumers. Research has also discovered that consumer motivation (in this case, to buy a product) is generated through pictorial and artistic information on packaging [31]. Particularly for primary packaging, any given brand has to win over consumers through the appearance of its packaging over other competitor brands, whose products may share the same shelf with its products.

On the other hand, packaging also plays a role as the communication medium that carries information about the product itself. If the information can be printed directly on the package, then the design team is able to specifically assign space for this information to be printed directly on the artwork on the packaging. In general, the artwork design should suppress information related to internal and logistical needs, which is irrelevant for consumers and is best hidden. Preferably, the packaging appearance should have a sense of beauty and be practical [31].

Internal information is usually online-based and hence can only be printed on a label during the production process when it gets stuck on the packaging. The problem arises when this label covers the artwork. It gets worse if the label needs to be placed in a specific area, like say, the center of the package, due to the needs and requirement of the automation system with several autonomous scanning devices. Hence, having a label is mandatory for production and logistical purposes, at the cost of packaging appearance. However, a report from an industry partner has suggested that this label has a negative FMOT impact for consumers.

This issue has led to these research questions:

1. Is it possible to eliminate or at least reduce the label size on the primary packaging?
2. What are the possible label alternatives that can be implemented?
3. What is the best label alternative for primary packaging, which can be applicable in the FMCG environment?

1.5. *Insights from the FMCG industry*

This work in answering the research questions was initiated by the FMCG industry partner. As one of the biggest FMCG industry, it has granted us access that enabled us to study and understand its vast supply chain network, and the logistic flow that caters to customers like the retail stores, and ultimately consumers across Europe during a short period of time. Besides tackling the research problem regarding labels and packaging, the way the company carried the project as a new initiative was also learnt. The findings and insights are then compared with the theoretical background.

The industry partner uses two main drivers to run new project initiatives: quality improvement and/or cost reduction. These drivers match with theoretical generic strategies to gain competitive advantage that firms in a broader target competitive scope must be in the cost leadership or have differentiation [34]. Cost leadership means offering the lowest price to consumers, and this can be achieved with cost reduction in all the processes but still maintaining the quality of the product. While quality improvement translates into a firm offering consumers the best and most innovative product, it includes packaging that differs from its competitors.

To reduce cost in the packaging area, the firm tried to reduce the number of customizations as much as possible to yield economies of scale. In several cases, when product variation within the same platform existed, packaging was customized accordingly within the boundary to maintain product quality and efficiency of the overall packaging. All production processes are also set to a global standard to reduce learning and maintenance effort. These were claimed to be an effective and efficient manufacturing process that delivers the same high-quality product across the region.

In the midst of quality improvement for the product brands, Procter & Gamble's former CEO, A. G. Lafley, argued that the best brand must win two moments of truth: the FMOT, and the second moment, which happens at home when consumers use the brand whether they are satisfied or not [17]. To win at the Second Moment of Truth (SMOT), the product must firstly win at the FMOT, where consumers choose to buy the product. In this case, packaging plays a significant role in winning the FMOT [15]. With this understanding, the packaging team in the industry partner has put a heightened effort to always deliver the best packaging quality for the ultimate FMOT experience.

Taking one step backward in the supply chain, FMOT will not be present if the product itself is not available on the shelf. For FMCG consumers, promotion and out-of-stock problem have become the second- and the third-most highlighted issue, respectively [35]. Moreover, from the suppliers' point of view, the Out of Stock (OOS) issue contributed to 54% of consumers deciding not to buy the specific brand and switched to other brands rather than try to find the same brand in other stores [36]. It means that On-Shelf Availability (OSA) is critically important not only to solve the OOS directly but also work as a promotion to capture consumers through the FMOT.

Procter & Gamble, for instance, tackles the OOS problem by integrating information digitally across the supply chain to gain the required agility to maintain OSA. From a theoretical perspective, IT integration has been proven to impact the supply chain flexibility directly by giving an increase in agility, which in the end provides a higher competitive business performance [23]. Once again, labels play a role in carrying the information as part of the integration. The main question then is: how can a firm deliver the best FMOT experience for consumers, through OSA and packaging appearance, while also having an effective and efficient production and logistical process with labeling that works across the supply chain?

2. Methodology

There are three major tasks—gap analysis, environment analysis and evaluation analysis—that can answer the research questions in the previous section, in assessing and acquiring a certain alternative from the available technologies (either services or products) as proposed by Tugrul Daim [6]. The gap analysis helps to define the problem or process improvement by providing the minimal requirements. It is followed by the environment analysis to see what the available options are and to identify all related technologies. Lastly, the evaluation analysis is used to evaluate and assess each available alternative with the help of the HDM [6].

In addition, the Political, Economic, Socio-cultural and Technological (PEST) analysis is also used as the basis framework to observe and narrow the large external business environment for learning in a specific organizational system [37, 38]. It was originally used to understand the position of a specific organization [39] and to study the viability of solutions for a general management in a business environment [38]. It is also deemed to be useful to provide analysis for strategic decision-making, marketing planning, organizational change, and product development [40]. The PEST analysis consists of political factors analysis that covers several forms of government involvement, economic factors, social factors including cultural and demographic, and technological factors over the external environment [41].

For this particular work regarding labels on primary packaging, political factors such as country-specific regulation must be applicable to all the alternatives, in order to rule it out in further analysis. From the study conducted in the industry partner, external economic factors that relate to the packaging are negligible compared to the internal change. Hence, the economic analysis is based only on the internal factor. Social factors are analyzed from all stakeholders within the supply chain, which consists of the producer, customers and consumers, who deal directly with the labels. At last, the technological factors are vital in determining the possible label alternatives.

The found PEST analysis is, therefore, to be modified and integrated with the gap analysis and the environment analysis. A portion of technology forecasting was also carried on to show the trends in labeling and technology diffusion in the market as an approach for the environment analysis to give a broader overview about the alternatives [6].

2.1. Gap analysis

The gap analysis was started with understanding the problem definition. Next, the detailed needs to implement the alternatives or the process improvement, for instance, technical parameter of the target values, shall be listed. In parallel, an internal audit is carried out to understand the availability of the infrastructure and capabilities within the system. The gap, which is the basis for the assessment criteria, is defined as the differences between the list of the needs and current capabilities from the internal audit [6].

In this research paper, the main problem was to find the best label alternatives for primary packaging in FMCG products. With the help of the PEST framework, there are three main aspects found as the gap between current capabilities within the whole supply chain for the particular consumer goods product and possible label alternatives requirement. They are:

1. Technical factors;
2. Economic factors;
3. Social factors.

Details for each factor was the result of an internal study from the production plant in the industry partner and throughout its supply chain network across Europe. In term of social factors, labels on primary packaging impact three categories of groups—producer, customers and consumers. From the thorough study, and to answer the first question research, it was found that labels on primary packaging couldn't be eliminated due to the needs of online data, which is used by producer and customers. It can, however, be hidden or at least minimized from the packaging appearance.

2.1.1. Technical gap

The technical factors consist of three parts—labeler machine, scanning devices, and artwork design (Tables 1–3). For the labeler machine and the scanning devices, the important gap is about capabilities and requirement. The artwork design, on the other hand, has a concern about integration between information on the label, which is referred to as a code, and the picture of the artwork. It deals with the position and orientation of the code as such that it can be readable and used throughout the supply chain. In the meantime, the current labeling machine and scanning device requirement in the industry partner is based on a self-adhesive thermal label. Therefore, the found parameters are suitable for the code applied in the specific label type. Other labeling types are out of scope from the gap analysis.

With the given current labeler machine, the most critical parameter lies in the labeling speed and the minimum label size that the labeler can stick onto the product due to the high-speed process of the production line. As for the label size, it is even desired that the label be removed totally, which currently cannot be achieved and is only limited to the smallest size the machine can handle. The minimum resolution is the smallest dot or cell of a code that the labeler can print. A change of label design means uploading and transferring data from the design template and online information such as batch number and order number. Equipment change is sometimes necessary to re-adjust and recalibrate during daily maintenance and the effort spent should be minimal.

Regarding the scanning device, the important points to take note of are the scanning and decoding speed, minimum resolution of the code, and the number of quiet zones. This reading speed is highly influenced by the conveyor belt speed that transports products throughout the production plant. The minimum resolution is the smallest dot or cell of the code that the scanner can read in the given configuration, and the number of quiet zones representing the distance between the edge

Table 1. Technical gap on the labeler machine.

Requirement	Minimum Parameter	Desired Parameter
Labeling speed	15 labels/min	>15 labels/min
Minimum label size	30 × 20 mm	15 × 15 mm
Minimum resolution	200 dpi	300 dpi
Label design change	2 minutes	Seamless
Equipment change	<30 minutes	<15 minutes

Table 2. Technical gap on the scanning device.

Requirement	Minimum Parameter	Desired Parameter
Scanning speed (via conveyor speed)	80 m/min	100 m/min
Minimum resolution	0.80 mm/cell	0.62 mm/cell
Quiet zone	6 cells	3 cells

Table 3. Technical gap on artwork design.

Requirement	Minimum Parameter	Desired Parameter
Code integration	Dedicated space	Seamless
Printing quality	On carton	On plastic bag and carton
Scanning quality	HRI	HRI, Machine

of the label to the edge of the code. As for the current system and equipment, it is only possible to work with 1D barcode and/or 2D codes. Therefore, from the producers' point of view, other codes cannot be implemented in the short to medium term.

The last part of the technical gap lies in the design of the artwork for the primary packaging. In principle, primary packaging is not made-to-order on site. Therefore, any online information can only be printed in very limited circumstance. It can also be caused by the high-speed production process that hinders proper online printing on the packaging. Hence, the connection between the information or code integration and the artwork design is very distant. However, several methods can be applied to avoid the perception of the label covering the artwork, one of which is providing a dedicated empty space for the label to be placed. However, the ultimate goal desires that all labels on primary packaging are to be either eliminated or integrated seamlessly with the artwork and that online information can directly be printed on the packaging bag.

2.1.2. *Economic gap*

The internal factors vastly outnumbered the external factors of the economic values, hence only the internal aspects were analyzed. Several studies have pointed out that the Net Present Value (NPV) is the most popular calculation for capital budgeting and decision-making [42–44]. This tool compares the total investment cost with the estimated cash inflow for the given time period with an appropriate discount rate. Assuming other variables are held constant, the total cost impact of implementing or changing labels on primary packaging lies within the total investment required to enable the alternative and also the future operating cost of using the alternative. In addition, the agile nature of an FMCG industry demands rapid innovation, and development time is of the essence, which is also confirmed by an observation made by the industry partner. Thus, the economic factors that are taken into analysis are development time, investment cost and future operating cost.

During the development time of a certain new technology for an online production system in the industry partner, it first requires a solid kick-off to allocate resource and working framework identification. It is followed by a study of technology feasibility and implementation. Once this knowledge is obtained, it is mandatory to do a proof of concept test. If this test is successful, then the technology may be applicable to the system, otherwise, it is concluded that the technology cannot be adopted at this moment. A validation test is then necessary to validate the capability of the new technology with a bigger test sample, longer test duration and higher success rate. If the technology passes the validation requirement, then the technology can be rolled out and implemented.

Unfortunately, there are no specific numbers that can be used as either a minimum parameter or a desired value. This is because the time frame for the development is varied with the urgency of the changes, business condition, the importance for the business, and also the risk of the new technology. For this label on a primary packaging issue, several experts from the company suggested that the development time window for the whole complete processes must be within three to 24 months.

For the investment cost, the important points in adopting a new technology from starting up the project until implementation and running the production lines are:

- Staff charter and/or training;
- Testing cost, for instance, product scrap, test material and third-party involvement;
- Equipment.

From those points, taking into account that an NPV calculation is often used to justify certain decisions of a project, this investment cost is then to be compared with the estimated future operating cost, assuming the label alternative is already implemented. The additional operating cost that label implementation may influence are:

- New staff for operation;
- Maintenance;
- Raw material, in this case, the label itself.

Nevertheless, the exact number to compare is not available as it varies drastically from project to project and with the same reason as of determining the development time. However, it is estimated that the cost of the packaging in an FMCG environment is varied and topped to a maximum of 10% of the overall product cost [4, 45] and a maximum of five years NPV period, which is favorable according to the industry partner. Therefore, the maximum limit of the total investment cost and the future operating cost can be estimated. However, the estimation is highly diverse due to the complexity of the packaging system across each product platform, for example, package type (primary, secondary and tertiary), forms (bag, bundle and box), and hence is not discussed.

Staffing decision also contributes to the total cost, specifically for packaging. From this observation, third party staffing is sometimes cheaper than hiring employees or training them into experts or specialists working on a specific task. In the scope of labeling, the industry partner has trained a couple of employees to be experts. This shows the importance of the label throughout the whole production system.

2.1.3. *Social gap*

The players—producer, customers and consumers—involved in the labels on primary packaging contribute social factors. The nature of this social gap analysis is more of a qualitative measure that explains the kind of experience that each label user across the complete supply chain is getting. These measures were found out during the observation in the industry partner's production plant and its supply network.

From the producer's point of view, the important aspects are the reliability of a certain technology, the compatibility to integrate the technology into the system, and the ease to implement the changes. Moreover, a mitigation plan for manual handling and operation for every technology shall also be available. As in the case of failure, the produced product must not be rejected only because of an error in the label on the packaging. Thus, all alternatives must abide by the standard operating procedure and regulations, according to the quality assurance and country specific regulations, if applicable.

For customers, the imperative factors are alignment in the usage of information on the label and trends in the labeling technology. Alignment is critical as the information in the labeling system is digitally integrated and therefore must be compatible throughout the supply chain—from producer to the end consumer—in order to avoid logistical errors. The complexity also increases because there are different customer groups that are geographically divided across Europe. Moreover, each group may probably also cater to several market segments, such as retail stores, hypermarkets and drug stores, and also e-commerce. Each and every group and segment sometime contribute with their own specific needs and regulation that have to be fulfilled.

Lastly, consumers consider the label as a part of the packaging. Therefore, the complete FMOT experience is important to attract consumers. Besides, labels may also be used to engage consumer participation, for example, through a website link in the code. There are no justified numbers found about consumers' engagement through that link. However, since the label becomes the part of the primary packaging, then it must also be useful for consumers in addition to the internal needs for the coded information. In the end, the ultimate goal is to improve consumers' willingness to buy the product with a good FMOT experience.

2.2. *Environment analysis*

The purpose of an environment analysis is to identify all related technologies and their maturity through market diffusion that will represent the alternative candidate for the technology assessment and acquisition [6]. For the purpose of this research paper, there are numerous automatic identification technologies available in the market. These technologies can be categorized into two groups: physical or visual contact medium, and contactless medium. Contact medium technology consists of all variants of barcodes that require either physical or visual contact, which is often referred to as the "line of sight". On the other hand, radio-frequency identification (RFID) is an example of a contactless medium, where the information can be transmitted wirelessly through radio frequency. The next section briefly discusses both groups of automatic identification technologies as the alternatives for the HDM.

2.2.1. Barcodes

Barcode is the most used and mature technology that has been adopted in supply chain management [46]. Barcode technology was adopted as early as the 1970s, and has since then improved data management and handling efficiency throughout the supply chain in companies [47]. The technology diffusion in the market is humongous as can be seen from the fact that approximately 80% to 90% of Fortune 500 firms have automated systems in their warehouses with barcode systems [48].

Barcode technology leads the automatic identification technology market due to its affordability, ease of use and reliability [46]. This humble technology requires only an empty label and a labeler to print the code. It is also easy to use—online information can be encoded and printed whenever required without manual data entry, and it only needs a barcode scanner device to retrieve the information [49]. Nowadays, there are even various free barcode reader applications for smartphones with camera. The reliability of the barcode is also undisputable as it has the highest successful scanning rate (of less than three misreads in ten thousand samples) in the industry. Of these misreads, they all fall under circumstances where there is misorientation of the product, wrongly printed information, and other errors that may have no relation at all with the code.

The first barcode technology that appeared in the market consisted of several bars (or parallel lines of different widths) of encoded numbers, which is today referred to as 1D barcoding [50]. As the carried information has grown massively over time, the 2D barcode was invented in the 1980s. There are several 2D barcodes variants that have been developed and are being used. Several examples of such a barcode are Code 49, QR Code, PDF417 and Data Matrix [49]. However, in addition to the 1D barcode, GS1 only standardizes QR Code and Data Matrix for 2D barcodes usage in a business application [25]. Most recently, there is an underdeveloped technology called Fiducial, which comes with a barcode shaped like a circle, and it is claimed that this barcode is be able to have thousands of different codes within [51]. However, the Fiducial barcode application is only applicable in augmented reality and has not been seen on consumer goods yet.

In addition to the generic version of the barcode, there have been numerous technologies invented to improve its performance in a specific industry. Some basic techniques were merely cosmetic by making the label appearance look nicer, for example, by transforming the basic rectangular form of the barcode into other preferred shapes. Other technologies improved the functionality of the barcodes. For instance, a barcode printed with special ink that is excitable only by specific wavelengths of incident radiation can reduce counterfeiting [52]. Other inventions tried to tackle the "line of sight" problem; an optical management system for a synthetic barcode module consisting of light management and a controller module was invented [53].

2.2.2 *Radio-Frequency Identification*

RFID is an information and communication technology that enables automatic identification of objects, locations and individuals to the computing system in a wireless environment without the need for manual intervention. In general, the technology consists of two components: the RFID tag and RFID reader. The RFID tag contains a unique identifier and is attached or embedded to the target object. The RFID reader is an electronic device that searches for a target tag. When such a tag is available in the vicinity, it reads the tag identification and retrieves the information in several ways depending on its application [54].

In business applications, RFID technology increases a significant positive impact of stock availability by 45.4% and improves supply performance by 36.3% [55]. However, for the labeling application on primary packaging, the RFID system is deemed too expensive because the total cost is not recoverable, even after sales improvement and cost savings made by RFID implementation [47, 56]. Several studies have tried to figure out a solution to optimize the economic benefit of RFID in several industries through cost sharing and collaborative practices throughout the supply chain stakeholders [55–57]. Moreover, a model for RFID application in consumer units shows that the costs and benefits of the technology are distributed in an asymmetric way [58]. This explains why the adoption of RFID technology is very limited due to the complexity of the supply network and the cost burden for the producers [47].

2.3. *Hierarchical Decision Model*

There are two methods of learning proposed by Thomas Saaty. The first method is through examining and studying the object and all of its properties, synthesizing the analysis, and then making conclusions from the whole observation. The second method is to learn the entity relative to other similar objects and then relating them through comparisons [59]. The argumentation is that complex mathematics is not needed to make a decision for a certain problem; it just requires the right mathematical formula. He then developed a mathematically based technique to analyze sophisticated situations through simplification. This technique has become famous as the Analytic Hierarchy Process (AHP), and has been very helpful for decision makers to structure and analyze a wide range of problems [60].

The AHP has been widely used, modified and improved over time. Dundar Kocaoglu is also one of the pioneers in developing multiple decision models such as the AHP and is also the creator of the HDM. The HDM is a mission-oriented technique to evaluate and/or select alternatives. The tool can be used to consider a wide range of alternatives from the business environment to daily life application. It is claimed that the HDM approach has been implemented in various industrial sectors, the education system and has even influenced government planning [61].

As the HDM has been widely used and researched across studies for assessment and selection in decision-making problems, this method is relevant for this research paper. Thus, we implemented the HDM here to answer our main research question: what are the best label alternatives on the primary packaging that are applicable in the FMCG environment?

The HDM follows closely the AHP in its process. The whole process consisted of four main steps. The first step is defining the problem and knowing which kind of knowledge will be observed. The second step is then to structure the hierarchy with the mission on the top, followed by the objectives in a wider perspective, through intermediate levels, and lastly the alternatives on the lowest level. The third step is finding the global weight for each element by having a pairwise comparison for each factor in the same cluster in each level. The last step is to score the alternatives, based on priorities or other normalized scales against each direct upper element. The score is then multiplied with the global weight and summed. In the end, the alternative that yields the highest value from the sum of the multiplication shall be the best alternative.

2.3.1. *Constructing the HDM*

The very first step in the HDM construction is to define the problem, which is the research question itself. The mission is to find the best label alternative that is applicable in the FMCG environment, meaning that the best alternative must also be suitable throughout the complete supply chain. Knowledge to be found generally lies in the study of the labels and the study of its application in the FMCG environment, which is partly fulfilled in the gap analysis.

The second step consists of analyzing the possible hierarchy based on the objectives in a broader spectrum. Carrying the main goal to find out the best label alternatives, the objectives found in the industry partner to reach the goal are based on the technical factor, economic factor and user experience. Through several rounds of interviews with the employees in middle management, the essential objectives were found for assessing the project to deliver a technically sound solution with the best economic benefit and also to satisfy all stakeholders. Interestingly, these objectives also matched with the conducted gap analysis. Then, the intermediate levels are constructed based on the study and observation of the needs and relation to the labels on the primary packaging.

The first objective in the technical factor consists of the technicalities of the label on primary packaging, which relate to the labeler machine and scanning device. Both these devices have a direct relation with the label—they are its main "creator" and "user". The exception for the HDM is that this technical factor is only capable of assessing labels that are printed on self-adhesive labels with normal ink. Other labeling technologies like RFID cannot be assessed, because the equipment and infrastructure for this technology were not available for analysis.

Inside the labeler machine, its printing attributes and stickability are important features for minimizing the label size. The printing attributes refer to the printing resolution, quiet zone and other HRI information or text that needs to be printed, which directly impacts the label size. The stickability criteria relates to the mechanism on how the labeler machine can handle and stick the smallest label to the product. There is a limit and threshold to a label's size that the machine can hold through vacuum technology and place on the product with enough sticking power.

For the scanning devices, their key parameters are scanning attributes and boundary conditions for the scanning process. The scanning attributes refer to the scanning resolution and quiet zones that translate into the total code size. Meanwhile, the boundary conditions are that of scanning orientation and product variation. These conditions have nothing to do with the label directly but have a significant impact on the scanning rate. Orientation means that the scanner can read a code or label that is not in the right scanning position, while the product variation, which makes a difference in a product's width, translates the distance between the label and the scanner to read the label. In this case, the very same label, which sticks to the minimum and maximum product variation, must also be readable by the scanning devices throughout its journey.

The second objective is about the economic factor that optimizes the economic benefit. There are three factors to reach this objective: development time, investment cost and future operating cost. The nature of an FMCG environment demands a rapid go-to-market solution in which speed is the essence. Hence, the development time, which is the time that the whole process needs from initiation until the rolling out of the technology that is used to deliver product to the end consumers, is important. All the costs during development are incurred in the investment cost. This includes the cost for product scrapping for live testing and new equipment when needed. The last part of the objective is the future operating cost, after the alternative is selected for implementation in the production system.

The third objective is about the users' user experience satisfaction and their interaction with the labels on primary packaging. There are three main users: producer, customers and the end consumers. Each user has its own impact and needs with the label on primary packaging.

The producer first needs to have a reliable label alternative that works with a product throughout its production process and distribution until it reaches the customer. Since there are many processes, the label alternative also has to be compatible with all the systems it has to pass through. One other concern is its ability to adapt in case of new technology adoption. In the previous section, we discussed that OSA is important. Therefore, any change that obstructs product delivery to the customer must be minimized.

The customer is the one who sells the product to the end consumers. In regard to labels on primary packaging, the essential factor for customers is information alignment when they are using the label. As the label functions as a communication medium, information that needs to be communicated must be aligned and available so that customers in all networks can use them. Secondly, it may be necessary to maintain the relevancy of labels to ensure that they remain aligned with present and future label trends.

Last but not least are the end consumers. For consumers, it has been proven that FMOT is decisive. Therefore, the negative impact from a label that covers the artwork on the primary packaging that it is stuck on must be minimized. On the other hand, since the label is already visible to consumers, it may be used to engage them by providing them with information about the product through the product webpage or other means of communication technology. All of these factors are then structured in the HDM as shown in Figure 3.

2.3.2. *The alternatives*

There are numerous labeling alternatives that were found in the environment analysis. However, several of them are not commonly used in consumer goods or customized for any specific industry and were hence eliminated from the analysis. The remaining alternatives taken into account for the HDM assessment, are: barcode, QR Code, Data Matrix, transparent label, invisible ink and RFID. However, the last three alternatives are not feasible because the producer and customer do not have

Organizational Transformation: Consumer Goods 549

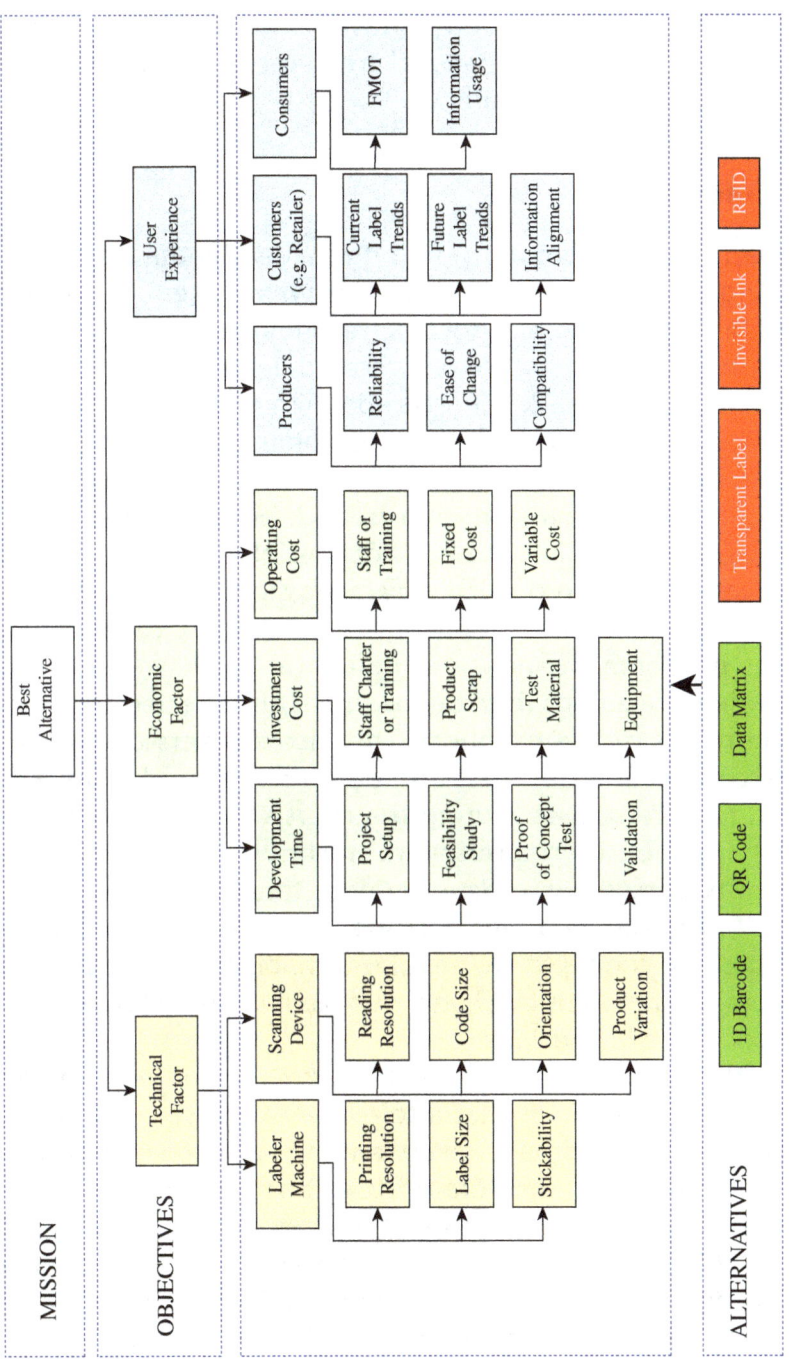

Figure 3. HDM for assessing label alternatives on primary packaging in FMCG.

the supporting system and infrastructure. Therefore, these alternatives cannot be analyzed from the technical factor of the HDM as shown in Figure 3.

2.4. *Summary*

The gap analysis is prolific in that it can identify the main problem and provide a bigger picture of what is required in the environment to adopt a specific technology. In this research paper, the gap analysis also answered the question of whether labels on primary packaging can be eliminated or reduced in size. The key answer to that is that online data information still needs to be available on the product, and the only way to do so is through a label that is stuck on the primary packaging. Therefore, elimination is not possible in the short term. However, label size reduction may be possible with other different techniques to the limit that the infrastructure can handle.

The environment analysis is useful in finding available technology in the market, which can be sorted and selected according to the needs of the producer. This also answers our second research question. Journals and patents are good sources to search for alternatives. During the course of our work on this research paper, there are several technologies found in the coding: 1D barcode, 2D codes (QR Code, Data Matrix, Code49 and PDF417), Fiducial, synthetic barcodes and RFID. In combination with barcode coding, there are also technologies in the physical medium—transparent label and invisible ink—in addition to the common self-adhesive label.

Lastly, the HDM is developed to assess and select the best alternatives. The tool is useful to evaluate and select different alternatives with a broader overview and perspective from various objectives that are related to the main mission. Expert opinions from different fields and levels in the industry partner were obtained to assign a weight for each element in the HDM according to their expertise through pairwise comparisons. However, not all alternatives found in the environment analysis were used against the HDM as some of them are not standardized by GS1 and thus not compatible with the infrastructure. Hence, the main analysis focusses only on the 1D barcode, QR Code and Data Matrix. For the HDM analysis, a software tool developed by Portland State University is used. The tool is accessible here: http://research1.etm.pdx.edu/HDM2/. The results of the HDM are discussed thoroughly in the following section.

3. Results

Fourteen employees from different fields of expertise and levels were asked to give their opinions based on their expertise and management levels. The experts gave opinions through a pairwise comparison for each element on the same level for each objective, respectively. From this comparison input, the software automatically computes the weight assigned by each expert and provides the global weight for each element after all the inputs have been obtained. This global weight will then be multiplied by the score for each alternative (Table 4).

In the HDM, a higher weight means that the element is more important, or in this work, the element with the higher weight also has a higher concern in terms of project execution. From each cluster of factors and objectives, all elements on the lowest level can be sorted according to their weight as shown in Figure 4.

The expert opinion has proven that FMOT is the most important factor in the FMCG industry, at least for the project relating to labels on primary packaging, as it has the highest weight and significantly stands out from the rest of the other factors. Ironically, the Ease of Change criterion that may enable a quick and rapid solution is shown to be the least significant in the project, regardless of the agile environment in the FMCG industry. Several similar arguments from the experts explain why this is so: Ease of Change is less important when compared to Reliability and Compatibility. However,

Table 4. Global weight for the HDM by experts' opinions.

	Factors		Weight
Technical Factor	Labeler machine	Printing resolution	0.025
		Label size	0.027
		Stickability	0.031
	Scanning device	Reading resolution	0.065
		Code size	0.029
		Orientation	0.042
		Product variation	0.056
Economic Factor	Development time	Project setup	0.015
		Feasibility study	0.015
		Proof of concept test	0.012
		Validation	0.016
	Investment cost	Staff charter	0.018
		Product scrap	0.014
		Test material	0.013
		Equipment	0.017
	Operating cost	Staff/Training	0.056
		Fixed cost	0.053
		Variable cost	0.051
User Experience	Producer	Reliability	0.034
		Ease of change	0.011
		Compatibility	0.028
	Customer	Current trend	0.041
		Future trend	0.026
		Information alignment	0.056
	Consumer	FMOT	0.224
		Information usage	0.025

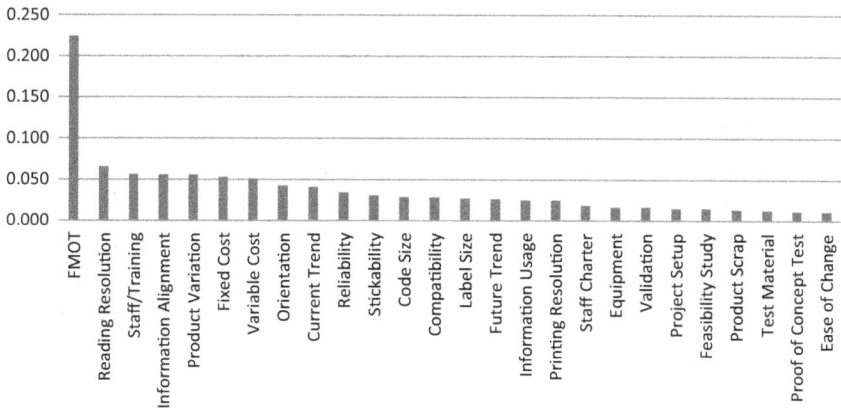

Figure 4. Global weight for the HDM by the experts' opinions.

throughout the execution, the project was still rushed as quickly as possible.

When taking expert opinion into account, it was found that all the experts were consistent with their opinions, in that their computed inconsistency values were always below 0.1. This meant that their argumentations were valid and taken into account. However, interestingly, this value is above 0.1 in the objectives section. The reason for this could be that the opinions gathered for the objectives section were obtained from middle management employees. The engineers, technicians and operational workers also have a slight disagreement value of between 0.01 and 0.1 when providing their opinions in the practical level. However, most shared similar opinions. All of the results from the experts are available in Appendix A.

From observations and the experts' feedback, there might be several reasons for the disagreement in opinion. Disagreement itself is a case of misunderstanding and/or dissent due to a failure to share common ground [62]. Firstly, some experts faced difficulties when quantifying and assigning numbers in the pairwise comparisons when comparing factors. Secondly, the disagreement was probably also caused by work differences and their views toward the project.

The first problem acknowledged by the experts that raised disagreement lies in the technicalities of using the tool. Most had a similar opinion that while a factor was more important than the other, they could not quantify the difference. This is clearly shown in the given opinion in the intermediate level of the Technical Factor and User Experience Objectives. For instance, within Technical Factor, all the experts agreed that the scanning device was more important than the labeling machine, but they had a slight disagreement with regard to the weight given in the pairwise comparison.

The second problem might come from a management point of view, as the disagreement raised in the Objectives level, which was filled by middle management experts who worked in different fields of expertise. Moreover, the seating plan that was designed in a functional centric field meant that they were also physically separated. A study showed that there is a 10% chance of interaction with someone who sits only two rows away [63]. This also justified recent changes by the management to reorganize the seating plan into a product platform-centric to boost interaction to share a common ground for employees to work together for specific product platform.

3.1. *Scoring calculation*

The scoring for the alternatives is based on a normalized scale in numerical values whenever data is available. For instance, in calculating the variable cost, it was found out the estimated variable cost for 1D barcode, QR Code and Data Matrix in the industry partner was €0.03, €0.0235 and €0.0215, respectively. Data Matrix obtained a maximum score of 1 (one) because it was the cheapest option and had the lowest cost. On the other hand, while the cost of 1D barcode was the highest it was not the most expensive; there was a more expensive option (€0.0425) on the same label type. Hence, that option received a score of 0 (zero). The score of the variable cost for the used 1D barcode and QR Code is then calculated in between the linear scale.

Several qualitative factors in the user experience cannot be directly quantified. In this case, a list of specific tasks in each element is used to determine the score. For instance, if an element has four tasks that need to be fulfilled and the alternative satisfies three out of these four, then the alternative score is ¾ or 0.75. The calculations for all the alternatives against the HDM are available in Appendix B.

From the results in Table 5, it shows that the 2D Data Matrix in the circumstances of the FMCG industry partner is the most suitable option for the label on primary packaging, followed by the QR Code and then the 1D barcode. Moreover, had RFID been taken into consideration, it would have shown that barcodes options are still more favorable than RFID. RFID technology may improve FMOT drastically by not requiring a label on the primary packaging, but the needed cost to adopt the technology is far too expensive compared to the barcode alternatives. Furthermore, even if the technology is already well developed and ready for operational use in the production plant, the investment and the operating costs negate the overall benefit and prove that RFID technology is not suitable for FMCG application, specifically from the selling unit's point of view (Table 6).

3.2. Sensitive analysis

The FMCG industry partner has confirmed that the FMOT is the most important element with regard to labels on primary packaging. Though, this is clearly a quality driven project to improve consumers' FMOT experience, the project also needs to be economically justified. Hence, a sensitive analysis was carried out to see if the best alternative changes if the most important element is now the Economic Factor instead of the FMOT. The sensitive analysis is done by reducing the global weight of the FMOT to a similar weight as that of the second highest. The weight saved is then distributed equally into all elements in the Operating Cost, which was chosen because according to the experts, this element

Table 5. Alternatives scoring against the HDM.

	Barcode		QR Code		Data Matrix		Comment
	Score	Sum	Score	Sum	Score	Sum	
Printing Resolution	0.23	0.006	0.23	0.006	0.44	0.011	—
Label Size	0.52	0.014	0.79	0.022	0.95	0.026	—
Stickability	1	0.031	0.66	0.020	0.32	0.010	—
Reading Resolution	0.78	0.051	0.78	0.051	1	0.065	—
Code Size	0.52	0.015	0.79	0.023	0.95	0.027	Code size is almost proportional to label size
Orientation	0.71	0.030	0.43	0.018	0.43	0.018	—
Product Variation	1	0.056	1	0.056	1	0.056	All label must be applicable to all product
Project Setup	1	0.015	1	0.015	1	0.015	This stage is done, unknown value to compare
Feasibility Study	1	0.015	1	0.015	1	0.015	This stage is done, unknown value to compare
Proof of Concept Test	1	0.012	1	0.012	1	0.012	This stage is done, unknown value to compare
Validation	1	0.016	1	0.016	0	0.000	This stage is not done, unknown value to compare

Organizational Transformation: Consumer Goods 557

Staff Charter	1	0.018	1	0.018	1	0.018	Unknown value to compare
Product Scrap	1	0.014	1	0.014	1	0.014	Unknown value to compare
Test Material	0.6	0.008	0.91	0.012	1	0.013	—
Equipment	1	0.017	1	0.017	0.63	0.010	From current status to the needs and requirement
Staff/Training	1	0.056	1	0.056	1	0.056	Unknown value to compare
Fixed Cost	1	0.053	1	0.053	1	0.053	No change in fixed cost/maintenance
Variable Cost	0.6	0.031	0.91	0.046	1	0.051	—
Reliability	1	0.034	1	0.034	1	0.034	—
Ease of Change	1	0.011	0.5	0.006	0.25	0.003	—
Compatibility	1	0.028	0.8	0.023	0.8	0.023	—
Current Trend	1	0.041	0.8	0.033	0.8	0.033	Author's observation
Future Trend	0.8	0.021	1	0.026	1	0.026	Author's observation
Information Alignment	1	0.056	1	0.056	1	0.056	All labels must be aligned
FMOT	0.52	0.117	0.79	0.177	0.95	0.213	—
Information Usage	0.25	0.006	0.5	0.012	0.5	0.012	—
Total		0.771		0.836		0.871	

Table 6. RFID scoring against the HDM.

	Weight	Score	Sum	Comment
Printing Resolution	0.025	1.00	0.025	N/A
Label Size	0.027	1.00	0.027	
Stickability	0.031	1.00	0.031	Different technology, printing technology is outweighed
Reading Resolution	0.065	1.00	0.065	
Code Size	0.029	1.00	0.029	
Orientation	0.042	1.00	0.042	N/A
Product Variation	0.056	1.00	0.056	No "Line of Sight" Needed
Project Setup	0.015	1.00	0.015	Assuming it is already done, else it will be zero and reduce the total score
Feasibility Study	0.015	1.00	0.015	
Proof of Concept Test	0.012	1.00	0.012	
Validation	0.016	1.00	0.016	
Staff Charter	0.018	1.00	0.018	
Product Scrap	0.014	1.00	0.014	Same amount needs to test
Test Material	0.013	0.00	0.000	Out of chart when compared with barcodes
Equipment	0.017	0.00	0.000	
Staff/Training	0.056	1.00	0.056	Same amount of operator
Fixed Cost	0.053	0.00	0.000	Out of chart when compared with barcodes
Variable Cost	0.051	0.00	0.000	
Reliability	0.034	1.00	0.034	Must be reached
Ease of Change	0.011	0.00	0.000	N/A
Compatibility	0.028	0.00	0.006	Different technology
Current Trend	0.041	0.20	0.008	Author's observation
Future Trend	0.026	1.00	0.026	Author's observation
Information Alignment	0.056	1.00	0.056	Must be reached
FMOT	0.224	1.00	0.224	Removed from appearance
Information Usage	0.025	0.00	0.000	Nothing can be used
Total Sum			0.770	

Table 7. Global weight difference.

Original Global Weight		
Operating Cost	Staff/Training	0.056
	Fixed cost	0.053
	Variable cost	0.051
Consumer	FMOT	0.224
Global Weight for Sensitive Analysis		
Operating Cost	Staff/Training	0.109
	Fixed cost	0.106
	Variable cost	0.104
Consumer	FMOT	0.065

Table 8. Total score difference.

Original Total Score		
Barcode	QR Code	Data Matrix
0.771	0.836	0.871
Sensitive Analysis Total Score		
Barcode	QR Code	Data Matrix
0.826	0.864	0.878

is more than twice as important as compared to Development Time and Investment Cost (Table 7).

The alternatives were once again calculated with the same score against the different global weight between the FMOT and the operating cost (Table 8). Interestingly, the result still holds with Data Matrix as being the best alternative, followed by the QR Code and the Barcode. However, the Barcode yields the highest increase from the original total score. It proves that choosing Data Matrix is not only beneficial for quality improvement through the FMOT, but also in the Economic Factor, especially the Operating Cost that contributes accordingly.

4. Conclusions and Outlook

The role of packaging has evolved from being protection for products into becoming the "silent salesman" that sells the product to consumers [5, 14, 15]. To be the silent salesman, the visual appearance of the packaging is the product's most important aspect [29, 30], which is later referred to as FMOT by Procter & Gamble. Furthermore, the packaging is also responsible as a communication media that carries important information used throughout the supply chain [7, 12, 14].

These information needs have led to the existence of labels on the primary packaging, even though these labels may end up covering the artwork and has a negative impact on the FMOT. This creates a dilemma: should a label serve as a communication medium or as the silent salesman? This research paper answers this question by providing the best label alternative through a gap analysis, environment analysis and an HDM with expert opinions from employees in one of the leading FMCG industry.

Data Matrix is selected as the best alternative for primary packaging and self-adhesive labeling in an FMCG environment. The main driver according to the expert from the FMCG industry is that FMOT must be the ultimate goal. Therefore, there should ideally be no labeling on primary packaging as it has a negative impact on the appearance of the packaging. Based on the HDM, Data Matrix was chosen because it had the smallest option possible. It is not only the best option from the FMOT perspective but is also economically beneficial. This was proven by the sensitive analysis that spread the FMOT importance into the Economic Factors and again Data Matrix yielded a higher score compared to the QR Code and 1D Barcode. This result shows that using Data Matrix for the self-adhesive label on primary packaging in FMCG is the best alternative to improve the FMOT. It is also economically advantageous.

Additionally, the result also shows that RFID technology is not favorable for labeling primary packaging. Even with a significant improvement in the FMOT, the cost to adapt and implement the technology is too much to be borne by the producer. This also proves that RFID is not suitable for selling units application in FMCG industries [47, 56]. In the future when access to technology has become significantly cheaper and the infrastructure is available throughout the supply chain, RFID can be reconsidered as an alternative as a label on primary packaging.

The results of this research paper will soon be implemented in the production plant as the industry partner is going to replace the current QR Code label with a smaller Data Matrix label. This will be rolled out in the near future. The chosen best label alternative on primary packaging shall reflect the actual solution that is the most suitable option in the FMCG environment. However, it may also not be applicable to other FMCG companies, because the raised problem and the obtained experts' opinions exclusively come from the industry partner. Even though it is a one of the biggest in a multinational FMCG company, other firms may have their own perspective, views and even company culture that are used to analyze and solve a problem. Further research is needed to obtain expert opinions from other FMCG firms to gain more confidence in interpreting the HDM, and in the end, provide a conclusive solution for the complete FMCG environment. Besides, further study is also required to understand more about consumer behavior toward labels on primary packaging.

References

1. M. Gomez, D. M. Consuegra and A. Molina, "The importance of packaging in purchase and usage behaviour", *International Journal of Consumer Studies,* **39** (2015) 203–211.
2. M. A. Pinero, L. Lockshin, R. Kennedy and A. Corsi, "Distinctive elements in packaging (FMCG): an exploratory study", *Australian and New Zealand Marketing Academy,* 2010.
3. B. Rundh, "The multi-faceted dimension of packaging: marketing logistic or marketing tool?" *British Food Journal,* **107**, 9 (2005) 670–684.

4. G. Prendergast and L. Pitt, (1996). "Packaging, marketing, logistics and the environment: are there trade-offs?" *International Journal of Physical Distribution & Logistics Management*, 26 (1996) 60–72.
5. D. Judd, B. Aalders and T. Melis, *The Silent Salesman: Primer on Design, Production, and Marketing of Finished Package Good* (Singapore: Continental Press, 1989).
6. T. U. Daim, "Technology assessment and acquisition", *The Handbook of Technology Management*, (ed.) H. Bidgoli (Hoboken: John Wiley & Sons, Inc, 2010), pp. 315–332.
7. F. Chan, H. Chan and K. Choy, "A systematic approach to manufacturing packaging logistics", *International Journal of Advanced Manufacturing Technology*, 29, (2006) 1088–1101.
8. M. Deliya, "Consumer behavior towards the new packaging of FMCG product", *Abhinav National Monthly Refereed Journal of Research in Commerce & Management*, 1, **11** (2012) 199–211.
9. M. Johnsson, "Packaging logistics: a value added approach", Dissertation, Lund University, Department of Engineering Logistics, Lund, 1998.
10. D. Hellström and M. Saghir (2007). "Packaging and logistics interactions in retail supply chains", *Packaging Technology and Science*, **20** (2007) 197–216.
11. M. Saghir, "Packaging information needed for evaluation in the supply chain: the case of Swedish grocery retail industry", *Packaging Technology and Sciences*, **15** (2002) 37–46.
12. D. Lambert, J. Stock and L. Ellram, *Fundamentals of Logistics Management* (Singapore: McGraw-Hill, 1998).
13. M. Christopher, "The agile supply chain: competing in volatile markets", *Industrial Marketing Management*, 29, **1** (2000) 37–44.
14. R. L. Underwood and N. M. Klein, "Packaging as brand communication: effects of product pictures on consumer response to the package and brand", *Journal of Marketing Theory and Practice*, 10, **4** (2002) 56–68.
15. M. Löfgren, "Winning at the first and second moments of truth: an exploratory study", *Managing Service Quality: An International Journal*, 15, **1** (2005) 102–115.
16. C. Olsmats, *The Business Mission of Packaging as a Strategic Tool for Business Development Towards the Future* (Abo: Abo University, 2004).
17. J. Lecinski, *Winning the Zero Moment of Truth* (Google, 2011).

18. M. Saghir, "The concept of packaging logistics", in *2nd World Conference on POM and 15th Annual POM Conference*, 2004.
19. S. Srinivasan and W. F. Lu, "Development of a supporting tool for sustainable FMCG packaging designs", in *21st CIRP Conference on Life Cycle Engineering*, 2014, pp. 395–400.
20. A. White, "Labels for Packaging", *Packaging Technology Fundamentals, Materials and Processes* (Woodhead Publishing Limited, 2012), pp. 395–407.
21. M. Farley, *Label Market Trends* (London: Tarsus Group, 2008).
22. B. Lowson, R. King and A. Hunter, *Quick Response: Managing the Supply Chain to Meet Consumer Demand* (New York: John Wiley & Sons, 1999).
23. P. M. Swafford, S. Ghosh and N. Murthy, "Achieving supply chain agility through IT integration and flexibility", *International Journal of Production Economics*, 2008, pp. 288–297.
24. C. Jannsen (ed.), "GS1 logistics label guideline", Release 1.2, GS1, September 2017.
25. "GS1 General Specification", Version 18, GS1, January 2018; "How we got here", GS1, 2018. Accessed: 8 August 2018. www.gs1.org/about/how-we-got-here.
26. Keyence, *1D Code Handbook* (Itasca: Keyence Corporation of America, 2006).
27. P. Peretz, "Are UPC-A and EAN-13 the Same?", *Nationwide Barcode*, 3 January 2012. Accessed: 8 August 2018. www.nationwidebarcode.com/are-upc-a-and-ean-13-the-same/.
28. M. H. McDonald, L. de Chernatony and F. Harris, "Corporate marketing and service brands—moving beyond the fast moving consumer goods model", *European Journal of Marketing*, 35, 3/4 (2001) 335–352.
29. Z. Harith, C. Ting and N. Zakaria, "Coffee packaging: consumer perception on appearance, branding and pricing", *International Food Research Journal*, 21, **3** (2014) 849–853.
30. R. Wang and W. Chen, "The study on packaging illustration affect on buying emotion", in *Proceeding of the International Association of Society of Design Research* (Hong Kong: The Hong Kong Polytechnic University, 2007).
31. P. Balaban-Durdev and V. Maletic, "Visual impact of graphic information in the package", in *Proceedings of Informing Science & IT Education Conference (InSITE)*, 2011, pp. 33–46.

32. N. Draskovic, J. Temperley and J. Pavicic, "Comparative perceptions of consumer goods packaging", *International Journal of Management Cases*, 2009, pp. 154–163.
33. W. Olins, *The New Guide to Identity* (Aldershot: Gower Publishing Ltd, 1995).
34. M. Porter, (1998). *Competitive Strategy: Techniques for Analyzing Industries and Competitors* (New York: The Free Press, 1998).
35. Roland Berger Strategy Consultants, *Optimal Shelf Availability: Increasing Shopper Satisfaction at the Moment of Truth* (ECR Europe, 2003).
36. D. Corsten and T. Gruen, "Desperately seeking shelf availability: an examination of the extent, the causes, and the efforts to address retail out-of-stocks", *International Journal of Retail & Distribution Management*, 2003, pp. 605–617.
37. A. Gupta, "Environment and PEST analysis: an approach to external business environment", *International Journal of Modern Social Sciences*, 2, 1 (2013) 34–43.
38. G. C. Peng and M. B. Nunes, "Using PEST analysis as a tool for refining and focusing contexts for information system research", in *6th European Conference on Research Methodology for Business and Management Studies*, 2007, pp. 229–237.
39. D. Vrontis and C. Vignali, "Dairy milk in France: a marketing investigation of the situational environment", *British Food Journal*, 103, **4** (2001) 201–206.
40. M. Morrison, "PESTLE analysis", *Factsheet*, Chartered Institute of Personnel and Development, 1 September 2017. Accessed: 10 August 2018. www.cipd.co.uk/knowledge/strategy/organisational-development/pestle-analysis-factsheet.
41. J. K.-K. Ho, "Formulation of a systemic PEST analysis for strategic analysis", *European Academic Research*, 2, 5 (2014) 6478–6492.
42. K. E. Okpala, T. S. Afolabi and D. Adegbola, "Management accounting information system and effective business decisions: an evaluation of quoted FMCG manufacturing firms", *Imsu Business & Finance Journal*, 2018, pp. 161–174.
43. P. A. Ryan and G. P. Ryan, (2002). "Capital budgeting practices of the Fortune 1000: how have things changed?" *Journal of Business and Management*, 8, 4 (2002) 355–364.
44. K. Ward and L. Ryals, "Latest thinking on attaching a financial value to marketing strategy: through brands to valuing

relationship", *Journal of Targeting, Measurement and Analysis for Marketing*, **4** (2001), 327–340.
45. S. Gangar, "Role of packaging in sales of FMCG products", Bachelor of Management Studies, University of Mumbai, 2015.
46. L. McCathie and K. Michael, "Is it the end of barcodes in supply chain management?", in *Proceedings of Collaborative Electronic Commerce Technology and Research Conference LatAm*, Chile, 2005, pp. 1–19.
47. R. Balocco, G. Miragliotta, A. Perego, and A. Tumino, (2011). "RFID adoption in the FMCG supply chain: an interpretative framework", *Supply Chain Management: An International Journal*, 16, **5** (2011) 299–315.
48. N. Varchaver, "Scanning the globe the humble bar code began as an object of suspicion and grew into a cultural icon", *Fortune*, 149, **11** (2004) 144–56.
49. J. Gao, L. Prakash, and R. Jagatesan, "Understanding 2D Barcode technology and applications in M-Commerce—design and implementation of a 2D Barcode processing solution", in *31st Annual International Computer Software and Application Conference (COMPSAC)*, Beijing, 2007, pp. 49–56.
50. R. C. Palmer, *The Bar Code Book*, 3rd edition (Helmers Publishing, 1995).
51. L. Naimark and E. Foxlin, "Circular data matrix fiducial system and robust image processing for a wearable vision-inertial self-tracker", in *Proceedings of the International Symposium on Mixed and Augmented Reality (ISMAR'02)*, 2002, pp. 27–36.
52. W. Berson and J. D. Auslander, US Patent No. US5525789A, 1994.
53. R. McGill and P. F. Shadwell, "US Patent No. US8196833B2", 2005.
54. G. Roussos, "What is RFID?" *Networked RFID: Systems, Software and Service* (London: Springer Verlag, 2008).
55. I. P. Vlachos (2014). "A hierarchical model of the impact of RFID practices on retail supply chain performance", *Expert System with Applications*, 41, **1** (2014) 5–15.
56. M. Bertolini, E. Bottani, A. Rizze and A. Volpi, "The benefits of RFID and EPC in the supply chain: lessons from an Italian pilot study", *The Internet of Things: 20th Tyrrhenian Workshops on Digital Communications*, 2010, pp. 293–302.
57. G. Demiralp, G. Guven and E. Ergen, (2012). "Analyzing the benefits of RFID technology for cost sharing in construction

supply chains: a case study on prefabricated precast components", *Automation in Construction* **24** (2012) 120–129.
58. G. M. Gaukler, R. W. Seifert and W. H. Hausman, "Item-level RFID in the retail supply chain", *Production and Operations Management,* 16, **1** (2007) 65–76.
59. T. L. Saaty, "Decision making with the analytic hierarchy process", *International Journal Services Sciences,* 1, **1** (2008) 83–98.
60. B. L. Golden, E. A. Wasil and P. T. Harker, *The Analytic Hierarchy Process* (Springer Verlag Berlin Heidelberg, 1989).
61. T. U. Daim, *Hierarchical Decision Modeling* (Springer International Publishing, 2016).
62. J. Stalpers, "The expression of disagreement", *The Discourse of Business Negotiation*, (eds.) K. Ehlich and J. Wagner, (Berlin, 1995, pp. 275–290).
63. R. Feintzeig, "The new science of who sits where at work", *The Wall Street Journal*, 8 October 2013.

Appendix A: HDM Expert Opinion

This section provides the lists of the weights for each element in the HDM according to the expert opinion, which was obtained from pairwise comparisons. In order to respect the experts' confidentiality in providing their opinions, private information such as their names, personal IDs and fields of working is abstracted.

Best Alternative	Technical Factor	Economic Factor	User Experience	Inconsistency
B3M1	0.23	0.23	0.54	0
B3M2	0.25	0.38	0.38	0
B3M3	0.41	0.41	0.18	0
B4M1	0.21	0.1	0.68	0.03
Mean	0.28	0.28	0.45	
Minimum	0.21	0.1	0.18	
Maximum	0.41	0.41	0.68	
Std. Deviation	0.08	0.12	0.19	
Disagreement				0.125

Technical factor

Technical Factor	Labeler Machine	Scanning Device	Inconsistency
B2M1	0.2	0.8	0
B2M2	0.3	0.7	0
B2M3	0.4	0.6	0
Mean	0.3	0.7	
Minimum	0.2	0.6	
Maximum	0.4	0.8	
Std. Deviation	0.08	0.08	
Disagreement			0.067

Technical factor subcriteria

Labeler Machine	Printing Resolution	Label Size	Stickability	Inconsistency
B1M1	0.33	0.33	0.33	0
B1M2	0.18	0.41	0.41	0
B2M3	0.38	0.25	0.38	0
Mean	0.3	0.33	0.37	
Minimum	0.18	0.25	0.33	
Maximum	0.38	0.41	0.41	
Std. Deviation	0.08	0.07	0.03	
Disagreement				0.061

Scanning Device	Reading Resolution	Code Size	Orientation	Product Variation	Inconsistency
B1M1	0.36	0.16	0.29	0.19	0.02
B1M2	0.31	0.22	0.09	0.38	0.01
B2M3	0.34	0.08	0.28	0.31	0
Mean	0.34	0.15	0.22	0.29	
Minimum	0.31	0.08	0.09	0.19	
Maximum	0.36	0.22	0.29	0.38	
Std. Deviation	0.02	0.06	0.09	0.08	
Disagreement					0.066

Economic factor

Economic Factor	Development Time	Investment Cost	Operating Cost	Inconsistency
B2M1	0.14	0.11	0.75	0.01
B2M3	0.2	0.26	0.54	0
B2M4	0.29	0.29	0.43	0
Mean	0.21	0.22	0.57	
Minimum	0.14	0.11	0.43	
Maximum	0.29	0.29	0.75	
Std. Deviation	0.06	0.08	0.13	
Disagreement				0.087

Economic factor subcriteria

Development Time	Project Setup	Feasibility Study	Proof of Concept Test	Validation	Inconsistency
B1M3	0.3	0.3	0.2	0.2	0
B2M3	0.22	0.22	0.22	0.33	0
B2M4	0.27	0.25	0.18	0.3	0
Mean	0.26	0.26	0.2	0.28	
Minimum	0.22	0.22	0.18	0.2	
Maximum	0.3	0.3	0.22	0.33	
Std. Deviation	0.03	0.03	0.02	0.06	
Disagreement					0.035

Investment Cost	Staff Charter	Product Scrap	Test Material	Equipment	Inconsistency
B1M3	0.22	0.22	0.22	0.33	0
B2M3	0.25	0.28	0.23	0.25	0
B2M4	0.44	0.17	0.17	0.23	0
Mean	0.3	0.22	0.21	0.27	
Minimum	0.22	0.17	0.17	0.23	
Maximum	0.44	0.28	0.23	0.33	
Std. Deviation	0.1	0.04	0.03	0.04	
Disagreement					0.057

Operating Cost	Staff/Training	Fixed Cost	Variable Cost	Inconsistency
B1M3	0.38	0.38	0.25	0
B2M3	0.33	0.33	0.33	0
B2M4	0.33	0.29	0.38	0.02
Mean	0.35	0.33	0.32	
Minimum	0.33	0.29	0.25	
Maximum	0.38	0.38	0.38	
Std. Deviation	0.02	0.04	0.05	
Disagreement				0.036

User experience

User Experience	Producers	Customers	Consumers	Inconsistency
B2M3	0.18	0.36	0.47	0
B3M1	0.23	0.23	0.54	0
B4M1	0.09	0.24	0.67	0
Mean	0.17	0.28	0.56	
Minimum	0.09	0.23	0.47	
Maximum	0.23	0.36	0.67	
Std. Deviation	0.06	0.06	0.08	
Disagreement				0.066

User experience subcriteria

Producers	Reliability	Ease of Change	Compatibility	Inconsistency
B1M4	0.41	0.18	0.41	0
B2M3	0.53	0.17	0.3	0.02
B2M4	0.44	0.11	0.44	0
Mean	0.46	0.15	0.38	
Minimum	0.41	0.11	0.3	
Maximum	0.53	0.18	0.44	
Std. Deviation	0.05	0.03	0.06	
Disagreement				0.047

Customers	Current Trend	Future Trend	Information Alignment	Inconsistency
B1M4	0.26	0.2	0.54	0
B2M3	0.29	0.33	0.38	0.02
B2M4	0.44	0.11	0.44	0
Mean	0.33	0.21	0.45	
Minimum	0.26	0.11	0.33	
Maximum	0.44	0.33	0.54	
Std. Deviation	0.08	0.09	0.07	
Disagreement				0.078

Consumers	FMOT	Information Usage	Inconsistency
B1M2	0.8	0.2	0
B1M5	0.95	0.05	0
B2M4	0.95	0.05	0
Mean	0.9	0.1	
Minimum	0.8	0.05	
Maximum	0.95	0.2	
Std. Deviation	0.07	0.07	
Disagreement			0.067

Appendix B: Alternatives Scoring

This section provides the calculation for the alternatives scoring when the numerical value to compare is available. Several points are scored based on a checklist design.

Technical factors—labeler machine

Printing Resolution		
Labeler capability	0.127 mm	
Label print resolution (in mm)		Score
Barcode	0.8	0.23
QR Code	0.8	0.23
Data Matrix	0.62	0.44

Label Size		
Bag Size	12.000 mm²	
Label Size (in mm²)		Score
Barcode	5.750	0.52
QR Code	2.500	0.79
Data Matrix	600	0.95

Stickability (Due to Sticking Area)		
	Label size	Score
Barcode	5750 mm²	1.00
QR Code	2500 mm²	0.66
Data Matrix	600 mm²	0.32

Technical factors—scanning device

Reading Resolution		
Scanner capability		0.62 mm
Label print resolution (in mm)		Score
Barcode	0.8	0.78
QR Code	0.8	0.78
Data Matrix	0.62	1.00

Orientation			
	Barcode	QR Code	Data Matrix
Pitch	ok	ok	ok
Skew	ok	ok	ok
Tilt	ok	ok	ok
Bag wrapped x-axis	Problem	Problem	Problem
Bag wrapped y-axis	ok	Problem	Problem
Distorted print x-axis	Problem	Problem	Problem
Distorted print y-axis	ok	Problem	Problem
Score	0.71	0.43	0.43

User experience—producers

	Reliability		
	Barcode	**QR Code**	**Data Matrix**
Succesful rate requirement	0.9999	0.9997	0.9997

	Ease of Change		
	Barcode	**QR Code**	**Data Matrix**
Vacuum plate	no change	change	change
Vacuum pressure	no change	no change	change
Labeler mechanism position adjustment	no change	change	change
Scanning devices	no change	no change	no change
Score	1	0.5	0.25

	Compatibility		
	Barcode	**QR Code**	**Data Matrix**
1D scanner	yes	no	no
2D scanner	yes	yes	yes
Labeler (printing)	yes	yes	yes
Labeler (sticking)	yes	yes	yes
Network	yes	yes	yes
Score	1	0.8	0.8

User experience—consumers

	FMOT	
Bag Size	12.000 mm²	
Label Size (in mm²)		Score
Barcode	5.750	0.52
QR Code	2.500	0.79
Data Matrix	600	0.95

	Information Usage		
	Barcode	QR Code	Data Matrix
Batch#	HRI	HRI	HRI
Order#	no	no	no
Product#	no	no	no
Website	N/A	yes	yes
Score	0.25	0.5	0.5

CPSIA information can be obtained
at www.ICGtesting.com
Printed in the USA
LVHW082041090221
678838LV00001B/27